Sound and Structural Vibration

Radiation, Transmission and Response

Sound and Structural Vibration
Radiation, Transmission and Response

FRANK FAHY
Institute of Sound and Vibration Research
The University
Southampton, England

1985

ACADEMIC PRESS
(Harcourt Brace Jovanovich, Publishers)
London Orlando San Diego New York
Toronto Montreal Sydney Tokyo

ACADEMIC PRESS INC. (LONDON) LTD.
24–28 Oval Road
LONDON NW1 7DX

United States Edition published by
ACADEMIC PRESS, INC.
Orlando, Florida 32887

British Library Cataloguing in Publication Data

Fahy, F.J.
Sound and structural vibration : radiation,
transmission and response.
1. Structural dynamics 2. Sound
I. Title
624.1'71 TA654

ISBN 0-12-247670-0

Library of Congress Cataloging in Publication Data

Fahy, Frank J.
Sound and structural vibration.

Includes index.
1. Vibration. 2. Sound-waves. I. Title.
TA355.F34 1985 620.2 84-20439
ISBN 0-12-247670-0 (alk. paper)

PRINTED IN THE UNITED STATES OF AMERICA

85 86 87 88 9 8 7 6 5 4 3 2 1

This book is dedicated to the memory of my parents who made many sacrifices in order that I could pursue my studies.

Contents

2. Sound Radiation by Vibrating Structures

3. Fluid Loading of Vibrating Structures

4. Transmission of Sound through Partitions

5. Acoustically Induced Vibration of Structures

6. Acoustic Coupling between Structures and Enclosed Volumes of Fluid

7. Introduction to Numerically Based Analyses of Fluid–Structure Interaction

Preface

In writing this book my aim has been to present a unified qualitative and quantitative account of the physical mechanisms and characteristics of linear interaction between audio-frequency vibrational motion in compressible fluids and structures with which they are in contact. The primary purpose is to instruct the reader in theoretical approaches to the modelling and analysis of interactions, whilst simultaneously providing physical explanations of their dependence upon the parameters of the coupled systems. It is primarily to the engineering student that the book is addressed, in the firm belief that a good engineer remains a student throughout his professional life. A preoccupation with the relevance and validity of theoretical analyses in relation to practical problems is a hallmark of the engineer. For this reason there is a strong emphasis on the relationship of results obtained from theoretical analysis of idealised models and the behaviour of the less than ideal realities from which they are abstracted.

The teacher of analysis in any sphere of applied science is faced with a central dilemma: systems which can be modelled and analysed in a manner sufficiently explicit and direct to illustrate a principle are usually gross oversimplifications of the real world and are hence, to some extent, trivial; systems which are of practical concern are usually much too complex to offer suitable examples for didactic purposes. In attempting to grasp this nettle I hope I may be forgiven by any physicists and applied mathematicians who may pick up this book for sacrificing a certain amount of mathematical rigour for the sake of qualitative clarity.

In teaching mechanical engineering and engineering acoustics over a number of years it has struck me forcibly that an appreciation of structural vibration as a form of wave motion, a concept readily grasped by the student of physics, is often lacking in those reared on a diet of lumped elements and normal modes. One unfortunate effect is that the associated wave phenomena such as interference, scattering and diffraction are often believed to be the preserve of water and air, and the link between natural modes and frequencies of structures, and the component waves intrinsic to these phenomena, is not readily perceived. The subject of this book appeared to be the ideal vehicle for persuading students of the advantage to be gained by taking a dual view of vibrational motion in distributed elastic systems. Hence I have emphasized the wave "viewpoint" right from the start, in the hope of encouraging the reader to "think waves."

The three main categories of practical problems to which the material of this book is relevant are sound radiation from vibrating structures, sound transmission between adjacent regions of fluid media separated by an intervening solid partition, and the response of structures to excitation by incident sound fields. Much of the source material is only available (in English at least) in articles scattered throughout the learned journals of the world. In particular, fundamental analyses in acoustics textbooks of sound transmission through partitions tend to be restricted to highly idealised cases, and the complicating effects of finite panel size, non-homogeneous structures, cavity absorption and frames, and panel curvature are at best briefly and only qualitatively described. This is why Chapter 4 is the longest in the book.

Although the aim of the book is instructional, it is different from many textbooks in that it is not divided into neat, self-contained sections of analysis, which can be concluded with Q.E.D.; it also contains a large amount of descriptive text. The first feature is connected with the "dilemma" previously mentioned; the second stems from a desire to provide a text from which the reader can learn in the absence of a formal lecture course, although it is hoped that my prolixity will not deter a lecturer from using the book to complement his course.

The arrangement of questions in the book does not generally follow the conventional pattern of formalised quantitative examples at the end of chapters. The reader is challenged at various places within the text to think about the material which he is currently reading, while it is fresh in his mind. I hope, in this way, to solicit more active cooperation in the learning process, and to stimulate a questioning approach to the material, rather than passive acceptance. The questions at the ends of chapters are linked to specific sections in the text and range from straightforward numerical evaluation of quantities, intended to encourage a "physical feel" for their orders of magnitude, to rather open-ended questions, which can only be answered in qualitative terms. The absence of a large number of formal calculation exercises reflects both the nature of the subject and the fact that the readership is expected to have developed previously the facility for performing formal analyses of fundamental vibrational and acoustical problems.

Numerous references to other books, research publications, and reports are provided in the text. The list is clearly not comprehensive, but it is hoped that it will provide the reader with jumping-off points for further and deeper study. The omission of any particular relevant reference in no way constitutes a reflection of its value, any more than the inclusion of a reference implies that it is to be considered uniformly meritorious and correct.

It is my hope that this book, for all the faults which will no doubt emerge, will help at least a few people to understand more fully the fascinating interplay between sound and structural vibration, and thereby serve to increase their ability to control whatever aspect of the subject commands their attention.

Acknowledgements

This book is the product not only of my knowledge, research and teaching experience, but of numerous discussions, debates and joint endeavours with many colleagues and students over a period of twenty years. In particular, I would like to acknowledge the help given to me by Professor Phil Doak and Dr. Denys Mead, Dr. Maurice Petyt, Dr. Stewart Glegg, Dr. Philip Nelson and Dr. Chris Morfey of the University of Southampton. For her vital contribution in translating my vile scribble into an excellent typescript, I am greatly indebted to Jan Ward, and for the skillful conversion of my original figures into reproducible form I wish to thank Georgina Allan. For the unenviable task of proofreading I am indebted to my youngest son Tom and my wife Beryl. I also wish to acknowledge the help and guidance provided by the editorial and production staff of Academic Press in bringing this offspring of my labour into the world in the fine form you have before you. Finally, I have great pleasure in acknowledging the loving patience and practical assistance, accompanied by innumerable cups of coffee, with which my wife Beryl supported me during the three-year gestation period.

Introduction

> I think I shall never undertake to write a book again. If one were a scamp, the work would be easy enough, but for an honest man it is dreadful.
>
> John Tyndall, 1859

As you read these words you are almost certainly experiencing various manifestations of the process of vibrational interaction between fluids and solid structures. In all probability traffic noise is being transmitted through the windows of the room, the plumbing system may be announcing its operation, or perhaps the radio is providing background music for your pleasure. The first two examples represent the undesirable aspect of the phenomenon and suggest that study of its qualitative and quantitative aspects is of importance to those concerned with the control and reduction of noise. The third example shows that the process may be put to good use; vibrations of musical instruments, microphone diaphragms, and loudspeakers act as intermediaries in the creation of sound which is, at least for some listeners, the very antithesis of noise. The function of this book is to explain the physical process of interaction and to introduce the reader to various mathematical models and analyses of the behaviour of coupled fluid–structural systems.

Acoustic vibrations in fluids and solid structures essentially involve the propagation of wave motion throughout the supporting media, although explicit recognition of this fact is not always apparent in textbooks on mechanical vibration. Indeed, an emphasis on the work–energy approach to vibration analysis, which is fundamental to many modern computational techniques, tends to obscure the wave nature of the processes under analysis. In dealing with audio-frequency vibrations of systems involving coupling of compressible fluids with plate and shell structures, it is important to possess an appreciation of the "wave view" of vibration. There are three main reasons for this requirement: the first concerns the three-dimensional nature, and often very great extent, of fluid volumes, which effectively rules out the assumption of a limited number of degrees of freedom in describing the vibrational state of the medium; the second reason relates to the description of the interaction of sound waves with structural boundaries of diverse geometric and dynamic form, which is most appropriately framed in terms of the wave field phenomena of reflection, diffraction, and scattering; and the third reason is associated with the fact that frequencies of practical concern are usually far above the fundamental natural frequencies of the structures involved, and discrete modal models are not appropriate because of uncertainties in the modelling of detail and the very large number of degrees of freedom involved. Hence, vibration wave field models, analogous to those used in room acoustics, are more useful and effective.

For these reasons Chapter 1 introduces the reader to a unified mathematical description of temporal and spatial distributions of wave field variables and presents a partly qualitative account of the characteristics of waves in beam, plate, and shell structures. In particular, it shows how the concept of a dispersion relationship between wave speed and frequency forms a basis for categorising the regimes of interaction between waves of various type travelling in contiguous media. The phenomena of natural frequencies and characteristic modes of bounded elastic systems and the related phenomenon of resonance are explained qualitatively in terms of wave reflection and interference; and the roles of outgoing and returning waves in determining the input impedances of distributed elastic systems are illustrated by one-dimensional examples.

In Chapter 2 the mechanics of sound radiation from vibrating surfaces is explained in terms of the distribution of normal surface acceleration. Analyses of sound radiation from planar surfaces by means of far-field evaluation of the Rayleigh integral and in terms of travelling wave Fourier component synthesis are presented in such a way that the equivalence of these dual approaches can be appreciated. The utility of the latter approach is illustrated by application to the evaluation of the contributions to radiated power made by locally applied forces and by reaction forces arising from the presence of local constraints. The chapter closes with brief treatments of sound radiation from orthotropic and sandwich plates and from circular cylindrical shells.

The problem of evaluating the reaction forces applied by a fluid to a vibrating structure is addressed in Chapter 3. The Kirchhoff–Helmholtz integral equation is introduced and the concept of complex acoustic radiation impedance illustrated by some elementary examples. The value of the concept of wave impedance in analysing wave propagation in coupled fluid–structure systems is demonstrated in the case of bounded and unbounded uniform plates; and the effect of fluid loading on plate natural frequencies is discussed. An elementary, one-dimensional example of fluid loading of an elastic structure by an enclosed volume of fluid is presented in order to demonstrate the existence of coupled system modes; this section is essentially a forerunner to a more comprehensive analysis of closed coupled systems presented in Chapter 6. Finally, brief mention is made of the effects of heavy fluid loading on the radiation of plates produced by locally applied forces.

Chapter 4 presents a detailed account of sound transmission through plane partitions of various forms, including single-leaf, double-leaf, and non-homogeneous constructions. A considerable amount of analytical detail is felt to be justified by the apparent lack of a unified, comprehensive treatment of the general problem in other English language textbooks on acoustics, much of the material appearing only in specialised papers and technical reports. The reader is led from elementary analyses of highly idealised systems, through a discussion of the relative importance of resonant and non-resonant transmission mechanisms, to an appreciation of the current state of the art with regard to understanding the behaviour of the more complex structures used in practice. A brief review is presented of the performance of enclosure structures and of stiffened, composite, and non-uniform panels. The final section consists of a fairly detailed treatment of the problem of sound transmission through thin-walled, circular cylindrical shells; unfortunately, this does not extend as far as the presentation of practical formulae because research has not yet produced definitive theoretical results and formulae.

Analysis of the vibrational response of thin-plate and shell structures to incident sound is the subject of Chapter 5. In practice, this topic is of importance in cases where the integrity of structures or attached systems is at risk because of excitation by very intense sound fields produced, for instance, by aircraft engines, rockets, or industrial plant components. The relationship between radiational and response characteristics of structures is strongly emphasized, since this property has considerable experimental significance; it is employed, for instance, in the underwater noise reduction of ships and is now being used in the study of the acoustics of stringed musical instruments. Among other practical applications of the theory are the optimisation of the performance of low-frequency panel absorbers and estimates of the absorbing properties of non-load-bearing structures in buildings.

In many cases of practical concern, structures totally or partially enclose fluid volumes. In such cases, the vibrational behaviour of two coupled energy storage systems is of interest; examples include vehicle cabins and fluid transport ducts. In Chapter 6, theoretical approaches to the analysis of coupled system behaviour are presented, and the conditions under which the coupling drastically modifies the vibrational characteristics of the components from those in the uncoupled state are explained. The final section analyses an elementary case of wave propagation in a waveguide comprising coupled fluid and structural components.

The development of efficient computational procedures for the analysis of the vibrational behaviour of systems described by large numbers of degrees of freedom has naturally led to their application to acoustic problems. Although such analyses are generally limited to the lower end of the audio-frequency range, they can be of considerable value, particularly where fluid-loading effects are strong, for instance, in the case of a sonar transducer arrays in which vibrations of the transducer support structure can significantly affect the array performance. Chapter 7 presents a necessarily brief introduction to such basic techniques as finite element analysis, the details of the modelling and computational techniques being too diverse to allow more than a superficial review to be presented herein. However, I trust that the brevity of the treatment will not discourage the reader from looking more closely into this fast-developing area of the subject, with the aid of the cited references.

F. J. Fahy

1 Waves in Fluids and Solid Structures

1.1 Frequency and Wavenumber

In this book we shall confine our attention largely to audio-frequency vibrations of elastic structures that take the form of thin flat plates, or thin curved shells, of which the thickness dimension is very much less than those defining the extent of the surface. Such structures tend to vibrate in a manner in which the predominant motion occurs in a direction normal to the surface. This characteristic, together with the often substantial extent of the surface in contact with a surrounding fluid, provides a mechanism for displacing and compressing the fluid: hence such structures are able effectively to radiate, and to respond to sound. In order to understand the process of acoustic interaction between solid structures and fluids, it is essential to appreciate the wave nature of the responses of both media to time-dependent disturbances from equilibrium, whether these be transient or continuous. In this chapter we shall take a look at the phenomena of natural frequency and resonance, and the impedance of simple structures, from a wave point of view.

A mechanical wave may be defined as a phenomenon in which a physical quantity (e.g., energy or strain) propagates in a supporting medium, without net transport of the medium. It may be characterised kinematically by the form of relative displacements from their positions of equilibrium of the particles of the supporting medium, that is to say the form of distortion, together with the speed and direction of propagation of this distortion. Wave disturbances in nature rarely occur at a single frequency; however, it is mathematically and conceptually more convenient to study single-frequency characteristics, from which more complex time-dependent behaviour can be synthesised mathematically if required.

Before we consider wave motion in particular types of physical systems, we shall discuss the mathematical representation of relationships between variations in time and space, which are fundamental to the nature of wave motion in general. Simple harmonic variations in time are most conveniently described mathematically by means of a complex exponential representation, of which there are two forms: in one, only positive frequencies are recognised; in the other, which is more appropriate to analytical and numerical frequency analysis techniques, both positive and negative frequencies are considered (Randall, 1977). In this book the former representation will be employed, because it avoids the confusion that can arise between signs associated with variations in time and space.

The basis of the representation is that a simple harmonic variation of a quantity with time, which may be expressed as $g(t) = A\cos(\omega t + \phi)$, where A symbolises amplitude and ϕ symbolises phase, can also be expressed as $g(t) = \mathrm{Re}\{\tilde{B}\exp(j\omega t)\}$, where $\tilde{B}$ is a complex number, say $a + ib$, and Re{ } means real part of. $\tilde{B}$ may be termed the complex amplitude. It will be seen that

$$A = (a^2 + b^2)^{1/2} \qquad a = A\cos\phi, \qquad b = A\sin\phi, \qquad \phi = \arctan(b/a).$$

Hence $g(t)$ may be represented graphically in the complex plane by a rotating vector (phasor) as illustrated in Fig. 1.

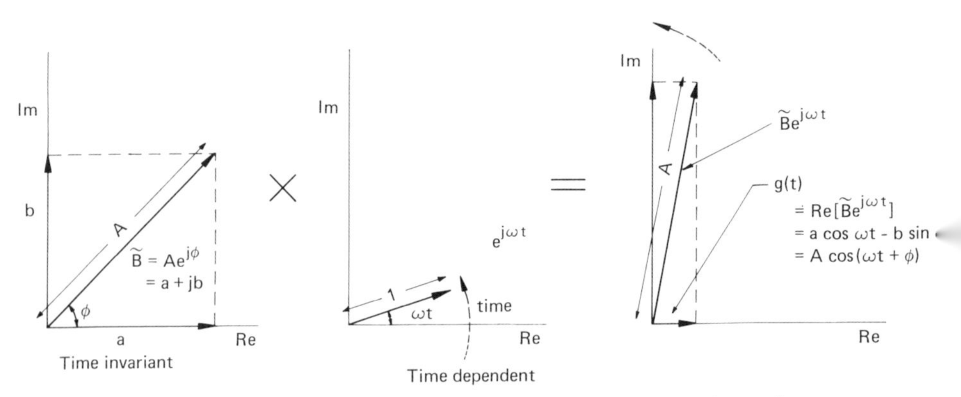

Fig. 1. Complex exponential representation of $g(t) = A\cos(\omega t + \phi)$.

In a wave propagating in only one space dimension x, of a type in which the speed of propagation of a disturbance is independent of its magnitude, a simple harmonic disturbance generated at one point in space will clearly propagate away in a form in which the spatial disturbance pattern, as observed at any one instant of time, is sinusoidal in space. This spatial pattern will travel at a speed c, which is determined by the kinematic form of the disturbance, the properties of the medium, and any external forces on the medium. As the wave progresses, the disturbance at any point in space will vary sinusoidally in time at the same frequency as that of the generator, provided the medium responds linearly to the disturbance. Suppose we represent the disturbance at the point of generation by $g(0, t) = \mathrm{Re}\{\tilde{B} \exp(j\omega t)\}$, as represented by Fig. 1. The phase of the disturbance at a point distance x_1 in the direction of propagation will *lag* the phase at 0 by an angle equal to the product of the circular frequency ω (phase change per unit time) of the generator and the time taken for the disturbance to travel the distance x_1: this time is equal to x_1/c. Hence, $g(x_1, t)$ may be represented by multiplying $\tilde{B} \exp(j\omega t)$ by $\exp(-j\omega x_1/c)$, i.e., $g(x_1, t) = \mathrm{Re}\{\tilde{B} \exp[j(\omega t - \omega x_1/c)]\}$. Thus we see that the quantity $-(\omega/c)$ represents phase change per unit increase of distance, in the same way as ω represents phase change per unit increase of time. Figure 2 illustrates the combined effects of space and time variations. The horizontal component of the rotating phasor represents the disturbance $g(x, t)$, which may or may not physically correspond to displacement in the x direction, depending upon the type of wave.

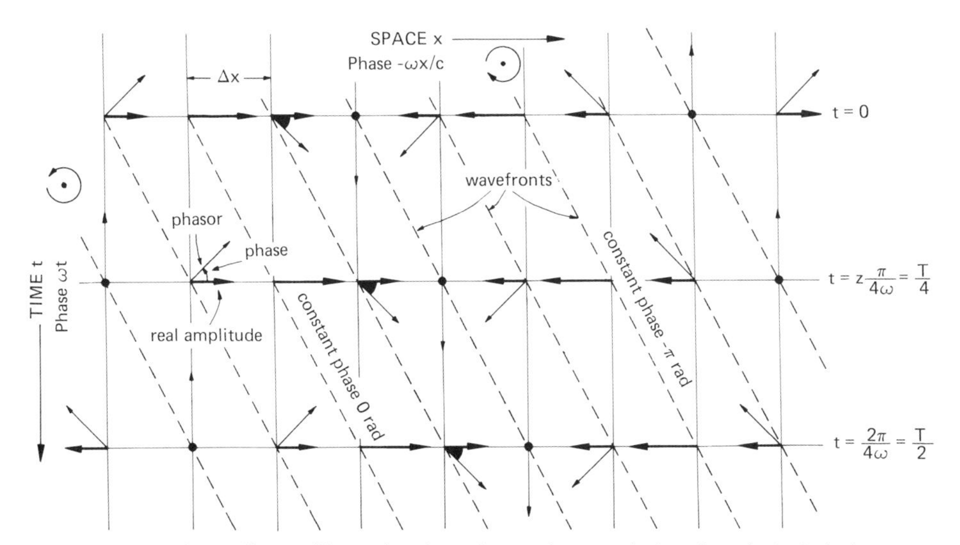

Fig. 2. Phasor diagram illustrating phase changes due to variation of t and x ($\omega\,\Delta x/c$ chosen to equal $\pi/4$ for clarity).

The general complex exponential form of a simple harmonic wave travelling in the positive x direction is seen to be $g^+(x, t) = \mathrm{Re}\{\tilde{B}\exp[j(\omega t - \omega x/c)]\}$. For a wave of the same amplitude travelling in the negative x direction, $g(x, t)$ would take the form $g^-(x, t) = \mathrm{Re}\{\tilde{B}\exp[j(\omega t + \omega x/c)]\}$.

The blocked-in phase angle shown in Fig. 2 illustrates why c is termed the *phase velocity* of the wave, which we shall in future symbolise as c_{ph}: an observer travelling in the direction of wave propagation at this speed sees no change of phase. The spatial period of a simple harmonic wave is commonly described by its wavelength λ. However, the foregoing exposition of the mathematical description of a wave suggests that spatial variations are better described by an associated quantity that represents phase change per unit distance and is equal to (ω/c_{ph}). This is termed the *wavenumber* and is generally symbolised by k. One wavelength clearly corresponds to an x-dependent phase difference of 2π: hence $\omega\lambda/c_{\mathrm{ph}} = k\lambda = 2\pi$. The following relationships should be committed to memory:

$$k = \omega/c_{\mathrm{ph}} = 2\pi/\lambda, \tag{1.1}$$

where k has the dimension of reciprocal length and unit reciprocal metre.

Wavenumber k is actually the magnitude of a vector quantity that indicates the direction of propagation as well as the spatial phase variation. This quality is of vital importance to the mathematical representation of two- and three-dimensional wavefields. The analogy between temporal frequency ω and spatial frequency k is illustrated by Fig. 3. Any form of spatial variation can be Fourier analysed into a spectrum of complex wavenumber components, just as any form of temporal variation (signal) can be analysed into a spectrum of complex frequency components.

Thus far no assumption has been made about the dependence of phase velocity on frequency; Eq. (1.1) indicates that this relationship determines

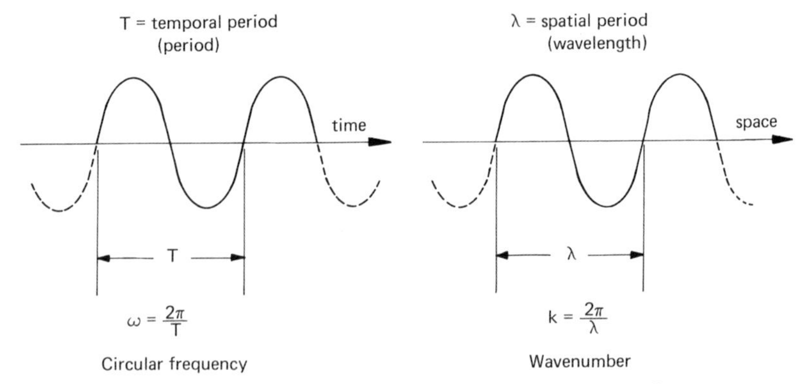

Fig. 3. Analogy between circular frequency and wavenumber.

the frequency dependence of wavenumber. The form of relationship between k and ω is termed the *dispersion* relationship; it is a property of the type of wave, as well as the type of wave-supporting medium. The dispersion relationship indicates, among other things, whether a disturbance generated by a process that is not simple harmonic in time will propagate through a medium unaltered in its basic spatial form (even if it is reduced in amplitude with distance traveled on account of two- or three-dimensional spreading of the wave). Only if the relationship between k and ω is linear will an arbitrary spatial form of disturbance not be subject to change as it propagates: such change is known as dispersion. Figure 4, which shows the progress of three different frequency components of a dispersive wave, illustrates the dispersion process. Any disturbance of finite time duration contains an infinity of frequency components and, if transported by a dispersive wave process, will be distorted as it propagates.

We shall see later that dispersion curves can be of great help in understanding the interaction between waves in coupled media, because where dispersion curves for two types of wave intersect, they have common frequency and wavenumber, and therefore common wavelength and phase speed. A dispersion curve can also tell us something about the speed at which energy is transported by a wave; this speed is called the group speed c_g. It is obtained from the dispersion curve by the relationship

$$c_g = \partial\omega/\partial k, \tag{1.2}$$

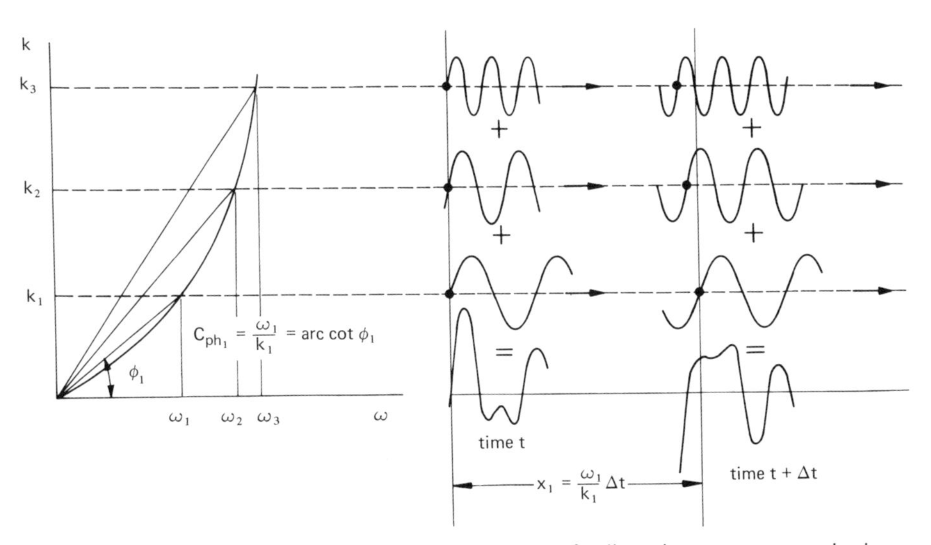

Fig. 4. Combination of three frequency components of a dispersive wave at successive instants of time.

which is the inverse slope of the curve. It is clear from Eqs. (1.1) and (1.2) that the phase and group speeds are only equal in non-dispersive waves, for which c_{ph} is independent of ω. Knowledge of group speed is useful in vibration analyses involving the consideration of wave energy flow, such as the analysis of the absorption of bending waves in plates by the boundaries (Heckl, 1962a).

1.2 Sound Waves in Fluids

The wave equation governing the propagation of *small* disturbances through a homogeneous, inviscid, compressible fluid may be written in rectangular Cartesian coordinates (x, y, z) in terms of the variation of pressure from the equilibrium pressure, the "acoustic pressure" p, as

$$\frac{\partial^2 p}{\partial x^2} + \frac{\partial^2 p}{\partial y^2} + \frac{\partial^2 p}{\partial z^2} = \frac{1}{c^2}\frac{\partial^2 p}{\partial t^2}. \tag{1.3}$$

The same equation governs variations in fluid density and temperature, and in particle displacement, velocity, and acceleration: c is the frequency-independent speed of sound given by $c^2 = (\gamma P_0/\rho_0)$, where P_0 is the mean fluid pressure and ρ_0 the mean density. The wave equation is derived from the linearised* forms of the continuity equation

$$\frac{\partial \rho}{\partial t} + \rho_0\left(\frac{\partial u}{\partial x} + \frac{\partial v}{\partial y} + \frac{\partial w}{\partial z}\right) = 0, \tag{1.4}$$

and momentum equations

$$\frac{\partial p}{\partial x} + \rho_0 \frac{\partial u}{\partial t} = 0, \tag{1.5a}$$

$$\frac{\partial p}{\partial y} + \rho_0 \frac{\partial v}{\partial t} = 0, \tag{1.5b}$$

$$\frac{\partial p}{\partial z} + \rho_0 \frac{\partial w}{\partial t} = 0, \tag{1.5c}$$

in which u, v, and w are the particle velocities in the x, y, and z directions, respectively: an adiabatic process is assumed.

The general solution of the wave equation may be expressed in terms of various coordinate systems; it is separable in all common forms such as

* Products of small quantities neglected.

rectangular Cartesian, cylindrical, spherical, and elliptical. In studying the interaction between plane structures and fluids, the most useful form of the equation is a two-dimensional form involving only variations in two orthogonal directions, in association with simple harmonic time dependence:

$$\frac{\partial^2 p}{\partial x^2} + \frac{\partial^2 p}{\partial y^2} = -\left(\frac{\omega}{c}\right)^2 p = -k^2 p. \tag{1.6}$$

The propagation of a plane wave in a two-dimensional space may be expressed as

$$p(x, z, t) = \tilde{p} \exp(-jk_x x - jk_y y) \exp(j\omega t). \tag{1.7}$$

Substitution of this expression into (1.6) yields the following wavenumber relationship:

$$k^2 = k_x^2 + k_y^2. \tag{1.8}$$

The interpretation of this equation is that only certain combinations of k_x and k_y will satisfy the wave equation at any specific frequency. The direction of wave propagation is indicated by the wave vector **k** as shown in Fig. 5. The locus of the wave vector **k** for fixed frequency ω and real values of k_x and k_y is clearly seen to be a circle of radius k. It is opportune at this point to warn the reader that only wave vectors, not wavelengths, should be directionally resolved.

The value of the linearised momentum equations (1.5) in the analysis of structure–fluid interaction is that they connect pressure gradient and particle

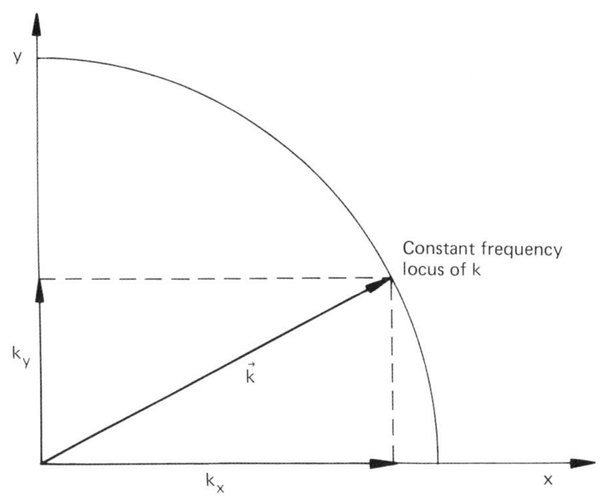

Fig. 5. Two-dimensional wave vector and components. Arrows over symbols indicate vectors.

acceleration; through the assumption of simple harmonic motion this becomes a relation between pressure gradient and particle velocity. Hence the transverse normal vibration velocity of a surface determines the value at the surface of the pressure gradient in the direction normal to the surface; e.g., from (1.5b), $(\partial p/\partial y)_{y=0} = -j\omega\rho_0(v)_{y=0}$, where the surface is assumed to lie in the x,z plane. This relationship is frequently used as a boundary condition in the analysis of sound radiation and scattering by vibrating surfaces. Assumption of a plane wave field yields, from (1.7),

$$jk_y(p)_{y=0} = j\omega\rho_0(v)_{y=0}$$

or

$$(p/v)_{y=0} = \omega\rho_0/k_y. \tag{1.9}$$

This ratio has the form of a specific acoustic impedance, which is defined to be the ratio of the complex amplitudes of pressure and particle velocity at a single frequency.

One of the applications of the impedance concept is in the evaluation of sound power radiated into a fluid by a vibrating surface. The time-averaged sound power radiated per unit area is given formally by

$$\bar{P} = \frac{1}{T}\int_0^T p(t)v(t)\,dt.$$

For single-frequency normal surface vibration of the form $v(t) = \tilde{v}\exp(j\omega t)$, which produces a local acoustic pressure $p(t) = \tilde{p}\exp(j\omega t)$, we may write the normal specific acoustic impedance of the fluid at the surface as $\tilde{z} = \tilde{p}/\tilde{v}$. Hence

$$\bar{P} = \tfrac{1}{2}\operatorname{Re}(\tilde{p}\tilde{v}*) = \tfrac{1}{2}|\tilde{v}|^2\operatorname{Re}(\tilde{z}) = \tfrac{1}{2}|\tilde{p}|^2\operatorname{Re}(1/\tilde{z}), \tag{1.10}$$

where $*$ indicates complex conjugate and $|\ |$ indicates magnitude (real amplitude). This equation is valid for all types of acoustic field. For the special case of plane-wave radiation assumed above, $\tilde{z}$ is given by (1.9): later we shall see that k_y is not always real!

1.3 Longitudinal Waves in Solids

In pure longitudinal wave motion the direction of particle displacement is purely in the direction of wave propagation; such waves can propagate in large volumes of solids. Two parallel planes in an undisturbed solid elastic medium, which are separated by a small distance δx, may be moved by dif-

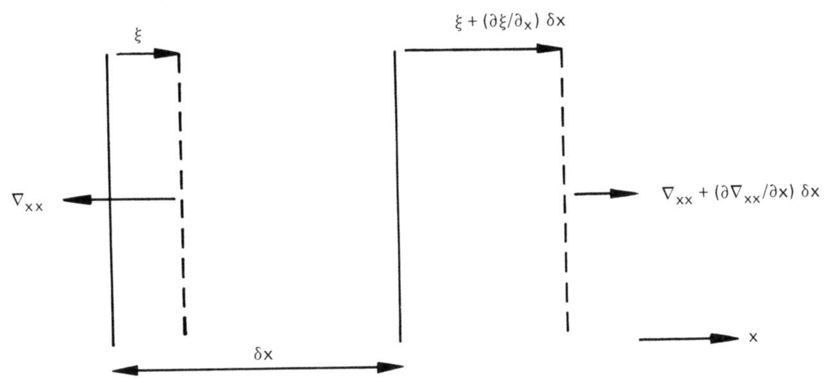

Fig. 6. Displacements from equilibrium and stresses in a pure longitudinal wave.

ferent amounts during the passage of a longitudinal wave, as illustrated in Fig. 6. Hence the element may undergo a strain ε_{xx} given by

$$\varepsilon_{xx} = \partial\xi/\partial x. \tag{1.11}$$

The longitudinal stress σ_{xx} is, according to Hooke's law, proportional to ε_{xx}. However, the constant of proportionality is not, as at first might be thought, the Young's modulus E of the material: Young's modulus is defined as the ratio of longitudinal stress σ_{xx} to longitudinal strain ε_{xx} in a *thin* bar under longitudinal tension (*ut tensio sic vis*). In this case the sides of the rod are under no constraint and, when longitudinal tension is applied, they move toward the axis of the rod, producing lateral strains ε_{yy} and ε_{zz}; of course, the lateral stresses σ_{yy} and σ_{zz} remain zero because no lateral constraints are applied. This phenomenon is termed *Poisson contraction*, Poisson's ratio being defined as the ratio of the magnitudes of the lateral strain to the longitudinal strain: $\nu = -\varepsilon_{yy}/\varepsilon_{xx} = -\varepsilon_{zz}/\varepsilon_{xx}$. In practice, ν varies between about 0.25 for glass and steel and 0.5 for virtually incompressible materials such as rubber.

When a one-dimensional pure longitudinal wave propagates in a solid medium, there can be no lateral strain, because all material elements move purely in the direction of propagation (Fig. 6). This constraint, applied mutually by adjacent elements, creates lateral direct stresses in the same way a tightly fitting steel tube placed around a wooden rod would create lateral compression stresses if a compressive longitudinal force were applied to the wood along its axis. In these cases, the ratio $\sigma_{xx}/\varepsilon_{xx}$ is not equal to E. Complete analysis of the behaviour of a uniform elastic medium shows that the ratio of longitudinal stress to longitudinal strain when lateral strain is prevented, is (Cremer *et al.*, 1973)

$$\sigma_{xx} = B\,\partial\xi/\partial x, \tag{1.12}$$

where $B = E(1 - \nu)/(1 + \nu)(1 - 2\nu)$. The resulting equation of motion of an element is

$$(\rho A\,\delta x)\,\partial^2\xi/\partial t^2 = [\sigma_{xx} + (\partial\sigma_{xx}/\partial x)\,\delta x - \sigma_{xx}]A = (\delta\sigma_{xx}/\partial x)\,\delta x\,A, \quad (1.13)$$

where ρ is the material mean density. Equations (1.12) and (1.13) can be combined into the wave equation

$$\partial^2\xi/\partial x^2 = (\rho/B)\,\partial^2\xi/\partial t^2. \quad (1.14)$$

The general solution of Eq. (1.14), which is of the same form as the one-dimensional acoustic wave equation, shows that the phase speed c_l is

$$c_l = (B/\rho)^{1/2}, \quad (1.15)$$

which is independent of frequency; these waves are therefore non-dispersive. A simple harmonic wave can therefore be expressed as $\xi^+(x,t) = \tilde{A}\exp j(\omega t - k_l x)$, where $k_l = \omega/c_l$. Because $B > E$, $c_l \gg (E/\rho)^{1/2}$: note that if $\nu = 0.5$, both B and c_l are infinite—no wonder solid rubber mats of large area are useless as vibration isolators!

1.4 Quasi-Longitudinal Waves in Solids

As we have noted, only in volumes of solid that extend in all directions to distances large compared with a longitudinal wavelength can pure longitudinal waves exist. A form of longitudinal wave can propagate along a solid bar, and another form can propagate in the plane of a flat plate. Because such structures have one or more outer surfaces free from constraints, the presence of longitudinal stress will produce associated lateral strains through the Poisson contraction phenomenon: pure longitudinal wave motion cannot therefore occur, and the term "quasi-longitudinal" is used. Elasticity theory shows that the relation between longitudinal stress and longitudinal strain in a thin flat plate is

$$\sigma_{xx} = [E/(1 - \nu^2)]\,\partial\xi/\partial x. \quad (1.16)$$

Hence the quasi-longitudinal wave equation for a plate is

$$\partial^2\xi/\partial x^2 = [\rho(1 - \nu^2)/E]\,\partial^2\xi/\partial t^2, \quad (1.17)$$

and the corresponding, frequency-independent phase speed is

$$c_l' = [E/\rho(1 - \nu^2)]^{1/2}. \quad (1.18)$$

The ratio of longitudinal stress to longitudinal strain in a bar is defined to be E. Hence the quasi-longitudinal wave equation is

$$\partial^2 \xi / \partial x^2 = (\rho / E)\, \partial^2 \xi / \partial t^2, \tag{1.19}$$

and the phase speed is

$$c_l'' = (E/\rho)^{1/2}. \tag{1.20}$$

The kinematic form of a quasi-longitudinal wave is shown in Fig. 7. Equations (1.16)–(1.20) are valid only for frequency ranges in which the quasi-longitudinal wavelength greatly exceeds the cross-sectional dimensions of the

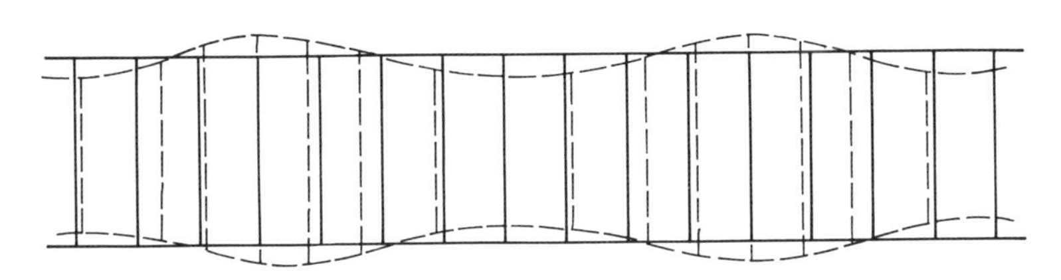

(a) Quasi-longitudinal wave
(transverse displacements exaggerated)

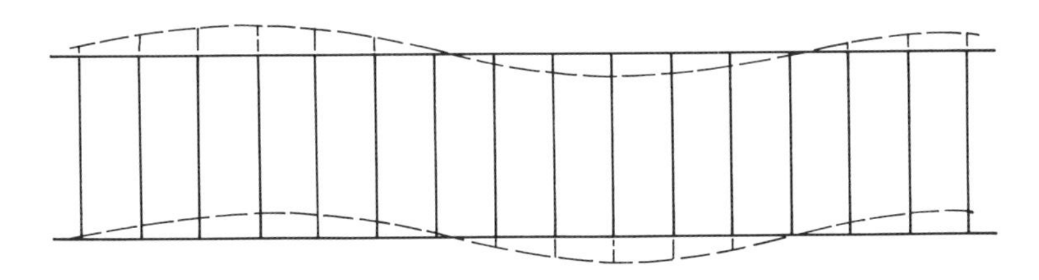

(b) Transverse shear wave

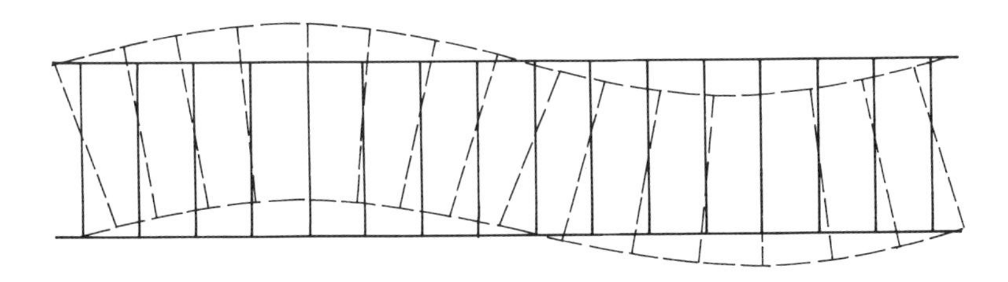

(c) Flexural (bending) wave

Fig. 7. Deformation patterns of various types of wave.

TABLE I

Material Properties[a] and Phase Speeds

Material	Young's modulus E (Nm^{-2})	Density ρ (kgm^{-3})	Poisson's ratio (ν)	c_l (ms^{-1})	c_l' (ms^{-1})	c_l'' (ms^{-1})	c_s (ms^{-1})
Steel	2.0×10^{11}	7.8×10^3	0.28	5900	5270	5060	3160
Aluminium	7.1×10^{10}	2.7×10^3	0.33	6240	5434	5130	3145
Brass	10.0×10^{10}	8.5×10^3	0.36	4450	3677	3430	2080
Copper	12.5×10^{10}	8.9×10^3	0.35	4750	4000	3750	2280
Glass	6.0×10^{10}	2.4×10^3	0.24	5430	5151	5000	3175
Concrete							
light	3.8×10^9	1.3×10^3				1700	
dense	2.6×10^{10}	2.3×10^3				3360	
porous	2.0×10^9	6.0×10^2				.1820	
Rubber							
hard	2.3×10^9	1.1×10^3	0.4	2120	1582	1450	867
soft	5.0×10^6	9.5×10^2	0.5			70	40
Brick	1.6×10^{10}	$1.9\text{–}2.2 \times 10^3$				2800	
Sand, dry	3.0×10^7	1.5×10^3				140	
Plaster	7.0×10^9	1.2×10^3				2420	
Chipboard[b]	4.6×10^9	6.5×10^2				2660	
Perspex[c]	5.6×10^9	1.2×10^3	0.4	3162	2357	2160	1291
Plywood[b]	5.4×10^9	6.0×10^2				3000	
Cork	—	$1.2\text{–}2.4 \times 10^2$				430	
Asbestos cement	2.8×10^{10}	2.0×10^3				3700	

[a] Mean values from various sources.
[b] Temperature sensitive.
[c] Greatly variable from specimen to specimen.

structure. This criterion will be satisfied by all practical structures and frequencies considered in this book: for example, at 10 kHz the wavelengths in bars of steel and concrete are approximately 500 and 350 mm, respectively. Values of E, ρ, ν, c_l, c_l', and c_l'' for common materials are presented in Table I.

1.5 Transverse Shear Waves in Solids

Solids, unlike fluids, can resist static shear deformation. Fluids in motion can generate shear stresses associated with laminar flow velocity gradients, but because the agent is molecular viscosity, which generates a dissipative, frictional type of force, shear waves generated in fluids decay very rapidly with distance and are therefore generally of little practical concern. The shear modulus G of a solid is defined as the ratio of the shear stress τ to the shear strain γ: shear deformation is represented in Fig. 8. The shear modulus G is related to Young's modulus E by $G = E/2(1 + \nu)$.

In *static* equilibrium under pure shear, the shear stresses acting in a particular direction on the opposite faces of a rectangular element are equal and

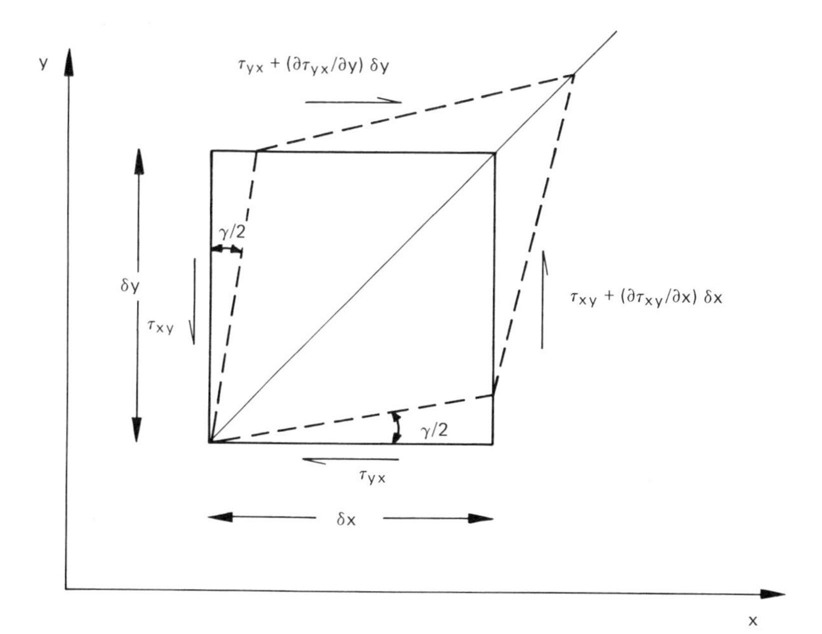

Fig. 8. Pure shear strain and associated shear stresses.

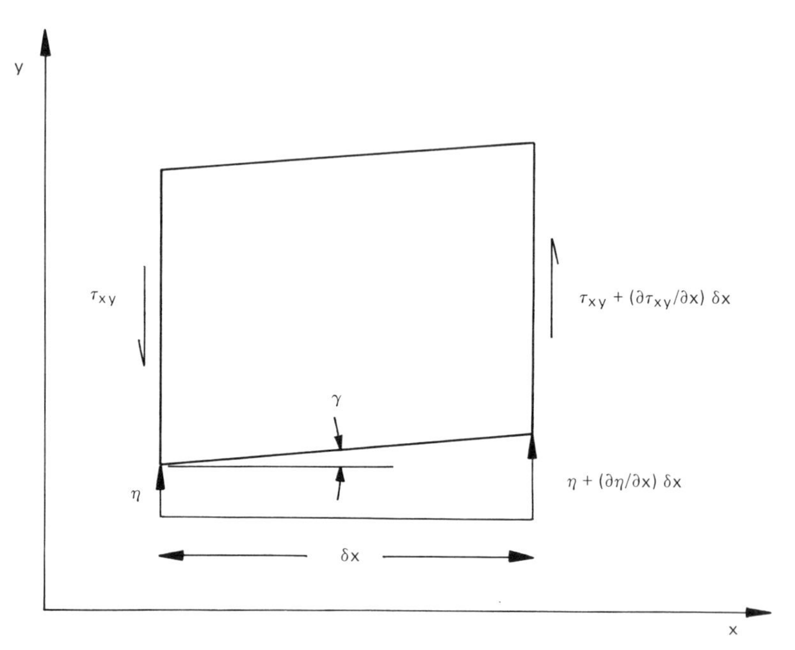

Fig. 9. Displacements, shear strain, and transverse shear stresses in pure transverse deformation.

opposite, because the direct stresses on these faces are zero, i.e., $\partial\tau_{xy}/\partial x = \partial\tau_{yx}/\partial y = 0$ and $\tau_{xy} = \tau_{yx}$ in Fig. 8. In a dynamic situation this is not necessarily so; the difference in the vertical shear stresses causes the vertical acceleration $\partial^2\eta/\partial t^2$ of the element (Fig. 9). Consideration of the other equilibrium equations indicates that rotation of the element must also take place. The equation of transverse motion of an element having unit thickness in the z dimension is

$$\rho\,\delta x\,\delta y\,\partial^2\eta/\partial t^2 = (\partial\tau_{xy}/\partial x)\,\delta x\,\delta y, \tag{1.21}$$

where η is the transverse displacement, and the stress–strain relationship is

$$\tau_{xy} = G\gamma = G\,\partial\eta/\partial x. \tag{1.22}$$

Hence the wave equation is

$$\partial^2\eta/\partial x^2 = (\rho/G)^{1/2}\,\partial^2\eta/\partial t^2. \tag{1.23}$$

The kinematic form of a shear wave is shown in Fig. 7. This takes the same form as the acoustic and longitudinal wave equations. The frequency-independent phase speed is

$$c_s = (G/\rho)^{1/2} = [E/2\rho(1 + \nu)]^{1/2} \tag{1.24}$$

The transverse shear wave speed is seen to be smaller than the quasi-longitudinal bar wave speed:

$$c_s/c_l = [1/2(1 + \nu)]^{1/2}. \quad (1.25)$$

This ratio is about 0.6 for many structural materials. Such shear waves can propagate in large volumes of solids and in the plane of extended flat plates of which the free surfaces have little influence, since in-plane direct stresses are negligible; hence the shear wave speed is very similar to that in a large volume. In general, in-plane shear waves are not easy to generate in plates by means of applied forces; however, they sometimes play a vital role in the process of vibration transmission through, and reflection from, line discontinuities such as *L*, *T*, and + junctions between structures such as those of buildings and ships (Kihlman, 1967a,b).

Torsional waves in solid bars are shear waves. The governing wave equation is

$$\partial^2\theta/\partial x^2 = (I_p/GJ)\,\partial^2\theta/\partial t^2, \quad (1.26)$$

where θ is the torsional displacement, I_p the polar moment of inertia per unit length of the bar about its longitudinal axis, and GJ the torsional stiffness of the bar, which is a function of the shape of the bar. The frequency-independent phase speed is $c_t = (GJ/I_p)^{1/2}$. Table II lists expressions for the torsional stiffness of solid bars of rectangular and circular sections. Calculation of the torsional stiffness of structures such as *I*-, *L*-, and channel-section beams is more complex, since such phenomena such as warping and bending-shear coupling must be taken into account (Argyris, 1954; Muller, 1983). The effects of constraint applied to such beams by connected structures such as plates have a substantial influence on their effective torsional stiffness: ship and aircraft structures commonly incorporate such beam-plate construction.

Shear-wave motion is of great importance in the vibration behaviour of sandwich and honeycomb plate structures, which have thin cover plates

TABLE II

Torsional Stiffness of Solid Bars

	h/b	J/b^3h
Rectangular ($b \times h$)	1	0.141
	2	0.230
	3	0.263
	4	0.283
	5	0.293
	10	0.312
Circular (radius a)	$J = \pi a^4/2$	

separated by relatively thick core layers; such cores are often designed to have low shear stiffness. In a mid-frequency range between low frequencies, at which overall section bending stiffness is dominant, and high frequencies, at which cover plate bending stiffness dominates, the wave speed is controlled primarily by the core shear stiffness. Advantage can be taken of this behaviour to design for optimum sound radiation and transmission characteristics. Analysis of wave motion in such structures is discussed in Chapter 4.

1.6 Bending Waves in Bars

Of the various types of wave that can propagate in bars, beams, and plates, bending (or flexural) waves are of the greatest significance in the process of structure–fluid interaction at audio frequencies. The reasons are that bending waves involve substantial displacements in a direction transverse to the direction of propagation, which can effectively disturb an adjacent fluid; and that the transverse impedance of structures carrying bending waves can be of similar magnitude to that of sound waves in the adjacent fluid, thereby facilitating energy exchange between the two media.

In spite of the fact that the transverse displacements of elements in structures carrying bending waves are far greater than the in-plane displacements of these elements, the bending stresses are essentially associated with the latter. Consequently, flexural waves can neither be classified as longitudinal nor as transverse waves. In pure bending deformation of a bar, cross sections are translated in a direction transverse to the bar axis, and are rotated relative to their equilibrium planes, as shown in Fig. 10. The transverse displacement η and rotation β are related approximately by

$$\beta = \partial\eta/\partial x. \tag{1.27}$$

The primary assumption of pure bending theory is that plane cross sections remain plane during bending deformation of the element. Let us examine this assumption as it relates to bending wave motion. Figure 11 shows an element of a bar undergoing pure bending deformation under the action of purely transverse forces: note that θ is assumed to be very small. The longitudinal strain ε_r, and hence σ_r, vary linearly with r according to our assumption. Because there is no applied force component in the direction of the bar axis, we can write

$$\int_{-r_0}^{h-r_0} \sigma(r)w(r)\,dr = 0,$$

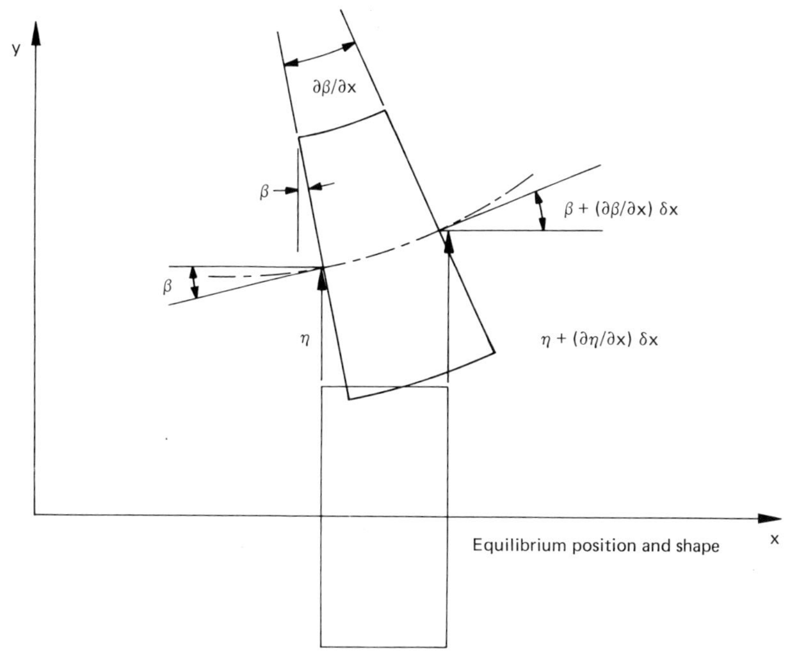

Fig. 10. Displacements and deformation of a beam element in bending.

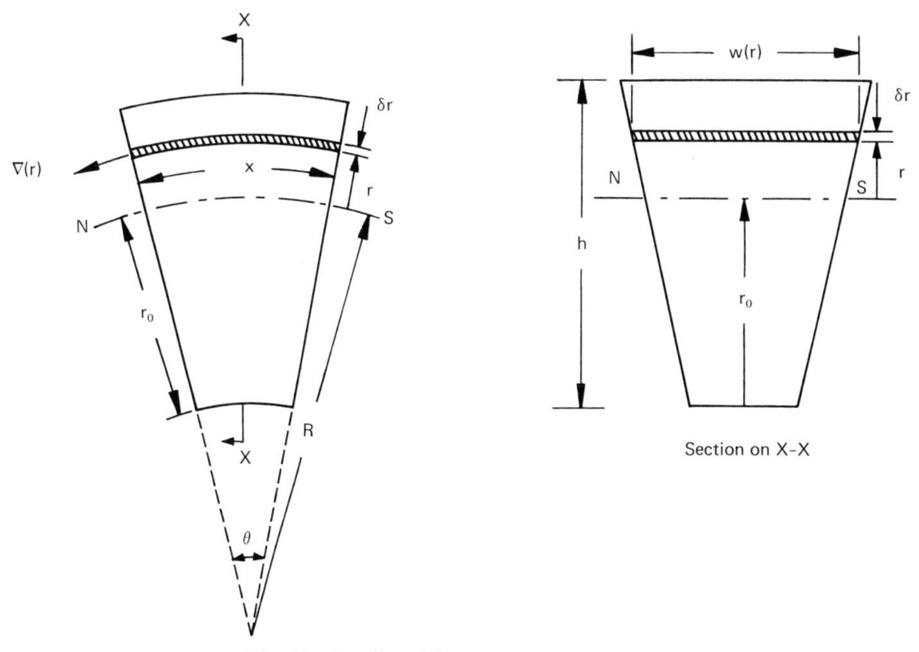

Fig. 11. Bending deformation of a bar element.

where $\sigma(r)$ is the longitudinal direct stress. This equation immediately indicates that $\sigma(r)$ must vary over both positive and negative values; hence $\sigma(r)$ is zero at some unique value of r. We denote the surface on which this occurs, called the "neutral surface," by $r = r_0$. The longitudinal strain on the deformed neutral surface is clearly zero, and the element retains its original length δx along this arc. Hence

$$\varepsilon(r) = [(R + r)\theta - \delta x]/\delta x. \tag{1.28}$$

We now assume that the relationship between $\varepsilon(r)$ and $\sigma(r)$ is the same as in longitudinal loading of a thin bar, i.e., zero lateral constraint: $\sigma(r) = E\varepsilon(r)$. The local radius of curvature R is related to θ and, from Fig. 10, to the slope and displacement of the bar axis, by

$$1/R = \theta/\partial x = -\partial\beta/\partial x = -\partial^2\eta/\partial x^2. \tag{1.29}$$

Hence

$$\sigma(r) = E\varepsilon(r) = -Er\,\partial^2\eta/\partial x^2. \tag{1.30}$$

In general, the bar curvature will vary with x, and so then will $\sigma(r)$. In a static situation the variation of $\sigma(r)$ in the axial direction must be balanced by other stresses so that the element is in longitudinal equilibrium; the corresponding force system on a section of bar above $r = r'$ is shown in Fig. 12.

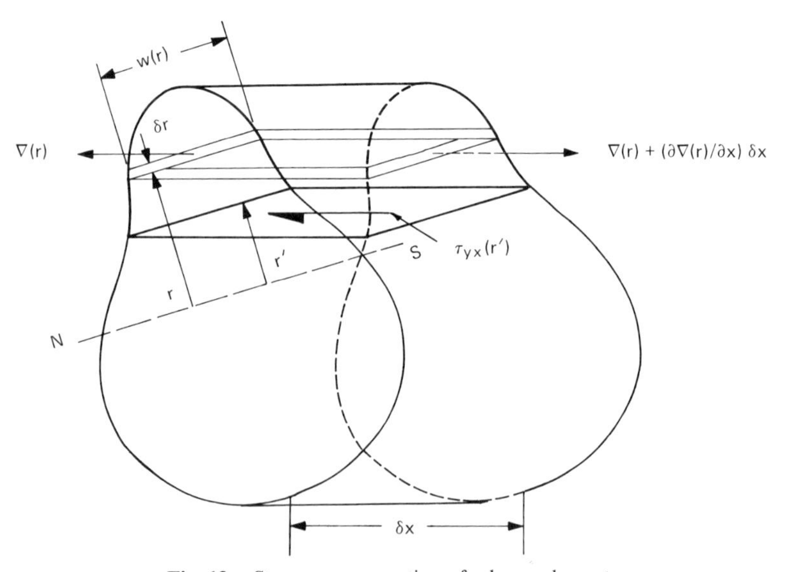

Fig. 12. Stresses on a portion of a beam element.

The equilibrium equation is

$$\tau_{yx}(r')w(r') = \int_{r'}^{h-r_0} \frac{\partial\sigma(r)}{\partial x} w(r)\,dr. \tag{1.31}$$

From Eqs. (1.30) and (1.31)

$$\tau_{yx}(r') = -\frac{E}{w(r')}\frac{\partial^3\eta}{\partial x^3}\int_{r'}^{h-r_0} rw(r)\,dr. \tag{1.32}$$

The integral can only be evaluated for specific variations of the bar width $w(r)$ with r. However, for a bar of uniform width, $w(r) = w$, and $r_0 = h/2$:

$$\tau_{yx}(r') = -\frac{E}{2}\frac{\partial^3\eta}{\partial x^3}\left[\left(\frac{h}{2}\right)^2 - (r')^2\right]. \tag{1.33}$$

This parabolic relationship only holds for bars of uniform width: however, the maximum shear stress occurs on the neutral surface $r' = 0$ in all cases.

It is vital to appreciate the role of this shear stress in opposing variation in the axial direction of longitudinal direct stresses in the beam, because where its action is destroyed, as in multi-laminated beams in which the adhesive fails, or where it can create large shear distortions because of a low shear modulus, as in the core of a sandwich structure, the fundamental assumption that plane sections remain plane must be invalid. The horizontal shear stress τ_{yx} is complemented by a vertical shear stress τ_{xy} of equal magnitude, which, of course, has the same distribution over the depth of the bar (Fig. 13). The

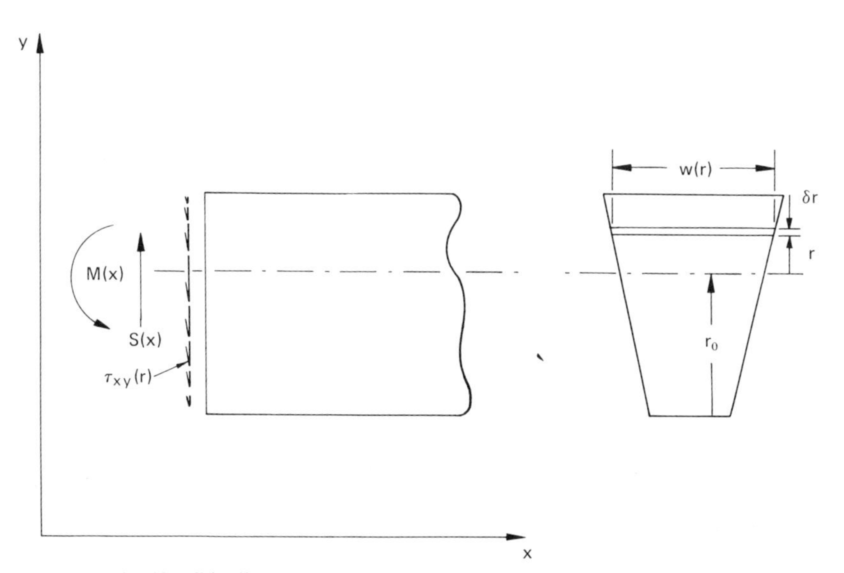

Fig. 13. Distribution of transverse shear stress in a beam in bending.

total vertical elastic shear force acting on any cross section is given by

$$S(x) = -\int_{-r_0}^{h-r_0} \tau_{yx} w(r)\, dr = E\frac{\partial^3 \eta}{\partial x^3}\int_{-r_0}^{h-r_0}\left[\frac{1}{w(r')}\int_{r'}^{h-r_0} rw(r)\, dr\right] w(r')\, dr'.$$

Integration of this equation by parts yields

$$S(x) = EI\frac{\partial^3 n}{\partial x^3}, \qquad \text{where} \qquad I = \int_{-r_0}^{h-r_0} w(r) r^2\, dr; \tag{1.34}$$

and I is defined to be the second moment of area of the cross section about the traverse axis in the neutral plane.

The bending moment acting on a section is reacted by the axial direct stresses $\sigma(r)$ acting about the neutral axis, and is given by

$$M(x) = -EI\, \partial^2\eta/\partial x^2,$$

where positive M acts to produce negative curvature. (Prove.)

The presence of shear stresses within the bar naturally produces shear distortion of the cross section, and it can be seen from Eq. (1.33) that the shear strain must vary with the distance r from the neutral axis and must be zero at the upper and lower surfaces. This form of deformation is incompatible with the assumption that plane sections remain plane, but in many cases the shear distortion of homogeneous bars and beams is rather small, and the contribution to transverse displacement η of the beam is also rather small. A rough assessment of the relative contributions to the transverse shear displacement can be obtained by considering the encastré beam, Fig. 14. For simplicity we assume that the vertical shear stress is uniform and equal to F/A. The vertical deflection of the tip of the beam due to shear is thus approximately equal to Fl/GA, where G is the shear modulus. Simple beam theory gives the bending deflection as $Fl^3/3EI$. Hence the ratio of the

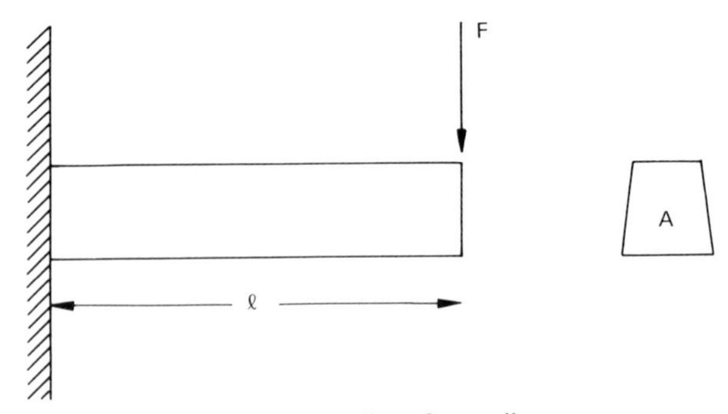

Fig. 14. Tip loading of a cantilever.

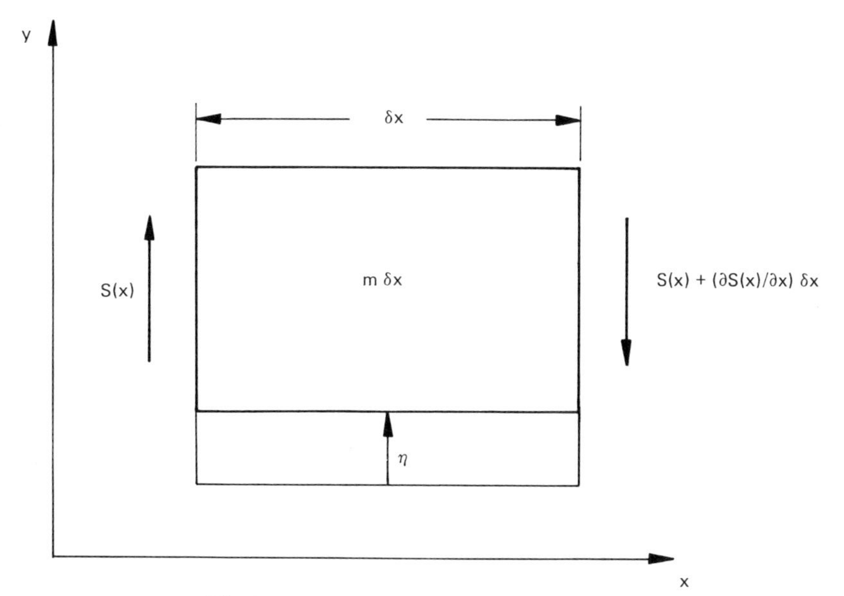

Fig. 15. Transverse forces on a beam element.

shear to bending deflection is approximately equal to $3EI/GAl^2$. For a rectangular section beam of depth h and width w, $I = wh^3/12$, and $A = wh$. Hence the ratio is approximately equal to $\frac{1}{2}(1 + \nu)(h/l)^2$. Thus shear deflections are significant if opposing transverse forces act on the beam at separation distances that are small compared with the depth of the beam. (Why doesn't this criterion apply to internal shear forces?)

The equation of transverse motion of an element of bar can be derived by reference to Fig. 15:

$$\partial S/\partial x = -m\,\partial^2\eta/\partial t^2, \tag{1.35}$$

where m is the mass per unit length of the bar.* The rate of change of S with x comes from Eq. (1.34); Eq. (1.35) can therefore be written

$$EI\,\partial^4\eta/\partial x^4 = -m\,\partial^2\eta/\partial t^2. \tag{1.36}$$

This is the bar wave equation, which is valid provided that the shear contribution to transverse displacement is negligible.

* The symbol m is used to represent mass per unit length or mass per unit area, as appropriate.

This equation differs radically from those governing all the previously considered forms of wave motion in that the spatial derivative is of fourth, and not second, order: the reason is that the bending wave is a hybrid between shear and longitudinal waves. Hence the phase speed of free bending waves cannot be deduced by inspection. Substitution of the complex exponential expression for a simple harmonic progressive wave $\eta(x,t) = \tilde{\eta}\exp[j(\omega t - kx)]$ yields

$$EIk^4 = \omega^2 m.$$

Hence $k = \pm j(\omega^2 m/EI)^{1/4}$ and $\pm(\omega^2 m/EI)^{1/4}$. The complete solution is thus

$$\eta(x,t) = [\tilde{A}\exp(-jk_b x) + \tilde{B}\exp(jk_b x) + \tilde{C}\exp(-k_b x) + \tilde{D}\exp(k_b x)]\exp(j\omega t), \tag{1.37}$$

where $k_b = (\omega^2 m/EI)^{1/4}$. The first two terms on the rhs of Eq. (1.37) represent waves propagating in the positive and negative x directions at a phase speed $c_b = \omega/k_b = \omega^{1/2}(EI/m)^{1/4}$. The second two terms represent non-propagating fields, the amplitudes of which decay exponentially with distance; their phase velocities are imaginary and they do not transport energy. They cannot strictly be called "waves."

The phase velocity c_b is seen to be proportional to $\omega^{1/2}$; bending waves in bars are therefore dispersive. The group speed $(c_b)_g = \partial\omega/\partial k = 2c_b$. (Prove.) The fact that the phase speed of bending waves varies with frequency has a profound influence on the phenomenon of acoustic coupling between structures and fluids, as will become evident later. The dispersive nature of bending waves also produces natural bending vibration frequencies of bars that are not harmonically related, in contrast to the harmonic progression associated with non-dispersive waves.

We need to estimate the contribution of shear distortion to transverse displacement in order to evaluate the range of validity of Eq. (1.36). The free bending wavelength is equal to $2\pi/k_b$. Hence points of maximum transverse displacement and acceleration are separated from points of zero displacement by a distance $\pi/2k_b$. In the d'Alembert view of dynamic equilibrium of bending waves, "inertia forces" may be considered to act in opposition to elastic shear forces. Maximum inertia forces act at points of maximum acceleration, and maximum shear forces act at points of zero displacement. (Prove this for yourself.) We may therefore replace the length l used in Fig. 14 by $\pi/2k_b$. The ratio of the contributions to transverse displacement of shear and bending is therefore given approximately by $\frac{1}{2}(2k_b h/\pi)^2$, where h is the bar depth. An appropriate condition for negligible shear contribution is $k_b h < 1$, or $\lambda_b > 6h$. For a bar of rectangular cross section this condition becomes $f < 4.6 \times 10^{-2}(c_l''/h)$ Hz: for I-sections the frequency is approximately 40%

lower. For example, this frequency for a 150-mm-deep rectangular section concrete bar is approximately 900 Hz. Above this frequency an extra, shear-controlled, term must be included in the equation of motion. The details can be found in more advanced books on the subject (Cremer *et al.*, 1973). The physical implication of the contribution of the shear distortion term is that at sufficiently high frequencies "bending" waves become very like pure transverse shear waves; the phase speed asymptotes to c_s, and does not go to infinity as the expression for c_b would suggest.

1.7 Bending Waves in Thin Plates

As far as the propagation of one-dimensional plane bending waves is concerned, a uniform flat plate of infinite extent is no different from a bar, except that the relationship between the longitudinal strains and longitudinal stresses [Eq. (1.30)] must be modified to that corresponding to Eq. (1.16), to allow for the lateral constraint that is absent in a finite width bar because of its free sides. Hence Eq. (1.36) becomes

$$\frac{EI}{(1-\nu^2)}\frac{\partial^4 \eta}{\partial x^4} = -m\frac{\partial^2 \eta}{\partial t^2}, \tag{1.38}$$

where m is now the mass per unit area of the plate and I the second moment of area per unit width: $I = h^3/12$ for a plate of thickness h. We may replace $Eh^3/12(1-\nu^2)$ by D, which may be termed the bending stiffness of the plate because the bending moment per unit width is given by $M = -D\,\partial^2\eta/\partial x^2$. The free-wave solution is the same as Eq. (1.37) with $k_b = (\omega^2 m/D)^{1/4}$. The phase speed $c_b = \omega^{1/2}(D/m)^{1/4}$. The dispersive character of plate bending waves may be observed by listening to the sound of a stone cast onto a sheet of ice on a pond.

Equation (1.38) is not, however, sufficient to describe two-dimensional bending wavefields in a plate in which propagation in the x and z directions may occur simultaneously. Derivation of the complete classical bending-wave equation, in which shear deformation and rotary inertia are neglected, is beyond the scope of this book and can be found, for example, in Cremer *et al.* (1973).

For a thin plate lying in the xz plane the bending-wave equation in rectangular Cartesian coordinates is

$$D\left(\frac{\partial^4 \eta}{\partial x^4} + 2\frac{\partial^4 \eta}{\partial x^2\,\partial z^2} + \frac{\partial^4 \eta}{\partial z^4}\right) = -m\frac{\partial^2 \eta}{\partial t^2}. \tag{1.39}$$

The free *plane-progressive* wave solution is the same for Eq. (1.38), and the condition for the neglect of shear deformation is rather similar to that for beams, namely, $k_b h < 1$. Derivation of the exact plate equations is extremely difficult. A rather complete approximate equation, which takes into account shear deformation and rotary inertia effects, has been published by Mindlin (1951).

Equation (1.38) describes plane-wave motion in the x direction only. However, plane bending waves may propagate in a plate in any direction in its plane. Consider a simple harmonic plane wave described by $\eta(x, z, t) = \tilde{\eta} \exp[j(\omega t - k_x x - k_z z)]$. Substitution into Eq. (1.39) yields

$$[D(k_x^4 + 2k_x^2 k_z^2 + k_z^4) - m\omega^2]\tilde{\eta} = 0,$$

or

$$D(k_x^2 + k_z^2)^2 - m\omega^2 = 0. \tag{1.40}$$

If we write $k_b^2 = k_x^2 + k_z^2$, we obtain

$$Dk_b^4 - m\omega^2 = 0, \tag{1.41}$$

which is the plane bending-wave equation for a wave traveling in the direction given by the vector sum of the wave vectors $\mathbf{k}_x$ and $\mathbf{k}_z$, i.e., in a direction at angle $\phi = \arctan(k_z/k_x)$ to the x axis. Hence $\mathbf{k}_b = \mathbf{k}_x + \mathbf{k}_z$ and $k_b = (\omega^2 m/D)^{1/4}$, as before.

1.8 Dispersion Curves

In analysing the coupling between vibration waves in solids and acoustic waves in fluids it is very revealing to display the wavenumber characteristics of the coupled systems on a common graph. This is done in Fig. 16 in the form of dispersion curves.

It should be recalled at this point that the phase velocity $c_{ph} = \omega/k$ and the group velocity $c_g = \partial\omega/\partial k$; hence the curves with lower slopes have higher group velocities. At the point marked C in Fig. 16 the phase speeds of the plate bending wave and of the acoustic wave in the fluid are equal. In wave-coupling terms this is called the *critical*, or *lowest coincidence*, frequency ω_c. (What is the ratio of the slopes of these two curves at ω_c?) The bending-wave curve is seen to approach the shear wave speed at high frequencies. The relative dispositions of the various curves depend, of course, upon the type and forms of material carrying the waves.

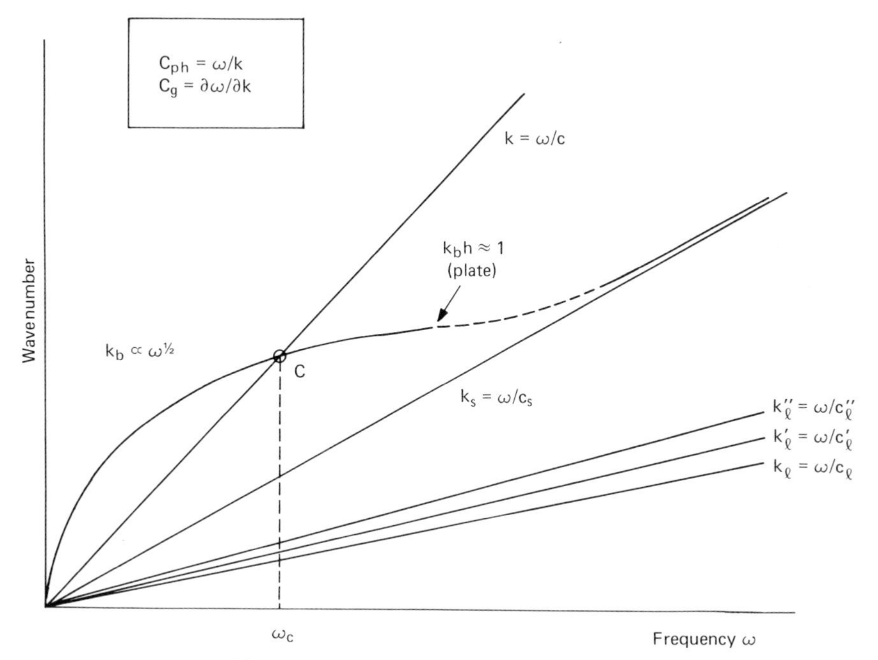

Fig. 16. Dispersion curves for various waves.

1.9 Waves in Thin-Wall Circular Cylindrical Shells

There are many structures that have a form approximating that of a thin-wall circular cylindrical shell: for example, pipes and ducts, aircraft fuselages, fluid containment tanks, submarine pressure hulls, and certain musical instruments. In many cases of long, slender cylinders having relatively thick walls, only the transverse bending, beamlike mode of propagation, in which distortion of the cross section is negligible, is of practical interest, particularly for low-frequency vibration of pipes in industrial installations such as petrochemical plants: Section 1.6 deals with such waves. However, if the ratio of cylinder diameter to wall thickness is large, as in aircraft fuselages, wave propagation involving distortion of the cross section is of practical importance even at relatively low audio frequencies. If the wall thickness is uniform, the allowable spatial form of distortion of a cross section must be periodic in the length of the circumference. The axial, tangential, and radial displacements of the wall must vary with axial position z and azimuthal angle θ as

$$u, v, w = [U(z), V(z), W(z)] \cos(n\theta + \phi), \qquad 0 \leqslant n \leqslant \infty. \tag{1.42}$$

The integer n is known as the circumferential mode order; at any frequency, *three* forms of wave having a given n may propagate along an *in vacuo* cylindrical shell: each has different ratios of U, V, and W, which vary with frequency (Leissa, 1973).

Whereas small amplitude bending waves in untensioned plates can be considered independently of longitudinal or in-plane shear waves, it is nearly always necessary to consider the three displacements u, v, w of a cylindrical shell. The principal reason is that a radial displacement of the wall of a cylinder creates tensile or compressive tangential and axial membrane stresses, depending upon whether the displacement is positive or negative, because the length of a circumference is proportional to its radius: in a flat plate the membrane stresses arising from bending deflections are always tensile (Fig. 17) and, if significant, produce non-linear vibrational behaviour.

A complete analysis of cylindrical shell vibration is beyond the scope of this book. However, a more detailed discussion appears in Chapter 4. A rather complete collation of theoretical and experimental data is presented by Leissa (1973). For the purposes of studying sound radiation and response behaviour of cylindrical shells that do not contain, or are not submerged in, dense fluids such as water or liquid sodium, it is only necessary to consider the vibrational waves in which the radial displacement w is dominant, the so-called flexural waves. The speed of propagation of these waves in the direction of the longitudinal axis of the cylinder is very dependent upon the circumferential mode order n because the relative contributions to strain energy of membrane strain and wall flexure depend upon this parameter. In addition, a cylindrical shell exhibits waveguide behaviour in that a mode of propagation having a given n cannot propagate freely below its "cutoff frequency," at which the corresponding axial wavenumber and group velocity are zero. A cutoff frequency corresponds to a natural frequency at which the modal pattern consists of a set of $2n$ nodal lines lying along equally spaced generators of a uniform shell of infinite length.

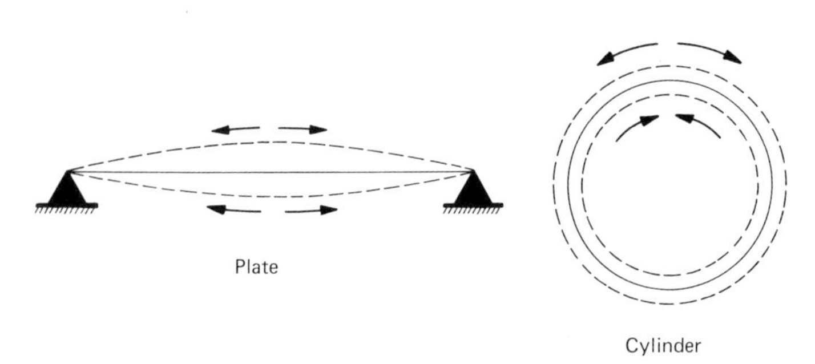

Fig. 17. Membrane stresses in transversely deflected plate and cylindrical shell.

The cutoff frequency of the $n = 0$, so-called breathing mode of a shell is termed the "ring frequency" and is given by

$$f_{\mathrm{r}} = c_l'/\pi d, \tag{1.43}$$

where d is the cylinder diameter. At this frequency, radial hooplike resonance occurs. In dynamic shell theory, frequency is usually made non-dimensional by dividing by f_{r}: $\Omega = f/f_{\mathrm{r}}$. Below $\Omega = 1$, the breathing mode cannot propagate, although axial and tangential $n = 0$ modes can. One result of the effect of membrane strain on flexural wave propagation speed in cylindrical shells is that the phase velocity of flexural waves in modes of low n can be much greater than the speed of sound in a surrounding fluid at frequencies well below the flat-plate critical frequency, as given by the particular combination of shell material, wall thickness, and speed of sound in the fluid. The effect is that such modes can rather effectively radiate and respond to sound. Cylinder radiation and sound transmission are discussed in Chapters 2–4.

1.10 Natural Frequencies and Modes of Vibration

Thus far we have considered the question of what types of wave can propagate naturally in forms of media that are uniform and unbounded in the directions of propagation. All physical systems are spatially bounded, and many incorporate non-uniformities of geometric form or material properties. Waves that are incident upon boundaries, or regions of non-uniformity, cannot propagate through them unchanged, and the resulting interaction gives rise to phenomena known as refraction, diffraction, reflection, and scattering. It is difficult concisely to define and to distinguish between each of these phenomena. However, broadly speaking, refraction involves veer in the direction of a wave vector due to spatial variations of phase velocity from place to place in a medium, or to wave transmission through an interface between different media; diffraction involves distortion of wave fronts (surfaces of uniform phase) due usually to the presence in a medium of bodies or obstacles of a highly disparate medium; reflection implies a reversal of the wave vector or a component thereof; and scattering refers to the redirection of wave energy flux, normally into diverse directions, due to the presence of localised regions of non-uniformity in a medium.

Although all these wave phenomena may occur in solid structures, the one having most practical importance is reflection. The reason for its importance is that it is responsible for the existence of sets of frequencies, and associated spatial patterns of vibration, which are proper to a bounded system. An infinitely long beam can vibrate *freely* at any frequency; a bounded beam can vibrate *freely* only at particular *natural* or *characteristic* frequencies. The elements of the beams which are not at boundaries satisfy the same equation of motion in both cases; they clearly only "know about" the boundaries because of the phenomenon of wave propagation and reflection. Wave reflection at boundaries also leads to the very important phenomenon of *resonance*. Note carefully that resonance is a phenomenon associated with *forced* vibration, generated by some input, whereas natural frequencies are phenomena of *free* vibration. Resonance is of very great practical importance because it involves large amplitude response to excitation, and can lead to structural failure, system malfunction, and other undesirable consequences.

In order to understand the nature of resonance we turn briefly to the consideration of the process of free vibration of an undamped, wave-bearing system. Mathematically, the question to be answered is: At what frequencies can the equation(s) of motion, together with the physical boundary conditions, be satisfied in the absence of excitation by an external source, and what characteristic spatial distributions of vibration are associated with these frequencies? This question can be interpreted in physical terms as follows: If the system is subjected to a transient disturbance, which frequencies will be observed to be present in the subsequent vibration, and what characteristic spatial distributions of vibration are associated with these frequencies?

It is possible to answer these questions without explicit reference to wave motion at all, which is a little surprising since any vibrational disturbance is propagated throughout an elastic medium in the form of a wave. However, it must be realised that an alternative macroscopic model to that of the elastic continuum is one consisting of a network of elemental discrete mass–spring systems. In fact, some of the earliest analysis of free vibration of distributed, continuous systems was based upon such a discrete element model (Lagrange, 1788). In this case, equations of motion can be written for each element, together with the coupling condition, thereby producing n equations of motion for n elements. Alternatively, statements can be made about the kinetic and potential energies of the elements and about the work done on them by internal and external forces: the energy approach forms the basis of most practical methods of estimation of natural frequencies of structures, and today it is most widely implemented as the *finite element method* (see Chapter 7). However, in this book we are concentrating upon the "wave picture" of vibration in fluids and solid structures, and hence we shall discuss natural vibrations of bounded systems mainly from this point of view.

In earlier sections we have seen that the wavenumbers and associated frequencies of waves propagating freely in unbounded uniform elastic systems are uniquely linked through the governing wave equation. Let us imagine what happens when a freely propagating wave meets an interface with a region of the system in which the dynamic properties are different from those of the uniform region previously traversed by the wave: in practice the interface could take the form of a boundary, a change of geometry or material, or a local constraint. The relationships between forces and displacements in this newly encountered region are different from those in the uniform region, and it is clear that the wave cannot progress unaltered: compatibility of displacements and equilibrium of forces must be satisfied at the interface and yet, if the wave were wholly transmitted past the interface, the forces associated with given displacements would be different on the two sides of the interface. [Consider two beams of different I joined together and refer to Eq. (1.34) with a progressive bending-wave form for transverse displacement.] Consequently a reflected wave must be generated that, in combination with the incident wave, is compatible and in equilibrium with the wave transmitted beyond the interface. As we shall see later, the wave reflection phenomenon may also be explained in terms of a change of wave impedance at the interface.

The amplitude and phase of a reflected wave relative to those of an incident wave depend *only* upon the dynamic characteristics of the region at and beyond the interface, which are manifested in the impedance at the interface. For simplicity we consider the case of a bending wave in an infinitely extended beam that is simply supported at one point. The incident wave displacement is

$$\eta_i^+(x,t) = \tilde{A} \exp[j(\omega t - k_b x)]. \tag{1.44}$$

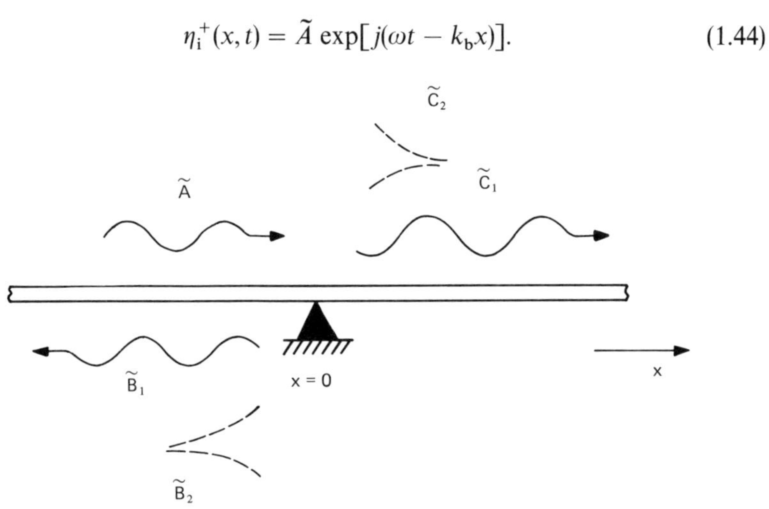

Fig. 18. Bending-wave reflection and transmission at a support.

The presence of the simple support completely suppresses transverse displacement and produces shear force reaction, but does not restrain rotational displacement and hence does not produce any moment reaction. The incident wave alone cannot satisfy the condition of zero transverse displacement at the support at all times; therefore, a reflected wave must be generated that in combination with the incident wave, does satisfy this condition. From Eq. (1.37) the general form for the propagating and non-propagating field components of negative-going, reflected bending waves is

$$\eta_{\rm r}^{-}(x,t) = [\tilde{B}_1 \exp(jk_{\rm b}x) + \tilde{B}_2 \exp(k_{\rm b}x)] \exp(j\omega t), \tag{1.45}$$

and for the positive-going, transmitted bending waves is

$$\eta_{\rm t}^{+}(x,t) = [\tilde{C}_1 \exp(-jk_{\rm b}x) + \tilde{C}_2 \exp(-k_{\rm b}x)] \exp(j\omega t). \tag{1.46}$$

We henceforth drop the time-dependent term and the superscripts indicating wave direction.

There are four complex unknowns to be related to $\tilde{A}$ through application of conditions of compatibility and equilibrium at $x = 0$:

(1) Compatibility

$$\begin{aligned} \eta_{\rm i}(0) + \eta_{\rm r}(0) &= \tilde{A} + \tilde{B}_1 + \tilde{B}_2 \\ &= \eta_{\rm t}(0) = \tilde{C}_1 + \tilde{C}_2 = 0. \end{aligned} \tag{1.47}$$

(2) Compatibility

$$\begin{aligned} \partial\eta_{\rm i}(0)/\partial x + \partial\eta_{\rm r}(0)/\partial x &= k_{\rm b}[-j\tilde{A} + j\tilde{B}_1 + \tilde{B}_2] \\ &= \partial\eta_{\rm t}(0)/\partial x = k_{\rm b}[-j\tilde{C}_1 - \tilde{C}_2]. \end{aligned} \tag{1.48}$$

(3) Equilibrium

$$\begin{aligned} EI[\partial^2\eta_{\rm i}(0)/\partial t^2 + \partial^2\eta_{\rm r}(0)/\partial x^2] &= EIk_{\rm b}^2[-\tilde{A} - \tilde{B}_1 + \tilde{B}_2] \\ &= EI\,\partial^2\eta_{\rm t}(0)/\partial x^2 \\ &= EIk_{\rm b}^2[-\tilde{C}_1 + \tilde{C}_2]. \end{aligned} \tag{1.49}$$

The solutions of Eqs. (1.47)–(1.49) are

$$\begin{aligned} \tilde{B}_1 &= -(\tilde{A}/2)(1 + j), \qquad & \tilde{B}_2 &= -(\tilde{A}/2)(1 - j), \\ \tilde{C}_1 &= (\tilde{A}/2)(1 - j), \qquad & \tilde{C}_2 &= -(\tilde{A}/2)(1 - j). \end{aligned} \tag{1.50}$$

It is interesting to note that, since the rate of transport of vibrational energy along a beam by a propagating bending wave is proportional to the square of the modulus of the complex amplitude, half the incident energy is reflected and half is transmitted by such a discontinuity. [Derive this result by considering the work done at a cross section by the shear forces and bending moments associated with internal stresses in the beam. Equation (1.54) should help.]

The displacement field on the incident side of the support ($x < 0$) is

$$\eta(x,t) = \tilde{A}[\exp(-jk_b x) - \tfrac{1}{2}(1+j)\exp(jk_b x) - \tfrac{1}{2}(1-j)\exp(k_b x)]\exp(j\omega t). \tag{1.51}$$

The final term in the square brackets represents the non-propagating component of the reflected field, which decays with distance from the support. It is not, strictly speaking, a wave component because it does not possess either a real phase velocity or a real wavenumber: its phase is independent of distance and, at a sufficiently large non-dimensional distance $k_b x$ from the support, it makes a negligible contribution to the vibrational displacement; hence it is termed the near-field component. The second term represents the true reflected wave, and its influence extends to $x = -\infty$. The first and second terms together represent an interference field that, following the graphical representation presented in Fig. 2, can be represented by Fig. 19. (Try to superimpose the nearfield term onto the figure: Is the support boundary condition satisfied?) Unlike the phasor that represents a pure progressive wave in Fig. 2, the phasor that arises from the addition of the two phasors representing the incident and reflected waves does not rotate $\pi/4$ rad per unit of distance Δx, where $k_b \Delta x = \pi/4$. The interference field is not a pure standing pattern, as the absence of any nodal point except the support demonstrates. (Why is this so?)

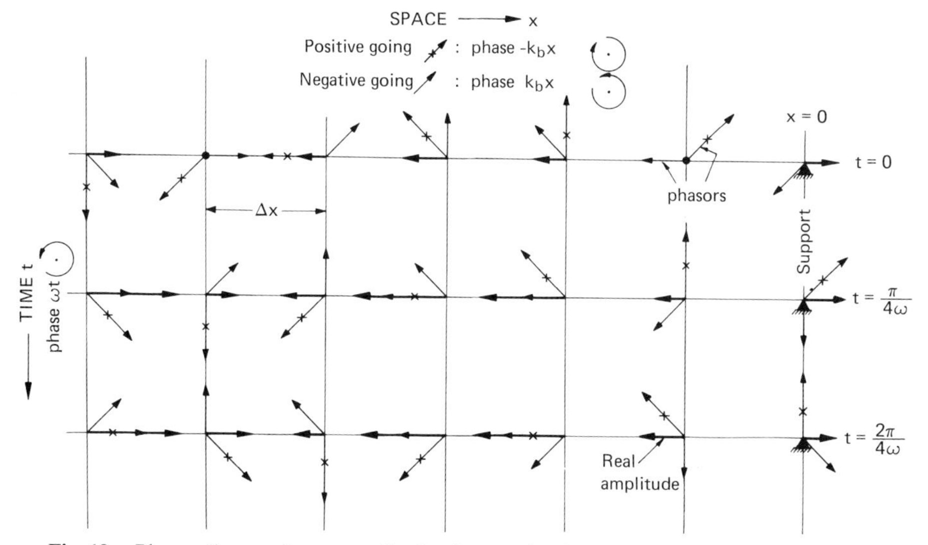

Fig. 19. Phasor diagram for wave reflection from a simple support (excluding the near field).

If the beam considered above had *terminated* at the simple support, the following coefficients would have been determined:

$$\tilde{B}_1 = -\tilde{A}, \qquad \tilde{B}_2 = 0. \tag{1.52}$$

(Try to obtain this result.) In this case, the beam displacement is

$$\eta(x, t) = \tilde{A}[\exp(-jk_b x) - \exp(jk_b x)] \exp(j\omega t) = 2j\tilde{A} \sin k_b x \exp(j\omega t). \tag{1.53}$$

Of course, all the incident wave energy is reflected. The resulting phasor diagram is shown in Fig. 20.

This interference field exhibits characteristics quite different from those of the pure progressive wave. The relative phase of the displacement at the various points on the beam takes only values of 0 or π; the points of zero and maximum amplitude are fixed in space; the field has the form of a *pure standing wave* in which the spatial and temporal variations of displacement are independent, as indicated by Eq. (1.53). Such forms of wave can only be produced by interference between two waves of the same physical type, having equal amplitudes and frequencies, and equal and opposite wave vector components in at least one direction.

If, instead of terminating the beam at a simple support, we chose to terminate it at a free end or a fully clamped support, or by a lumped mass or a springlike element, none of which dissipate or transmit energy, we would

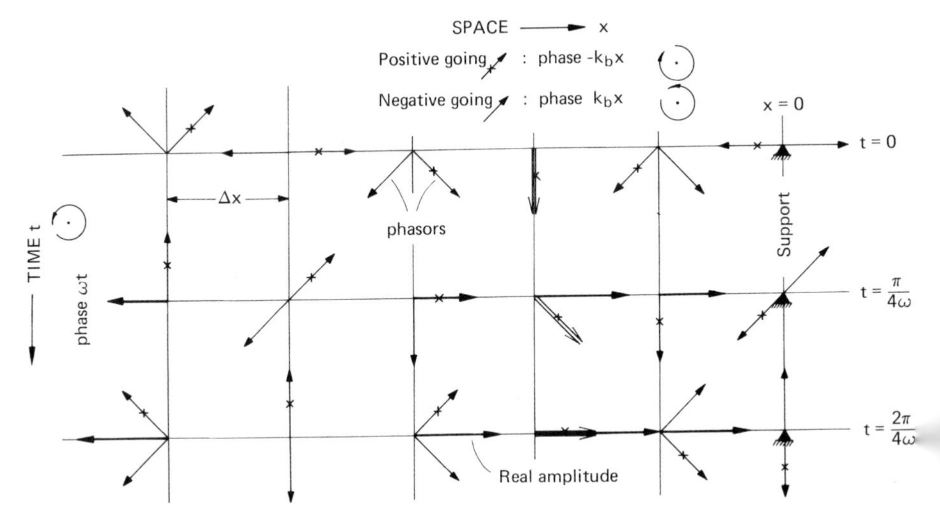

Fig. 20. Phasor diagram for wave reflection from a simple support termination ($k_b \Delta x$ chosen to equal $\pi/4$ for clarity).

find that the amplitude of the reflected wave would be of equal magnitude to that of the incident wave, although the near-field component amplitudes would be particular to the type of termination. Therefore, all such terminations produce pure standing-wave interference patterns.

Suppose now the beam is of finite length and supported at both ends by non-dispersive, non-transmitting terminations. Consider the progress of the wavefront of a *single-frequency* wave after reflection from the right-hand end. It is incident upon the left-hand end and perfectly reflected; it then returns along the path of the original incident wave. The phase of the returning wave, relative to that of the original incident wave, will depend upon the nature of both terminations, upon the frequency of the oscillation, through the associated free wavenumber, and upon the length of the beam. Certain combinations of these parameters will produce equality of phase, or phase coincidence,* of the original and returning waves; other combinations will not. The possibility of phase coincidence implies the possibility of existence of sustained, pure standing-wave fields in which original and returning waves are indistinguishable.

The frequencies at which such pure standing-wave fields can occur in a given system in free vibration are known as the characteristic or natural frequencies of the system. The associated spatial distributions of vibration amplitudes are known as the characteristic functions or natural modes of the system: These are clearly strongly related to the free wavenumbers associated with the natural frequencies. The natural frequencies of any continuously distributed elastic system are infinite in number, although lumped-element idealisations of such systems exhibit finite numbers of natural frequencies, e.g., a chain of lumped masses linked by discrete springlike elements.

A very useful device for evaluating the approximate distribution of natural frequencies of a one-dimensional bounded system of length l is to superimpose on the dispersion curves (such as those in Fig. 16) a set of horizontal lines representing phase coincidence, i.e., at intervals of k of size $n\pi/l$. Intersections with the dispersion curves indicate the approximate natural frequencies. (What does such a construction tell you about the natural frequency distribution of a bar in flexure?) Such a device can be extended to two- and three-dimensional systems to supply estimates of the average number of natural frequencies per unit frequency; this measure is called the modal density.

The question of the role of any near-field influence on phase coincidence is interesting, since a beam having different end conditions has different forms of near field at each end. In practice, this influence is only significant for the

* Note: The condition described as "phase coincidence" is also termed "phase continuity" by some authors.

lowest-frequency or fundamental modes of uniform systems, because the near field decays as $\exp(-k_b x)$, and $k_b l$ is about 3 at the lowest natural frequencies of beams of length l. However, near fields are much more important in systems incorporating multiple discontinuities, such as ribbed plates. It must be noted that the nature of a termination affects the phase of the reflected traveling wave relative to that of the incident wave, and this must be taken into account in any analysis based upon the phase coincidence criterion; for instance, both free and fully clamped beam ends produce a phase change of $-\pi/2$ rad, whereas a simply supported end produces a change of $-\pi$ rad.

Of course, all real systems possess some mechanisms of energy dissipation, or radiation, and hence waves cannot propagate to and fro forever during free vibration. Damping mechanisms are various: they include internal friction, interface friction, and radiation of energy into contiguous fluids or structures. In most practical cases the presence of damping mechanisms does not significantly change the phase velocity or the phase changes at reflection. Hence the above arguments relating to phase coincidence are unchanged, and the damped and undamped natural frequencies are almost equal.

In a case such as that of an infinite beam on two simple supports, where energy is partially transmitted (radiated) through boundaries into connected *unbounded* systems, phase changes upon reflection are affected by the presence of the sections of the beam beyond the supports; but natural frequencies exist because multiple wave reflection can occur, albeit with successive reduction of amplitude upon reflection due to energy transmission through each support. The effect is akin to that of damping described above.

Where further discontinuities occur in the beam extensions, reflections will occur there also, and then considerations of phase coincidence must include all such reflections. This observation highlights the fact, ignored by many vibration analysts, that there are no such entities as physical natural modes of individual parts of continuous structures, although they may exist in the minds of the theoreticians, or in models fed to the computer. However, in order not completely to disillusion the reader, by implying that the whole world must always be included in a vibration model, it must be said that, in many practical cases, discontinuities that divide one part of a continuous structure from the rest are often of such a form as to transmit very little energy; reflections return after a double transit of the discontinuity with so little amplitude as to alter very little the resultant phase of the motion in that part. Consequently a calculation made on the basis of an isolated subsystem with boundary conditions corresponding to an infinitely extended contiguous structure is often sufficiently accurate, at least for those models in which the vibration energy of the structure is contained largely in the particular region considered. An alternative view of ancillary structure effects is given by Bishop and Mahalingham (1981).

The condition of phase coincidence as an explanation of the existence of natural frequencies may, in principle, be extended to embrace two-, and three-dimensional elastic systems. However, as we shall see, it becomes increasingly difficult to apply purely physical arguments for its action as the complexity of the geometry of the system increases. The simplest two-dimensional structural example of practical engineering interest is that of a thin, uniform, rectangular plate supported on all sides by simple supports. Since phase coincidence must occur anywhere on the plate at a natural frequency, it is clear that if it is satisfied independently in the two orthogonal directions parallel to the sides, then it is satisfied everywhere. It turns out in this particular case that the wavenumber components in these directions at the natural frequencies correspond to those for simply supported beams of lengths equal to those of the plate sides, with the further condition that, in satisfaction of the bending-wave equation, their vector sum equals that of the free bending wavenumbers in the plate at that frequency, by analogy with the acoustic case expressed in Eq. (1.8) and Fig. 5.

The difficulty of extending the phase coincidence argument to more complex geometries is simply illustrated by imagining one corner of a simply supported rectangular plate to be removed! It seems, therefore, that we must trust to the remarkable fact, noted by Albert Einstein, that mathematical solutions to mathematical models, obtained by the application of mathematical rules developed in the abstract and having no apparent empirical basis, can provide solutions that correspond to the behaviour of real physical systems! Just how pure standing-wave fields can be created in any elastic system, by reflection of waves from boundaries of arbitrary geometry, is something of a mystery.

We have seen that the reflection of waves from boundaries or from other discontinuities creates an interference field that, if composed solely of waves of a frequency equal to a natural frequency of the system, takes the form of a standing-wave field in which the associated distribution of vibration amplitude is characteristic of the system and is known as a characteristic function, or natural mode shape, of the system. It consists of regions, or cells, of vibration of uniform phase, separated by zero-amplitude nodal lines from adjacent regions of vibration of opposite phase. A mode shape can be graphically represented most completely by contour plots of equal vibration amplitude, together with associated sections: however, it is more common simply to plot the nodal lines, especially for plates and shells of uniform thickness. For the purpose of analysing fluid–structure interactions, it is usually necessary only to plot vibration distributions normal to the interface between the media. For general formulas and numerical data the reader is referred to compendia of natural frequencies and mode shapes by Leissa (1969, 1973) and Blevins (1979).

1.11 Forced Vibration and Resonance

So far we have considered vibration frequencies that are proper to a system in the physical sense that they predominate in the frequency spectrum of free vibration observed subsequent to a *transient* disturbance from equilibrium, and in the mathematical sense that they are solutions of the equation of free motion. The word "predominate" is used because the spectrum of free motion of a damped system does not consist of discrete spectral lines since it is not periodic in the pure sense. The spectrum of a decaying "sinusoid" is spread around the frequency of the undamped sinusoid: the more rapid the decay, the broader the spread. (Check this by Fourier transform of a decaying oscillation.)

If a *linear* elastic system is subject to an unsteady disturbance that is continuous in time, the consequent vibration will contain *all and only* those frequencies present in the disturbance, not only the natural frequencies; indeed these may, in some cases, be completely absent. How can this be, when the natural frequencies have been established as those proper to the system? Perhaps a one-dimensional wave problem will help to clarify the nature of the process.

Imagine a tube with a movable piston in one end, a microphone at an arbitrary point in the tube, and a plug at the other end. If the piston is impulsively displaced inwards, an acoustic pulse will travel down the tube, reflect off the plug, travel back to the now stationary piston, reflect down the tube, and so on, theoretically endlessly. The microphone will register a succession of pulses having a basic period given by twice the length of the tube $2l$ divided by the speed of sound c. Frequency analysis of this periodic signal will produce a spectrum with discrete lines at frequencies given by $f_n = nc/2l$, where n is any positive integer. As explained in the previous section, these frequencies correspond to the acoustic natural frequencies of the tube, because the acoustic pressure phase shift at each wave reflection is zero: hence the round-trip phase shift is $k(2l) = 4\pi f_n l/c$. (What is the acoustic particle velocity phase shift?)

Imagine now that the piston is impulsively displaced periodically at intervals of time T. The pulse pattern period at the microphone will be T, and the frequencies in the received signal will be m/T, where m is any positive integer. In spite of the fact that *each* pulse does a round trip in the same time as before, namely $2l/c$, the natural frequencies will not appear in the signal spectrum unless $m/T = nc/2l$, or $T/(2l/c) = m/n$, in which case the exciting pulses will reinforce certain reflections of previously generated pulses, and strong signal components at frequencies m/T will be observed.

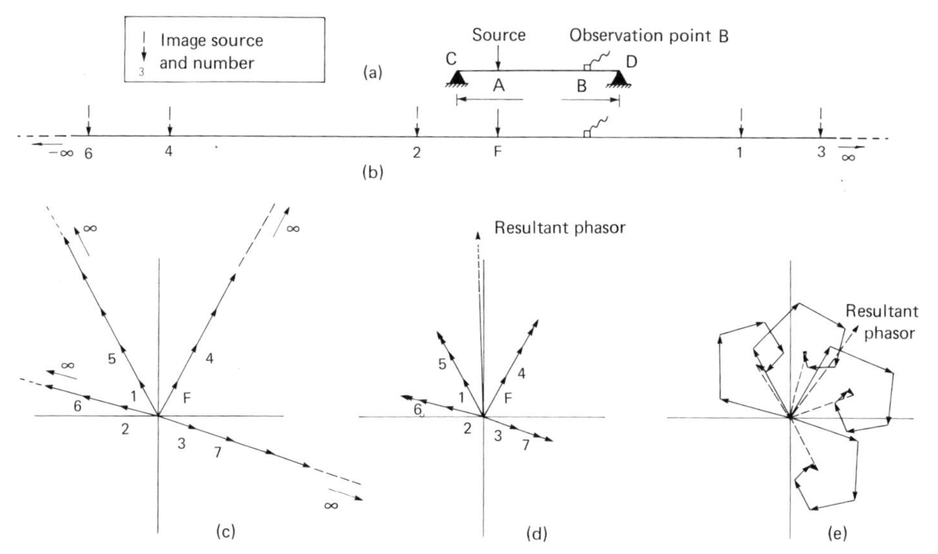

Fig. 21. Phasor diagram of direct and reflected waves: (a) system, (b) equivalent image system, (c) resonance frequency of undamped system, (d) resonance frequency of damped system, (e) damped system off-resonance.

(Check by graphical means.)* Such coincidence of excitation frequency and natural frequency, leading to response reinforcement, is known as resonance.

The phenomenon of resonance is usually associated with single-frequency excitation, but the above example shows that it occurs whenever a continuous input possesses a frequency component that corresponds to a natural frequency of a system. The response of a system excited at resonance is clearly unlimited unless some of the energy of the continuously generated and multiply reflected waves is dissipated: otherwise waves pile limitlessly upon waves, as can be demonstrated by sloshing the water in a bath tub up and down the length at the right frequency. Hence resonant response is *damping* controlled: the damping may take the form of material damping, interface damping, or radiation of energy into contiguous structures or fluids. (The bathroom floor may also be "damped" in a rather different manner!)

A system resonance has an associated bandwidth that is a measure of the range of frequencies around the natural frequency within which excitation will produce a significant response. Consider the one-dimensional wave-bearing system illustrated in Fig. 21a. A source of constant-frequency excitation acts at point A and an observation of response is made at point B.

* This form of argument cannot be directly applied to dispersive waves. (Why not?)

Waves generated by the source travel away to the boundaries, where we assume them to be perfectly reflected, in this case with zero phase shift. Perfectly reflecting boundaries of wavefields generated by point sources may be replaced by image sources, acting on equivalent unbounded systems as shown in Fig. 21b: such a device sometimes helps the analyst to visualise better the effects of reflections. We shall assume that the waves reflect without phase shift so that the images have the same phase as the real source: this assumption is not however necessary to the image concept. The total steady-state field is the sum of the outgoing waves plus the waves produced by all the images.

In Fig. 21 all the wave component amplitudes and phases are represented by phasors. The phasor labelled F is the component travelling directly from A to B; other components originate from the various images (or reflections) as labeled. The magnitude of the resultant response depends upon the positions of A and B in the system. (What happens as A and B are changed?) Wave components generated *earlier* than F, traveling in opposite directions, return repeatedly to B after travelling distances that are integer multiples of twice the system length $2nl$: alternatively it will be seen that the image sources provide these multiple wave passages past B. The phase shift during one complete return journey is $2kl$. If this is equal to $2m\pi$ and the system is undamped, the phasors representing the multiple reflection build up indefinitely (Fig. 21c), except at an observation point where mutual destructive interference occurs (nodal point). This condition is a resonance condition because the excitation frequency equals a natural frequency. (What can you conclude about the conditions on boundary phase shift for zero damping in the system?)

If the system is damped, but the damping does not affect the phase shift on reflection, the phasors will line up, but will sum to a finite value (Fig. 21d). In this case pure nodal points cannot occur. (Why?)

If now the frequency of excitation is altered slightly, the round-trip phase shift will not equal 2π, and the phasor diagram will appear as in Fig. 21e. It will be seen that the greater the damping the greater the off-resonance response relative to that at resonance, i.e., the greater the damping the greater the bandwidth. At certain frequencies intermediate to natural frequencies the phasor resulting from multiple reflection will oppose the first-passage phasor and produce negligible response: these are anti-resonances.

In terms of the complementary standing-wave, or modal, view of resonant vibration, the bandwidth of a mode is most commonly defined as the "half-power bandwidth," at the limits of which the rms modal response is $1/\sqrt{2}$ times the peak resonant response: this bandwidth is related to the modal loss factor η and modal quality factor $Q = \eta^{-1}$ by $\Delta f = \eta f_n = f_n/Q$. The measurement of the form of a modal frequency response curve around the resonance frequency is one technique for the estimation of modal bandwidth. It is only practicable when the frequency separation between adjacent resonance frequencies is significantly greater than one bandwidth.

In this and the previous section we have discussed in a qualitative manner the nature of natural modes of distributed elastic systems, and have seen how wave reflection from boundaries and other discontinuities is the origin of resonant behaviour. For clarity of exposition the wave reflection examples have been mainly one dimensional, but the principles are the same for bounded elastic systems of any kind. Most systems of practical interest comprise a number of components, each in isolation having rather different wave propagation and natural vibration characteristics. What happens when they are joined together? One approach to answering this question is based upon the concept of the vibrational impedance of components, which concerns the relationship between oscillating forces and velocities, and in terms of which the dynamic properties of components and assemblies of components may be quantified. The following sections introduce and explain this concept.

1.12 The Concept of Impedance

It is customary in books on the dynamic behaviour of structures to present analyses of response to applied simple harmonic forces; these may take the form of concentrated "point" forces and moments, or continuous spatial distributions of forces and moments. Extension to force systems having more complex variations with time can be synthesised from simple harmonic response characteristics by Fourier or Laplace transform techniques. Although this approach serves to illustrate the general frequency response characteristics of uniform structural idealisations such as beams or plates, the student is frequently uncertain how the results can be applied to a more complex system that consists of assemblies of subsystems, each of which has a different form or is constructed from a different material: this is particularly true when only the force(s) applied to one of the subsystems is known or can be estimated. As an example, we might wish to investigate the behaviour of a machine mounted on resilient isolators, which in turn are mounted upon a building floor that is connected to the supporting walls. Typically, only the forces generated within the machine by its operating mechanisms might be known. In such a case, the forces applied to the isolators, the floors, and the walls are not known a priori. However, we can assume that the motions of connected sub-structures at their interfaces are identical and that Newton's law of action and equal and opposite reaction applies at these interfaces. In vibration analysis it is convenient to represent these physical facts in terms of equality at the interfaces of linear or angular velocities, and forces and moments. By using the complex exponential representation of simple harmonic time dependence, the ratio of the complex amplitudes of the forces and velocities at any interface for a given frequency can be represented by a complex number, which is termed the *impedance* of the total system evaluated at that particular interface; it is sometimes more useful to use the inverse of impedance, termed the *mobility*.* Hence it is convenient to characterise the individual sub-structures by their complex impedances (or mobilities) evaluated at their points, or interfaces, of connection to contiguous structures. The response of the total system to a known applied force may then be evaluated in terms of the impedances (or mobilities) of the component parts.

The use of the concepts of impedance and mobility greatly facilitates the process of evaluating vibration energy or power flow through a complex system: at any interface the time-averaged power transfer for simple harmonic time dependences is given by

$$\bar{P} = \frac{1}{T}\int_0^T P(t)\,dt = \int_0^{2\pi/\omega} \mathrm{Re}\{\tilde{F}e^{j\omega t}\}\,\mathrm{Re}\{\tilde{v}e^{j\omega t}\}\,dt = \frac{1}{2}\,\mathrm{Re}(\tilde{F}\tilde{v}^*). \quad (1.54)$$

The impedance $\tilde{Z} = \tilde{F}/\tilde{v}$ and the mobility $\tilde{Y} = \tilde{v}/\tilde{F}$. Hence

$$\bar{P} = \tfrac{1}{2}|\tilde{F}|^2\,\mathrm{Re}\{\tilde{Z}^{-1}\} = \tfrac{1}{2}|\tilde{v}|^2\,\mathrm{Re}\{\tilde{Z}\} = \tfrac{1}{2}|\tilde{F}|^2\,\mathrm{Re}\{\tilde{Y}\} = \tfrac{1}{2}|\tilde{v}|^2\,\mathrm{Re}\{\tilde{Y}^{-1}\}. \quad (1.55)$$

[Why are the complex conjugate signs omitted in Eq. (1.55)?] The real part of $\tilde{Z}$ is called the *resistance* and the imaginary part the *reactance*. If we write $\tilde{Z} = R + jX$, then Eq. (1.55) can be written

$$\bar{P} = \tfrac{1}{2}|\tilde{F}|^2\,|R/(R^2 + X^2)| = \tfrac{1}{2}|\tilde{v}|^2\,R. \quad (1.56a)$$

Also

$$\tilde{Y} = (R - jX)/(R^2 + X^2) \quad (1.56b)$$

The practical advantage of this formalism, which is not immediately obvious, is that in many cases it is possible to make assumptions about the magnitude of $\tilde{F}$ or $\tilde{v}$ at an interface from a knowledge of the impedance characteristics of the structures joined thereat, together with an observation of the force and/or velocity at the interface when a contiguous structure is

* Strictly, these quantities should be termed the *mechanical impedance* and the *mechanical mobility* to distinguish them from acoustic impedances in later chapters.

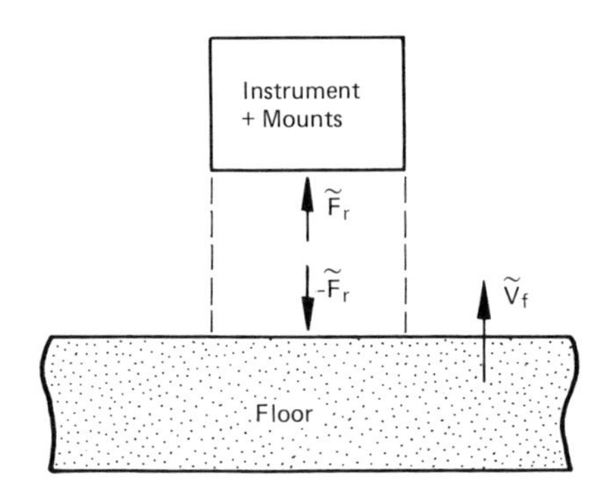

Fig. 22. Instrument mounted on vibrating floor.

disconnected. For example, consider the vibration of an instrument that is resiliently mounted upon a building floor, which itself is subject to vibration from an external source (Fig. 22). Suppose we have measured the vibration level of the floor and wish to estimate the vibration level of the instrument. When the instrument is placed on its mounts upon the floor, the vibration level of the floor changes—but by how much?

The reaction force upon the floor created by the presence of the mounted instrument is

$$\tilde{F}_{\mathrm{R}} \exp(j\omega t) = \tilde{Z}_{\mathrm{I}}\tilde{v}_{\mathrm{F}} \exp(j\omega t), \tag{1.57}$$

where $\tilde{Z}_{\mathrm{I}}$ is the impedance of the mounts plus instrument, and $\tilde{v}_{\mathrm{F}}$ the floor velocity in the presence of the instrument. It will be seen that the relationship between the floor velocity $\tilde{v}_0$ in the absence of the instrument, and the floor velocity $\tilde{v}_{\mathrm{F}}$ with the instrument mounted, is given in terms of the floor impedance $\tilde{Z}_{\mathrm{F}}$ as

$$\tilde{v}_{\mathrm{F}} = \tilde{v}_0 - \tilde{F}_{\mathrm{R}}/\tilde{Z}_{\mathrm{F}} = \tilde{v}_0 - \tilde{Z}_{\mathrm{I}}\tilde{v}_{\mathrm{F}}/\tilde{Z}_{\mathrm{F}}. \tag{1.58}$$

Hence

$$\tilde{v}_{\mathrm{F}} = \tilde{v}_0[1 + \tilde{Z}_{\mathrm{I}}/\tilde{Z}_{\mathrm{F}}]^{-1}. \tag{1.59}$$

The floor vibration velocity is seen to be altered by the presence of the instrument to a degree characterised by the ratio $\tilde{Z}_{\mathrm{I}}/\tilde{Z}_{\mathrm{F}}$ of mounted instrument to floor impedance. The modulus of this ratio will normally be much smaller than unity; hence it may often be assumed that the floor velocity is unaffected by the presence of a mounted object, and an estimate of the instrument vibration may be made on this basis. (Could $\tilde{v}_{\mathrm{F}}$ exceed $\tilde{v}_0$ under any circumstances?)

So far we have considered the concept of impedance as applied to lumped element models of structures, in which motions are described by single coordinates. For distributed, wave-bearing, elastic systems it is customary to define three main forms of impedance, namely, point impedance, line impedance, and wave impedance (introduction to the third form is delayed until the next chapter).

1.13 Point Force Impedance of an Elastic Structure

It is not possible physically to apply a transverse force to the surface of an elastic structure at a mathematically defined point. (Why?) However, in practice, it is possible to apply such a force over a very small area of surface. A force having simple harmonic time dependence will generate waves in the structure that will propagate away from their source at phase speeds dependent upon the wave type and, in the case of dispersive waves, upon the frequency. Small-amplitude waves in elastic structures are normally linear and, therefore the wave frequency will equal the forcing frequency.

The relationship between the applied force and the resulting velocity at the point of application, in other words the "driving point impedance," depends essentially upon two factors: (1) the properties, and hence wave-bearing characteristics, of the structure in the close vicinity of the driving point; (2) the amplitude and phase at the driving point of any waves reflected back to that point from discontinuities or constraints in the surrounding structure. These factors may be illustrated by analysis of the one-dimensional problem of point force excitation of a uniform bar or beam shown in Fig. 23.

The applied force is reacted by elastic shear forces in the beam; the equation of free motion of the *undamped* beam is Eq. (1.36):

$$EI\frac{\partial^4\eta}{\partial x^4} + m\frac{\partial^2\eta}{\partial t^2} = 0. \tag{1.60}$$

Note that the terms in this equation express the transverse elastic and inertia forces per unit length. The equation applies to all the points on the beam that are not subject to external forces such as at the driving forces or external constraints. At the driving point the transverse force per unit length is discontinuous, and the transverse shear force suffers a step change equal to the applied force, as it does in a static loading case. Mathematically such a con-

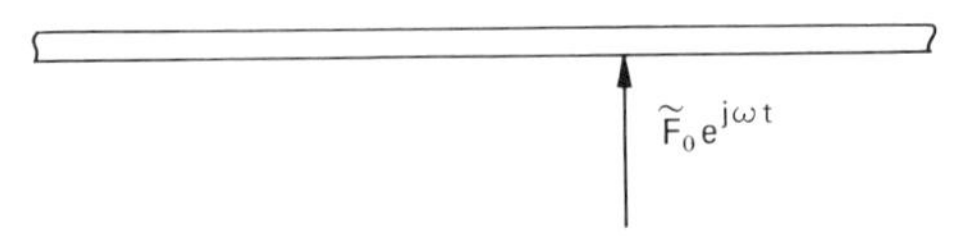

Fig. 23. Point force excitation of an infinite uniform beam.

centrated force applied at $x = a$ may be represented by a Dirac delta function representation of the distributed transverse loading $F'(x)$ *per unit length* thus:

$$F'(x) = F(x)\,\delta(x - a), \tag{1.61}$$

where $F(x)$ is the transverse applied force distribution. The Dirac delta function operates upon the transverse force distribution $F(x)$ such that $\int_{-\infty}^{\infty} F(x)\,\delta(x - a)\,dx = F(a)$, i.e., it "picks out" the force acting at $x = a$. Note that the dimension of the one-dimensional Dirac delta function is reciprocal length, as seen from Eq. (1.61). Since the loading per unit length is the derivative of the shear force with respect to x, it is seen that the Delta function is the derivative of a step function. These relationships are illustrated in Fig. 24.

Equations (1.60) and (1.61) may be combined:

$$EI\,\frac{\partial^4 \eta}{\partial x^4} + m\,\frac{\partial^2 \eta}{\partial t^2} = \tilde{F}(x)\,\delta(x - a)\exp(j\omega t). \tag{1.62}$$

In this case the force is simple harmonic, but any time history may be inserted.

Equation (1.62) is insufficient to describe the behaviour of any particular beam to a point force; boundary conditions at the ends of the beam are also required. Let us initially assume that the beam is infinitely extended. Equation (1.60) is equivalent to Eq. (1.62) at every point except that at which the force is applied. The solutions, given in Eq. (1.37), represent all the physically

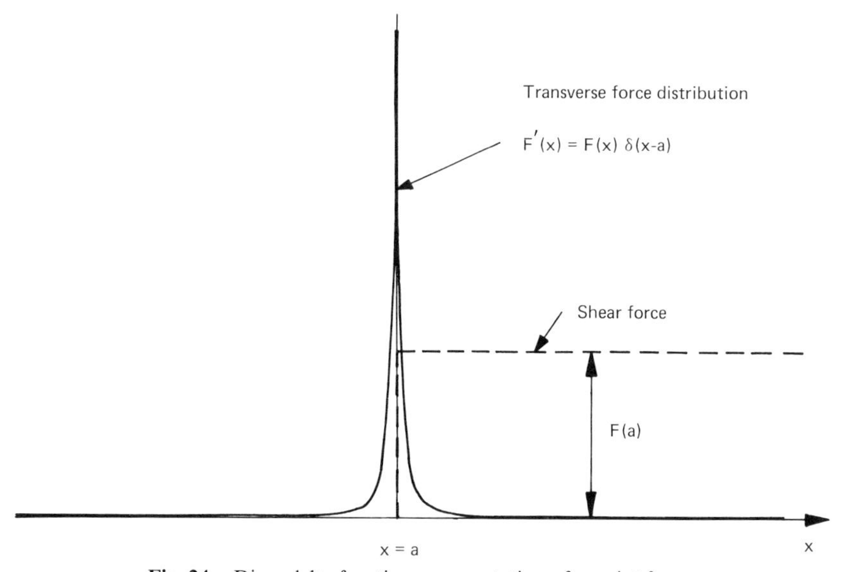

Fig. 24. Dirac delta function representation of a point force.

possible forms of free flexural vibration of the beam:

$$\eta(x,t) = [\tilde{A}\exp(-jk_b x) + \tilde{B}\exp(jk_b x) + \tilde{C}\exp(-k_b x) + \tilde{D}\exp(k_b x)]\exp(j\omega t), \quad (1.63)$$

where $k_b = (\omega^2 m/EI)^{1/4}$. Because the beam is infinite, $\tilde{A}$ and $\tilde{C}$ must be zero in the region $x < 0$, and $\tilde{B}$ and $\tilde{D}$ must be zero in the region $x > 0$. (Why?) The coefficients $\tilde{A}$, $\tilde{B}$, $\tilde{C}$, and $\tilde{D}$ may now be found by satisfying conditions of equilibrium immediately to the left and right of $x = 0$, i.e., at $x = 0^-$ and $x = 0^+$. According to Fig. 13 and Eq. (1.34), the elastic shear stresses produce a downward force of magnitude $EI\,\partial^3\eta/\partial x^3$ at the left-hand end of an elemental beam section, and an upward force of equal magnitude at the right-hand end of a section. Since the system is symmetrical about $x = 0$, this force at $x = 0^+$ is equal in magnitude and sign to that at $x = 0^-$, each being equal to $\frac{1}{2}\tilde{F}(0)\exp(j\omega t)$. If the applied force is considered to be positive in the positive y direction, then at $x = 0^+$,

$$\tilde{F}(0)/2 - EI(jk_b^3\tilde{A} - k_b^3\tilde{C}) = 0, \quad (1.64a)$$

and at $x = 0^-$,

$$\tilde{F}(0)/2 + EI(-jk_b^3\tilde{B} + k_b^3\tilde{D}) = 0. \quad (1.64b)$$

In addition, because of symmetry, the slope of the beam at $x = 0$ is zero. Hence

$$-jk_b\tilde{A} - k_b\tilde{C} = jk_b\tilde{B} + k_b\tilde{D} = 0. \quad (1.64c)$$

Equations (1.64) yield

$$\tilde{A} = \tilde{B} = j\tilde{C} = j\tilde{D}, \quad (1.65)$$

and

$$\tilde{A} = -j\tilde{F}(0)/4EIk_b^3. \quad (1.66)$$

Equation (1.63/1.37) can be written

$$\eta(0^+) = -[j\tilde{F}(0)/4EIk_b^3](1 - j) = \eta(0^-). \quad (1.67)$$

The point impedance is hence

$$\tilde{Z}_F(0) = \tilde{F}(0)/\{(\partial/\partial t)[\eta(0)]\} = 2(EIk_b^3/\omega)(1 + j). \quad (1.68)$$

The explicit dependence of $\tilde{Z}_F(0)$ on the beam parameters is seen by substituting for k_b:

$$\tilde{Z}_F(0) = 2EI(\omega^2 m/EI)^{3/4}(1 + j)/\omega = 2(EI)^{1/4}\omega^{1/2}m^{3/4}(1 + j). \quad (1.69)$$

For rectangular-section beams, having $I = wh^3/12$ and $m = \rho_s wh$, it is seen that the beam depth h has the strongest influence on the impedance:

$\tilde{Z}_F(0) \propto h^{3/2}$. However, the beam does not offer a stiffness reactance because the reactance has a positive sign and is therefore masslike, although the effective mass is frequency dependent. When a beam is elastically supported at a point a reasonance phenomenon can be created by the combined effects of local stiffness and local effective mass. The resistive component of the impedance, which represents the ability of the beam to carry energy away from the point of force application, is equal in magnitude to the reactance. Both components are seen to increase monotonically with frequency; resonance of the beam alone cannot occur at any frequency. Of course, Eq. (1.69) is only valid for the frequency range in which the simple bending equation is itself valid.

Real beams are not of infinite length, and therefore the impedance derived above is not valid if waves from boundaries return to the driving point with amplitudes significant compared with the outgoing waves. Suppose the beam is of length $2l$, is undamped, and is simply supported, i.e., the transverse displacement and bending moment are zero at each end. In this case all four coefficients $\tilde{A}$, $\tilde{B}$, $\tilde{C}$, and $\tilde{D}$ may be non-zero in both regions $x < 0$ and $x > 0$. They must take values that satisfy the boundary conditions at both ends. Suppose the force is applied at the midpoint of the beam. Then, at $x = -l$,

$$\eta(-l) = \tilde{A}_1 \exp(jk_b l) + \tilde{B}_1 \exp(-jk_b l) + \tilde{C}_1 \exp(k_b l) + \tilde{D}_1 \exp(-k_b l) = 0, \tag{1.70}$$

and

$$\begin{aligned} \partial^2\eta(-l)/\partial x^2 &= -k_b^2\tilde{A}_1 \exp(jk_b l) - k_b^2\tilde{B}_1 \exp(-jk_b l) \\ &\quad + k_b^2\tilde{C}_1 \exp(k_b l) + k_b^2\tilde{D}_1 \exp(-k_b l) \\ &= 0. \end{aligned} \tag{1.71}$$

At $x = l$,

$$\begin{aligned} \eta(l) = \tilde{A}_2 \exp(-jk_b l) + \tilde{B}_2 \exp(jk_b l) + \tilde{C}_2 \exp(-k_b l) \\ + \tilde{D}_2 \exp(k_b l) = 0, \end{aligned} \tag{1.72}$$

and

$$\begin{aligned} \partial^2\eta(l)/\partial x^2 &= -k_b^2\tilde{A}_2 \exp(-jk_b l) - k_b^2\tilde{B}_2 \exp(jk_b l) \\ &\quad + k_b^2\tilde{C}_2 \exp(-k_b l) + k_b^2\tilde{D}_2 \exp(k_b l) \\ &= 0. \end{aligned} \tag{1.73}$$

The subscripts 1 and 2 distinguish the coefficients appropriate to the two halves of the beam.

In addition, shear force equilibrium equations must be applied on each side of the face at $x = 0^-$ and $x = 0^+$, as before. Hence at $x = 0^-$,

$$\tilde{F}(0)/2 + EI[jk_b^3\tilde{A}_1 - jk_b^3\tilde{B}_1 - k_b^3\tilde{C}_1 + k_b^3\tilde{D}_1] = 0, \tag{1.74}$$

and $x = 0^+$,

$$\tilde{F}(0)/2 - EI[jk_b^3\tilde{A}_2 - jk_b^3\tilde{B}_2 - k_b^3\tilde{C}_2 + k_b^3\tilde{D}_2] = 0. \tag{1.75}$$

The bending moment is continuous through $x = 0$, because the applied force exerts no moment there. Hence

$$EI\frac{\partial^2\eta(0^+)}{\partial x^2} = EI\frac{\partial^2\eta(0^-)}{\partial x^2},$$

and

$$-k_b^2\tilde{A}_2 - k_b^2\tilde{B}_2 + k_b^2\tilde{C}_2 + k_b^2\tilde{D}_2 = -k_b^2\tilde{A}_1 - k_b^2\tilde{B}_1 + k_b^2\tilde{C}_1 + k_b^2\tilde{D}_1. \tag{1.76}$$

Finally, the slope of the beam is continuous through $x = 0$. Hence,

$$-jk_b\tilde{A}_1 + jk_b\tilde{B}_1 - k_b\tilde{C}_1 + k_b\tilde{D}_1 = -jk_b\tilde{A}_2 + jk_b\tilde{B}_2 - k_b\tilde{C}_2 + k_b\tilde{D}_2 = 0. \tag{1.77}$$

Equations (1.70)–(1.77) contain 16 unknowns, which can be obtained as a function of frequency or wavenumber. Because we have chosen a physically symmetric configuration, it is possible by physical reasoning to obtain certain relationships between the coefficients on the two halves of the beam. (What are they?) The solutions to these equations are as follows:

$$\tilde{A}_1 = \frac{j\tilde{F}_0[1 + \exp(-2jk_bl)]}{8EIk_b^3(1 + \cos 2k_bl)} = \tilde{B}_2, \tag{1.78a}$$

$$\tilde{B}_1 = -\tilde{A}_1\exp(2jk_bl) = \tilde{A}_2, \tag{1.78b}$$

$$\tilde{C}_1 = \frac{\tilde{F}_0}{4EIk_b^3[1 + \exp(2k_bl)} = \tilde{D}_2 \tag{1.78c}$$

$$\tilde{D}_1 = -\tilde{C}_1\exp(2k_bl) = \tilde{C}_2. \tag{1.78d}$$

At $x = 0$, the deflection η is given by

$$\eta(0, t) = \frac{F_0e^{j\omega t}}{4EIk_b^3}\left[\frac{\sin 2k_bl}{1 + \cos 2k_bl} + \frac{1 - \exp(2k_bl)}{1 + \exp(2k_bl)}\right]. \tag{1.79}$$

The velocity at $x = 0$ is $j\omega\eta(0, t)$. Hence the impedance is

$$\tilde{Z}_F(0) = -\frac{j4EIK_b^3}{\omega}\left[\frac{\sin 2k_bl}{1 + \cos 2k_bl} + \frac{1 - \exp(2k_bl)}{1 + \exp(2k_bl)}\right]^{-1}$$

$$= \frac{j4EIK_b^3}{\omega}(\tanh k_bl - \tan k_bl)^{-1}. \tag{1.80}$$

Comparison of this expression with that in Eq. (1.68) shows some similarity in its dependence upon the beam parameters, but it is very different in nature because it is purely imaginary. Equation (1.45) shows that no power can be transferred from the force to the beam; this makes physical sense because the beam has been assumed to possess no damping and therefore cannot dissipate power. Neither can power be transmitted into simple supports because they do not displace transversely, and there is no bending moment to do work through the beam end rotation.

Unlike the infinite beam impedance, that of the bounded beam varies with frequency over a wide range of positive and negative values (Fig. 25). The frequencies corresponding to $\text{Im}\{\tilde{Z}_F(0)\} = \tilde{Z}_F(0) = 0$ are resonance frequencies at which the theoretical model employed here predicts infinite driving point response to an applied force. The frequencies corresponding to maxima of $|\tilde{Z}_F(0)|$ are anti-resonance frequencies, at which the driving point response is a minimum. (Find expressions for these frequencies.) Note that, as frequency increases, the beam response between resonances is alternately masslike and stiffnesslike.

Of course, all real structures possess some degree of damping. The mechanisms are various; they include material or internal damping, friction at supports and joints, and transfer of vibrational energy to connected structures or fluids (radiation damping). It is found in practice that structural

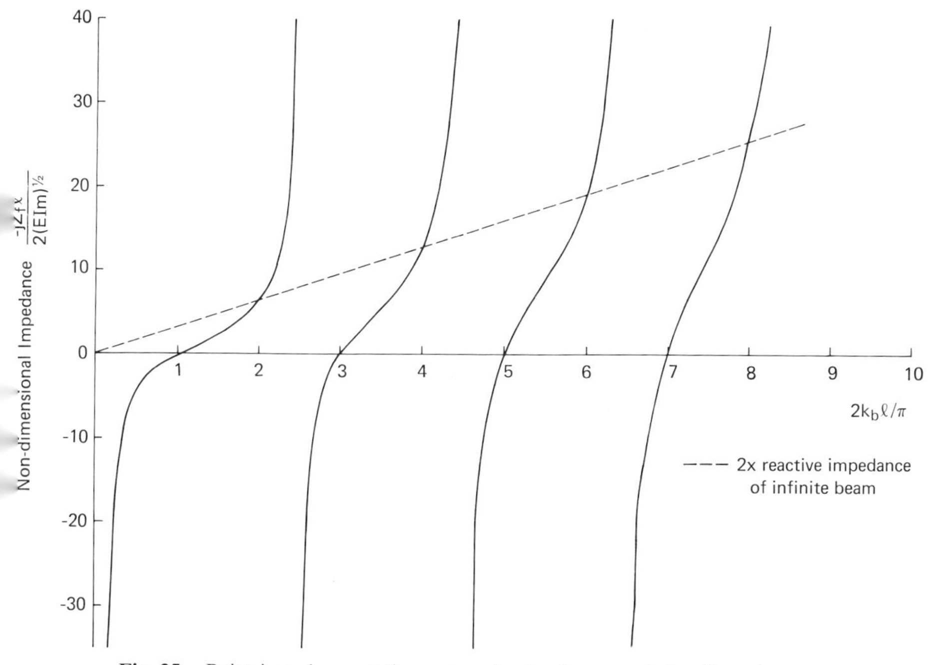

Fig. 25. Point impedance at the centre of a simply supported uniform beam.

damping may in many cases be reasonably represented mathematically by attributing a complex elastic modulus to the material: $E' = E(1 + j\eta)$, where η is termed the loss factor and is generally much smaller than unity. [The author hopes that no confusion will arise between the displacement $\eta(x)$ in the y direction and the loss factor η.] In practical structures that have not been specially treated with damping material, η usually has values in the range 5×10^{-4} to 5×10^{-2}; for most structures η tends to decrease with frequency, roughly as $\omega^{-1/2}$.

The inclusion of damping produces a modified form of bending wave equation (1.36). Unfortunately the complex elastic modulus model is not strictly valid in the time-dependent form of equation, because it can lead to non-causal transient solutions. However it is valid if restricted to steady-state, simple-harmonic, or multiple-harmonic vibration. Hence Eq. (1.36) may be written

$$E(1 + j\eta)I\,\frac{\partial^4 \eta(x)}{\partial x^4} = \omega^2 m\eta(x). \tag{1.81}$$

We now assume a one-dimensional wave solution for $\eta(x)$ and allow for energy dissipation by assuming a complex form for the wavenumber:

$$\eta(x, t) = \tilde{A}\exp[j(\omega t - k'x)], \tag{1.82}$$

where $k' = k(1 - j\alpha)$ and k is real. Substitution of this assumed solution into Eq. (1.71) yields

$$EI(1 + j\eta)k^4(1 - j\alpha)^4\eta(x) = \omega^2 m\eta(x). \tag{1.83}$$

In general, α is much less than unity, and binomial expansion of the term $(1 - j\alpha)^4$, together with neglect of small terms, yields

$$EI(1 + j\eta)k^4(1 - 4j\alpha) = \omega^2 m. \tag{1.84}$$

By separating real and imaginary parts, and ignoring the term $4\eta\alpha$ in comparison with unity,

$$EIk^4 = \omega^2 m \tag{1.85a}$$

and

$$\alpha = \eta/4. \tag{1.85b}$$

Hence $k = \pm k_b$, $\pm jk_b$, where $k_b = (\omega^2 m/EI)^{1/4}$, as in the case of the undamped beams. The complete solution for $\eta(x, t)$ corresponding to Eq. (1.37) is

$$\begin{aligned}\eta(x, t) = {} & \tilde{A}\exp(-jk'_b x) + \tilde{B}\exp(jk'_b x) \\ & + \tilde{C}\exp(-k'_b x) + \tilde{D}\exp(k'_b x)]\exp(j\omega t),\end{aligned} \tag{1.86}$$

where $k'_b = k_b(1 - j\eta/4)$.

It is not so simple to identify the propagating and non-propagating (near-field) wave components, as in Eq. (1.37). Application of the boundary conditions at the ends of the beam and at the point of force application leads to exactly the same form of relationship between the coefficients as given by Eqs. (1.78a–d), but with k_b replaced by k_b' and E replaced by E'. Hence the driving-point impedance has the same form as that given by Eq. (1.80) with k_b substituted by k_b' and E substituted by E':

$$\tilde{Z}_F(0) = -\frac{j4E'IK_b'^3}{\omega}\left[\frac{\sin 2k_b'l}{1+\cos 2k_b'l} + \frac{1-\exp(2k_b'l)}{1+\exp(2k_b'l)}\right]^{-1}. \qquad (1.87)$$

The traveling-wave energy is proportional to $\omega^2\eta(x)^2$. Equation (1.82), with substitution $k' = k(1 - j\alpha)$, takes the form

$$\eta(x, t) = \tilde{A}\exp[j(\omega t - kx)]\exp(-k\alpha x).$$

Hence the traveling-wave energy is proportional to $\exp(-2k\alpha x)$, that is, it suffers a fractional decrease of $\exp(-2k\alpha)$ per metre. The energy of waves generated at the point of force application decreases by a factor $\exp(-2k_b\alpha l)$ during the passage of the wave to a beam boundary and back. If the factor $2k_b\alpha l$ is sufficiently large, the presence of the beam boundaries, as "made known" at the driving point by the return thereto of reflected wave, will not significantly affect the driving-point impedance: the beam will therefore "appear to the force" to be unbounded and resonant and anti-resonant behaviour will disappear. The appropriate criterion for this to occur is $2k_b\alpha l \gg 1$, or $\eta \gg 2/k_b l$. Since $k_b = (\omega^2 m/EI)^{1/4}$, the value of the loss factor necessary to produce this condition decreases with frequency for a given beam and increases weakly with beam bending stiffness.

If it is assumed that $2k_b\alpha l \gg 1$ in Eq. (1.87), the driving-point impedance is approximately

$$\tilde{Z}_F(0) \approx 2(EI)^{1/4}\omega^{1/2}m^{3/4}(1 + j), \qquad (1.88)$$

which is the infinite-beam impedance [Eq. (1.69)]. (Prove.)

For comparison, the point force impedance of an infinite plate of thickness h and mass per unit area m'' is

$$\tilde{Z}_F = 8[Eh^3/12(1 - \nu^2)]^{1/2}(m'')^{1/2}. \qquad (1.89)$$

It is purely real, exhibiting no local reactive behaviour, unlike the beam. The physical reason for the difference lies in the essential difference between one- and two-dimensional wave radiation. Somewhat surprisingly, $\tilde{Z}_F$ for a plate is independent of frequency. Care should be taken in attempting to apply Eq. (1.89) to relatively thick plates, since the bending-wave equation on which the equation is based is not valid in the vicinity of locally applied transverse forces, and local reactive effects can occur (Timoshenko and Goodier, 1951).

Problems

1.1 Derive an expression for the speed of longitudinal waves in a uniform circular-section bar composed of a core of material of diameter d_1, elastic modulus E_1, and density ρ_1, sheathed in a different material of outer diameter d_2, elastic modulus E_2, and density ρ_2. Neglect the Poisson effect and assume zero slip between the two materials. Check your expression by considering the extreme values of the ratio d_1/d_2.

1.2 Why must rotation of an element occur in a transverse shear wave?

1.3 Prove that the ring frequency of a circular cylindrical shell corresponds to the natural frequency of breathing ($n = 0$) motion by deriving the equation of radial motion of an elemental circumferential segment of shell.

1.4 Derive an expression for the power flow between a vibrating floor and a mounted object in terms of the quantities used in Section 1.12. By assuming a simple model of a mass mounted on an infinite flat plate via a parallel spring/dashpot isolator, derive an expression for the ratio of the velocity of the mass to that of the floor in the absence of the mounted system, in terms of non-dimensional parameters involving the mass, spring stiffness, dashpot coefficient, floor thickness, density, and elastic modulus [Eq. (1.89) applies]. At what frequency is this ratio a maximum?

1.5 Why is the effective mass at the driving point on an infinite uniform beam said to be frequency dependent? A transverse simple harmonic force, of frequency-independent amplitude $\tilde{F}$, is applied to an infinite uniform beam that is supported at the driving point by a spring of stiffness K. At what frequency will maximum power be transmitted to the beam? Equation (1.69) applies.

1.6 Derive expressions for the bending-wave field in a uniform beam reflected at a termination consisting of (a) a point mass M, and (b) a spring of stiffness K. Check your expressions for travelling-wave energy reflection, and for free and simply supported end conditions.

1.7 Show that longitudinal waves in a system of equal masses connected by equal springs are dispersive. Derive expressions for the group velocity and the maximum frequency of free-wave propagation.

1.8 Derive an expression for the power flux in a simple harmonic free progressive bending wave in a uniform beam. Note that both transverse force and bending moment are involved. Hence find the power transmission coefficient for a simple support applied to an infinite beam.

1.9 By considering the dispersion curves for torsional waves in a thin flat strip of width h and thickness b, and bending waves in a plate of the same material and thickness, find an expression for the critical frequency of strip torsion with the plate, onto the edge of which the strip is welded at right angles along its centreline. What happens in the plate when the strip is excited in torsion below this frequency? Equations (1.26) and (1.41) and Table II apply.

2 Sound Radiation by Vibrating Structures

2.1 The Importance of Sound Radiation

A large proportion of all sources of sound radiate energy through the action of vibrating solid surfaces upon surrounding fluid. Some of these sources, such as pianos, radio loudspeakers, and church bells, are generally desirable; many, such as internal combustion engine blocks, punch presses, and train wheels, are not. There exist, as well, many sources of sound that do not radiate through the action of vibrating solid surfaces. (Can you think of any?)

The subject of sound radiation from vibrating structures is of great practical importance. It is imperative that designers of loudspeakers understand the mechanism of sound radiation so that they can improve the quality of the product. Designers of industrial machinery must take into account the widespread operation of industrial and community noise limitation regulations and therefore must understand the mechanisms of sound generation that operate in their machines, and also how most effectively and economically to eliminate or suppress them. The noise control engineer has to understand the nature and methods of control of sound radiation from vibrating structures. The designer of military ships and fishing vessels needs to reduce the radiation of sound from hull structures in order to minimise, respectively, the chances of detection or the disturbance of fish.

Structures that radiate sound through vibration are extremely diverse in their geometric forms, material properties, and forms of construction: Contrast a violin with a marine diesel engine! It should not be thought that the process of theoretical estimation of sound radiation from practical structures is simple: in general, it is quite impossible theoretically to evaluate the detailed form of a radiation field in terms of amplitude and phase, although the advent of computer-based numerical techniques is beginning to expand the possibilities in this direction. However, in many cases, it is only required to estimate the total sound power radiated by a structure, together with some indication of its frequency distribution. If so, then analytical methods of evaluation are more easily applied.

Although it may be said that the mechanism of generation of sound by surface vibration, namely the acceleration of fluid in contact with the surface, is common to all such sources, the effectiveness of radiation in relation to the amplitude of vibration varies widely from source to source. In order for a vibrating surface to radiate sound effectively, it must not only be capable of compressing or changing the density of the fluid with which it is in contact, but must do so in such a manner as to produce significant density changes in fluid remote from the surface.

Consider a volume of fluid contained in a short closed tube fitted with a plunger. Pushing the plunger inwards will necessarily change the volume and hence the density and pressure of the fluid. Compression will occur whatever the time dependence of the displacement of the plunger, although the pressure–density relationship may depend upon the speed. (Why and how?) Suppose now that the end of the cylinder opposite to the plunger is opened; How much compression will now occur? The intuitive answer is "none," because the fluid will escape freely through the hole. However, that answer is incorrect. The degree of initial compression depends upon the acceleration of the plunger, because the fluid has inertia, a property by which it resists acceleration, and because it applies a reaction force to any agent attempting to accelerate it. Hence, in place of the end of the container, which completely constrains the motion of the fluid, there is the alternative but less effective inertial constraint imposed by the fluid at the mouth of the short container on the fluid in contact with the plunger. The greater the plunger acceleration, the greater the initial compression of the fluid in contact with it.

Perhaps the nearest subjective impression of the process that we can envisage is that of being barged in the back while standing in a queue.

Rather like a fluid particle, one's body generates reaction by a form of distortion—to which one is sensible! Undoubtedly this impression is in proportion to the acceleration imparted to one from behind. Why acceleration and not velocity? Well, as we know, a queue can surge slowly to quite a high mean velocity without anyone being substantially "compressed"; had that same velocity been achieved by one body very rapidly, the mechanical reaction of his neighbour would be very much greater—to say nothing of his emotional reaction!

This analogy, although not exact, can be carried further to some advantage. If one is in a two-dimensional crowd, rather than in a one-dimensional crowd, there is a limit to the change of speed and direction, relative to those of one's neighbours, that one can achieve without causing serious discomfort to oneself and one's temporary neighbours. The accommodation of the crowd to one's action is limited by the rate at which reaction to it can propagate among the surrounding persons. If they cannot adjust their positions sufficiently quickly to avoid a local change of packing or of "density," there will occur an associated generation of stresses within the body. If they can, the local changes in density will be small with respect to the average in the undisturbed state, and discomfort will be minimised. The speed of propagation of disturbances from equilibrium in fluids is the speed of sound. Hence, if local disturbances of particle position due to surface vibration occur sufficiently slowly, the adjacent region of fluid *may* be able to accommodate to the changes by virtually incompressible motion. The qualification is necessary because the ability to adjust "incompressibly" depends very much upon the spatial distribution of the disturbance. If persons in adjacent positions in a line in a crowd interchange positions, local interaction and adjustment can occur with little disturbance to surrounding regions; if, on the other hand, persons in a line all attempt simultaneously to move off in the same direction normal to the line, a compressible disturbance will be propagated throughout the whole crowd. This analogy must be treated with caution because individual persons can move independently, whereas fluid particles accelerate under the action of pressures in the surrounding fluid.

On the basis of this qualitative picture of the generation of sound by a vibrating surface we can conclude that the acceleration of the surface normal to the surface, and the spatial distribution of that acceleration, significantly influence the effectiveness of fluid compression and, hence, of sound radiation. The spatial factor must be considered in relation to the time taken for a disturbance to propagate between regions of surface motion having accelerations of opposite sign. We shall now consider the physical characteristics and methods of mathematical analysis of sound radiation from planar and circular cylindrical surfaces.

2.2 The Volume Source

Surfaces vibrating in contact with fluids displace fluid volume at the interface. Consequently it is sensible initially to investigate the sound field generated by the fluid volume displacement produced by a small element of a vibrating surface. By the principle of superposition one would expect to be able to construct the field by summation of the fields from elementary sources distributed over the entire surface. Although such an exercise seems simple at first, it is generally not so, because the field generated by an elementary source depends upon the geometry of the whole surface of which it is a part, and upon the presence of any other bodies in the fluid. However, there are many cases of practical importance to which a relatively simple theoretical expression applies with reasonable accuracy.

It is shown in books on acoustics (Kinsler *et al.*, 1982) that the pressure field generated in free field (no reflections) by the uniform radial pulsation of a sphere of radius a at frequency ω is

$$p(r,t) = \frac{1}{1+jka}\frac{j\omega\rho_0\tilde{Q}}{4\pi r}\exp\{j[\omega t - k(r-a)]\}, \tag{2.1}$$

where r is the radial distance from the centre of the sphere and $\tilde{Q}$ the volume velocity of the source. The volume velocity equals the rate of displacement of fluid volume; if the normal displacement of the surface of the sphere is $\xi = \tilde{\xi}\exp(j\omega t)$ then $\tilde{Q} = j\omega 4\pi a^2\tilde{\xi}$. Equation (2.1) indicates that, as indicated previously, the surface acceleration is important in determining the radiated sound field. The significance of the rate of change of volume flow dQ/dt may be demonstrated by bringing the cupped hands together at different rates: the sound is created when the rate of volume displacement suddenly changes as the hands meet. (Draw a graph showing how you imagine Q to vary with time.)

Figure 26 illustrates the spherical source. It is clear, on account of symmetry, that the presence of a rigid plane AB in Fig. 26 in no way alters the sound field. (How do you know that the tangential component of particle velocity is zero?) If the source dimension is made very small in comparison with a wavelength, $ka \ll 1$, Eq. (2.1) reduces to

$$p(r,t) = j\omega\rho_0\frac{\tilde{Q}}{4\pi r}\exp[j(\omega t - kr)]. \tag{2.2}$$

The half-source on one side of the plane AB may represent an elementary volume source operating in an otherwise *infinite, rigid, plane* surface. To an observer confined to one side of the plane, the volume velocity source

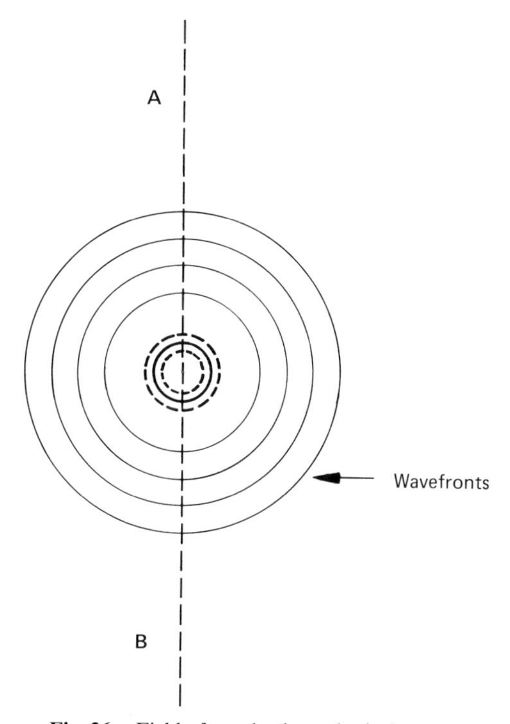

Fig. 26. Field of a pulsating spherical source.

strength appears to equal $\tilde{Q}/2$; the pressure, however, is still given by Eq. (2.2). Let the normal surface velocity of an elemental surface area δS be $v_n(t) = \tilde{v}_n \exp(j\omega t)$; then $\tilde{Q}/2 = \tilde{v}_n \, \delta S$, and we can rewrite Eq. (2.2) as

$$p(r, t) = j\omega\rho_0 \frac{2\tilde{v}_n \, \delta S}{4\pi r} \exp[j(\omega t - kr)]. \tag{2.3}$$

Carefully note the conditions under which this equation is valid.

In writing $\tilde{Q} = 2\tilde{v}_n \, \delta S$ we have tacitly assumed that the field produced by a small volume velocity source is independent of the detailed form of distribution of volume velocity over the source surface. This is, in fact, true in the limit of vanishingly small sources, and is true except very close to the source region even for sources of finite extent, provided the typical dimension d of the source region satisfies the condition $kd \ll 1$. Equation (2.3) may be applied to extended *plane* surfaces by summation or integration over the elementary sources. The resulting integral formulation was derived by Lord Rayleigh (1896):

$$p(\mathbf{r}, t) = \frac{j\omega\rho_0}{2\pi} e^{j\omega t} \int_S \frac{\tilde{v}_n(\mathbf{r}_s) e^{-jkR}}{R} \, dS, \tag{2.4}$$

where $\mathbf{r}$ is the position vector of the observation point, $\mathbf{r}_s$ the position vector of the elemental surface δS having normal velocity $\tilde{v}_n(\mathbf{r}_s)$, and R the magnitude of the vector $\mathbf{r} - \mathbf{r}_s$: $R = |\mathbf{r} - \mathbf{r}_s|$. In the following sections, Eq. (2.4) is applied to the analysis of radiation from baffled pistons and vibrating plates.

2.3 The Baffled Piston

The model of a rigid circular disc vibrating transversely to its plane in a coplanar rigid baffle is amenable to mathematical analysis and also constitutes a reasonable representation of a loudspeaker cone in a baffle or cabinet. In addition, this model has application to the design of cylindrical sonar transducers, which are usually mounted in baffles. The model is shown in Fig. 27. As with most sound radiation problems, it is difficult to evaluate the field at a distance from the source surface comparable with, or much less than, a typical source dimension. The problem of evaluating the field on the vibrating surface is particularly difficult and will be treated in Chapter 3 on

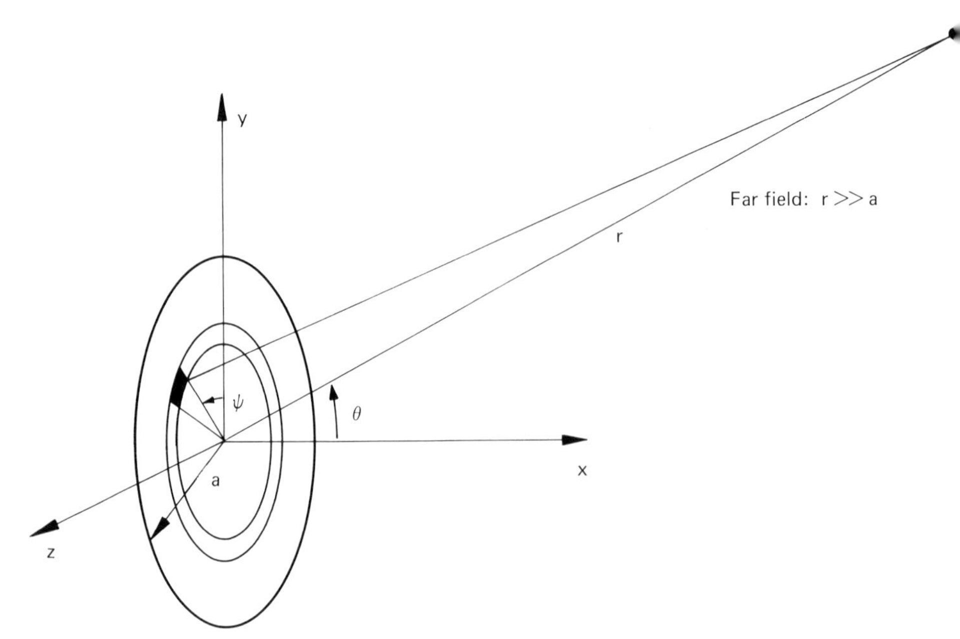

Fig. 27. Piston element and field coordinate system.

fluid loading of vibrating structures. The difficulty in evaluating the integral in Eq. (2.4) for distances not great compared with the source dimensions is associated with the fact that the distance R between elementary source and the source point is generally a rather complicated function of the coordinates of the two points. However, if the observation point is at a distance great compared with the source dimensions, R in the denominator of Eq. (2.4) may be approximated by a constant that is an average distance. On the other hand, the phase variation with $\mathbf{r}_s$, given by $\exp(-jkR)$, is a rapidly varying function of R: hence this term must be left within the integral.

In terms of the coordinate system and variables shown in Fig. 27 the integral becomes

$$\begin{aligned} p(r,t) &\simeq \frac{j\omega\rho_0\tilde{v}_n e^{j\omega t}}{2\pi r}\int_0^a y\,dy \int_0^{2\pi} \exp(-jky\sin\theta\cos\psi)\,d\psi \\ &= \frac{j\omega\rho_0\tilde{v}_n e^{j\omega t}}{r}\int_0^a J_0(ky\sin\theta)y\,dy \\ &= j\rho_0 cka^2\tilde{v}_n e^{j\omega t}\left[\frac{2J_1(ka\sin\theta)}{ka\sin\theta}\right]\frac{e^{-jkr}}{2r}. \end{aligned} \tag{2.5}$$

The functions J_0 and J_1 are Bessel functions of the first kind of the first and second order (Watson, 1966). In the limit $ka \to 0$, $J_1(ka\sin\theta)/ka\sin\theta \to \frac{1}{2}$ and the result corresponds to Eq. (2.2), with $\tilde{Q} = 2\pi a^2\tilde{v}_n$. The term containing the Bessel function is a far-field directivity term, which can be explained physically by interference between the fields radiated by the distributed elemental sources. Briefly, when $ka \ll 1$, $p(r)$ is nearly independent of θ, and the radiation is omnidirectional; when $ka \gg 1$ the sound field is much stronger on and near the polar axis, and there is very little radiation in the lateral direction [details are given in various acoustics books such as Kinsler *et al.* (1982)].

Before the vibrational characteristics of loudspeaker cones had been studied in detail, cone radiation was frequently modelled as piston radiation. Recent theoretical and experimental work using optical transducers (Bank and Hathaway, 1981) has shown that at frequencies well within the audio frequency range, loudspeaker cones exhibit radial wave patterns in which annular rings of alternate phase appear in the outer region of the cone (Fig. 28); the number of the rings increases with frequency and consequently the effective radiating radius decreases with frequency. Hence loudspeaker directivity does not increase as rapidly with frequency as piston theory would suggest. The fluid loading on the piston due to its motion is studied in Chapter 3.

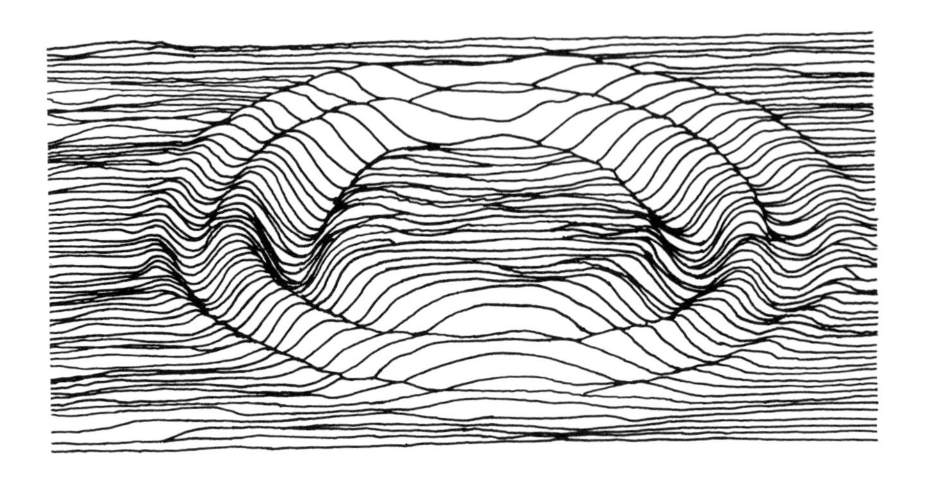

(a) New bass unit: 4 kHz

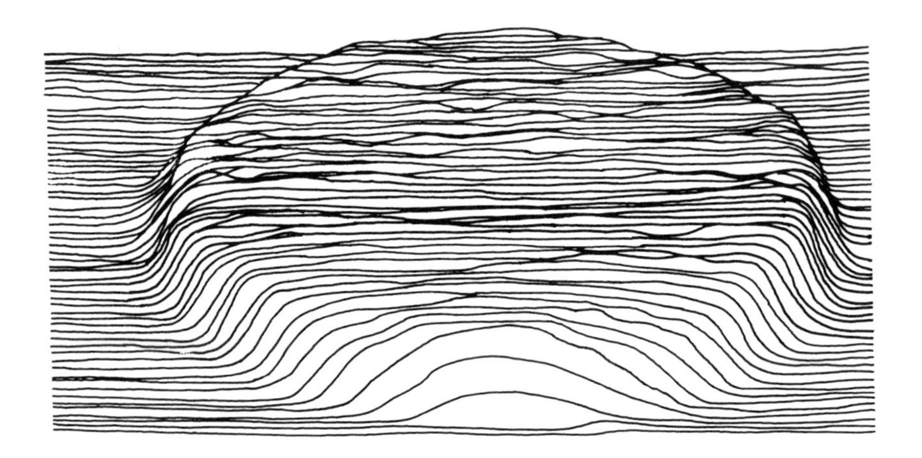

(b) Soft dome tweeter: 8 kHz

Fig. 28. Vibrational displacement distribution on loudspeakers (Bank and Hathaway, 1981). (a) New bass unit, 4 kHz; (b) Softdome tweeter, 8 kHz. Courtesy of Celestion International Ltd.

2.4 Sound Radiation by Flexural Modes of Plates

For the purpose of estimating their sound radiation characteristics, many structures of practical interest may be modelled sufficiently accurately by rectangular, uniform, flat plates. Examples include walls and floors of buildings, factory machinery casings, parts of vehicle shells, and hulls and bulkheads of ships. The natural modes of vibration of such plates vary in shape and frequency with their edge conditions and, as has already been observed, it is not strictly correct to consider the modes of isolated panels when they are dynamically coupled to contiguous structures, except in a purely mathematical sense as component modes of larger systems. However, the isolated rectangular panel forms a useful starting point for the illustration of flexural-mode radiation behaviour, and the effects of structural complications can often be estimated reasonably accurately once an understanding of the basic physics of modal radiation has been achieved.

Flexural-mode patterns of rectangular panels take the general form of contiguous regions of roughly equal area and shape, which vary alternately in vibrational phase and are separated by nodal lines of zero vibration. For simply supported edges, the normal vibration velocity distribution is

$$\tilde{v}_{\mathrm{n}}(x, z) = \tilde{v}_{pq} \sin(p\pi x/a) \sin(q\pi z/b), \qquad \begin{cases} 0 \leqslant x \leqslant a \\ 0 \leqslant z \leqslant b. \end{cases} \tag{2.6}$$

Radiation from lightly damped ($\eta < 5 \times 10^{-2}$), mechanically excited structures is usually associated with resonant vibration of the modes, i.e., around their natural frequencies. However, we can evaluate the radiation from a modal vibration distribution at any *arbitrary* frequency ω, whether it is vibrating at resonance or not. Equation (2.4) becomes

$$p(x', y', z', t) = \frac{j\omega\rho_0\tilde{v}_{pq}e^{j\omega t}}{2\pi} \int_0^a \int_0^b \frac{\sin(p\pi x/a)\sin(q\pi z/b)e^{-jkR}}{R}\, dx\, dz. \tag{2.7}$$

As already stated, this integral does not admit an analytical solution for arbitrary observer points x', y', z'. However, in the far field, where R is much greater than the source size as defined by the larger edge of the two panel dimensions a and b, Wallace (1972) has produced an analytical solution, using the coordinate system shown in Fig. 29, and Eq. (2.7) in the form*

$$p(r, \theta, \phi, t) = \frac{j\omega\rho_0\tilde{v}_{pq}e^{-jkr}e^{j\omega t}}{2\pi r} \int_0^a \int_0^b \sin\frac{p\pi x}{a} \sin\frac{q\pi z}{b} \times \exp\left[j\left(\frac{\alpha x}{a}\right) + j\left(\frac{\beta z}{b}\right)\right] dx\, dz, \tag{2.8}$$

where

$$\alpha = ka \sin\theta \cos\phi, \qquad \beta = kb \sin\theta \sin\phi. \tag{2.9}$$

* Wallace has $\exp(jkr)$ instead of $\exp(-jkr)$ because he uses the time exponent $-j\omega t$.

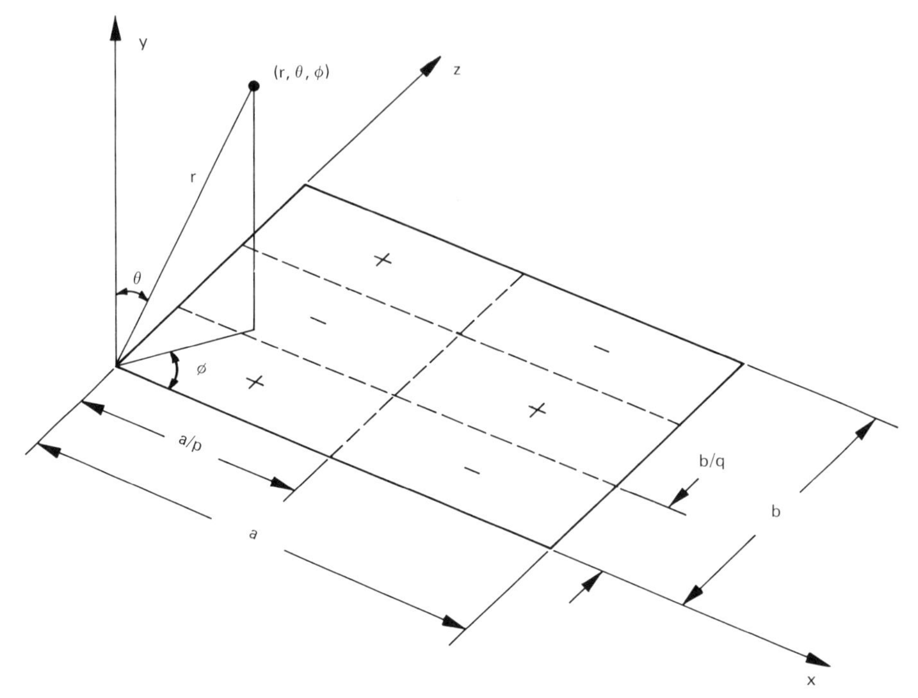

Fig. 29. Coordinate system, nodal lines and phases of a vibrating rectangular panel.

Comparison of Eqs. (2.7) and (2.8) shows that R, the distance between a surface element at (x, y) and the observation point, is related to r, the distance of the observation point from the coordinate origin, by the approximate relationship

$$R \simeq r - x \sin\theta \cos\phi - z \sin\theta \sin\phi. \tag{2.10}$$

Provided that $R \gg a$ and $R \gg b$, this relationship is sufficiently accurate for the purpose of evaluating the integral in Eq. (2.7). The solution to Eq. (2.8) is

$$\tilde{p}(r, \theta, \phi) = j\tilde{v}_{pq} k \rho_0 c \frac{e^{-jkr}}{2\pi r} \frac{ab}{pq\pi^2} \left[\frac{(-1)^p e^{-j\alpha} - 1}{(\alpha/p\pi)^2 - 1}\right]\left[\frac{(-1)^q e^{-j\beta} - 1}{(\beta/q\pi)^2 - 1}\right]. \tag{2.11}$$

Note that the distance dependence of $\tilde{p}$ is characteristic of the far field: $\tilde{p} \propto (1/r)\exp(-jkr)$.

We are not normally interested in the phase relationships in the far field, but more in the sound intensity and sound pressure level, both of which are functions of mean square pressure $\overline{p^2} = |\tilde{p}|^2/2$. The intensity is

$$\frac{|\tilde{p}(r,\theta,\phi)|^2}{2\rho_0 c} = 2\rho_0 c|\tilde{v}_{pq}|^2\left(\frac{kab}{\pi^3 rpq}\right)^2\left\{\frac{\begin{matrix}\cos\\ \sin\end{matrix}\left(\frac{\alpha}{2}\right)\begin{matrix}\cos\\ \sin\end{matrix}\left(\frac{\beta}{2}\right)}{[(\alpha/p\pi)^2-1][(\beta/q\pi)^2-1]}\right\}^2, \quad (2.12)$$

where cos $(\alpha/2)$ is used when p is an odd integer, and $\sin(\alpha/2)$ is used when p is an even integer; $\cos(\beta/2)$ is used for even q and $\sin(\beta/2)$ for odd q. Because Eq. (2.12) expresses far-field intensity, the intensity vector, of which Eq. (2.12) expresses the magnitude, has the same direction as the radial coordinate r.

Equation (2.12) reveals a great deal about the characteristics of panel radiation. The form of the denominator suggests infinite intensity when $\alpha = p\pi$ or $\beta = q\pi$. However this is not so. (Why?) The equation for the maximum intensity at a given frequency $\omega = kc$ is

$$\frac{|p(r,\theta,\phi)|^2_{\max}}{2\rho_0 c} = \rho_0 c\frac{|\tilde{v}_{pq}|^2}{2}\left(\frac{kab}{8\pi r}\right)^2, \quad (2.13)$$

where $\alpha = p\pi$ and $\beta = q\pi$, the ϕ direction of maximum radiation being given by $\tan\phi = (\beta/b)/(\alpha/a) = (q/b)/(p/a)$. This condition cannot occur at all frequencies because the substitution of $\alpha = p\pi$ and $\beta = q\pi$ in Eq. (2.9) gives

$$p\pi/ka = \sin\theta\cos\phi, \quad (2.14a)$$

$$q\pi/kb = \sin\theta\sin\phi. \quad (2.14b)$$

Since the right-hand sides of Eqs. (2.14) cannot exceed unity, the condition corresponds to $k > p\pi/a$ and $k > q\pi/b$; hence the frequency must be sufficiently high to make the acoustic wavelength $\lambda = 2\pi/k$ smaller than the traces of the structural wavelength in both the x and y directions, which are $2a/p$ and $2b/q$, respectively.

When the frequency is such that the acoustic wavelength greatly exceeds both the structural trace wavelengths, i.e., $ka \ll p\pi$ and $kb \ll q\pi$, the terms $\alpha/p\pi$ and $\beta/q\pi$ in the denominator of Eqs. (2.11) and (2.12) are small compared with unity, and can be neglected. The maximum value of intensity is

$$\frac{|p(r,\theta,\phi)|^2_{\max}}{2\rho_0 c} = 2\rho_0 c|\tilde{v}_{pq}|^2\left(\frac{kab}{\pi^3 rpq}\right)^2. \quad (2.15)$$

This value is of the order of $(pq)^{-2}$ times that given by Eq. (2.13). It is instructive to compare this intensity with that which would be radiated by a single modal cell of area ab/pq in the absence of any other surface motion. Since the dimensions of the cell have been assumed to be far smaller than an acoustic wavelength, the single cell may be modelled by a point source of volume velocity given by

$$\tilde{Q} = 2\tilde{v}_{pq} \int_0^{a/p} \int_0^{b/q} \sin(p\pi x/a) \sin(q\pi z/b)\, dx\, dz$$

$$= \tilde{v}_{pq}(8ab/\pi^2 pq). \tag{2.16}$$

Equation (2.2) applies, and the far-field intensity at all points a distance r from the source is

$$|\tilde{p}(r)|^2/2\rho_0 c = 2\rho_0 c|\tilde{v}_{pq}|^2(kab/\pi^3 rpq)^2. \tag{2.17}$$

By comparing Eqs. (2.17) and (2.15) we see that the intensity that would be produced at low frequencies by one cell equals the maximum intensity generated by all the pq cells acting together; and because the intensity produced by one cell is independent of θ and ϕ, whereas the intensity produced by the complete panel is less in all other directions than the maximum given by Eq. (2.15), the total sound power radiated by a single isolated cell would exceed that radiated by the whole panel.

A more useful measure of the effectiveness of sound radiation by vibrating surfaces than that of the maximum intensity is the total radiated sound power normalised on the panel area and the velocity of surface vibration. The radiated power can be obtained by integrating the far-field intensity over a hemispherical surface centred on the panel. Since the far-field intensity is directed radially, the time-average power may be written as

$$\bar{P} = \int_S I(\theta, \phi)\, dS = \int_0^{\pi} \int_0^{2\pi} I(\theta, \phi) r^2\, d\theta\, d\phi. \tag{2.18}$$

A measure of the velocity of vibration is the space-average value of the time-average normal vibration velocity $\langle \overline{v_n^2} \rangle$ defined by

$$\langle \overline{v_n^2} \rangle = \frac{1}{S} \int_S \left[\frac{1}{T} \int_0^T v_n^2(x, y, t) \right] dS, \tag{2.19}$$

where T is a suitable period of time over which to estimate the mean square velocity $\overline{v^2(x, y)}$ at a point (x, y), and S extends over the total vibrating surface: $\langle \overline{v_n^2} \rangle$ is sometimes known as the "average mean square velocity." For the modal distribution of velocity given by Eq. (2.6),

$$\langle \overline{v_n^2} \rangle_{pq} = |\tilde{v}_{pq}|^2/8. \tag{2.20}$$

A radiation efficiency, ratio, or index,* as it is variously termed, is defined by reference to the acoustic power radiated by a uniformly vibrating baffled

* Logarithmic measure = $10 \log_{10} \sigma$.

piston at a frequency for which the piston circumference greatly exceeds the acoustic wavelength: $ka \gg 1$. Use of Eq. (2.5) to evaluate the far-field intensity of a piston, together with integration over a hemisphere, yields the following expression for radiated power in this case:

$$\bar{P} = \tfrac{1}{2}\rho_0 c\pi a^2 |\tilde{v}_n|^2. \tag{2.21}$$

As a definition of reference radiation efficiency σ, we write

$$\bar{P} = \sigma\rho_0 cS\langle\overline{v_n^2}\rangle, \tag{2.22}$$

where σ is unity in the reference case.

Hence the radiation efficiency is defined as the ratio of the average acoustic power radiated per unit area of a vibrating surface to the average acoustic power radiated per unit area of a piston that is vibrating with the same average mean square velocity at a frequency for which $ka \gg 1$: note also that the fluid impedances $\rho_0 c$ should be equal. There is no physical reason why the radiation efficiency of a vibrating surface should not exceed unity, and therefore the term "radiation efficiency" is rather misleading. However, in most practical cases the radiation efficiency is either below or very close to unity.

The time-average vibration energy of an oscillator or set of oscillators in temporally stationary resonant or broad-band excited vibration is equal to twice the time-average kinetic energy. (Prove.) For a panel, this gives

$$\bar{E} = \int_S m(x, y)\overline{v_n^2(x, y)}\, dx\, dy, \tag{2.23}$$

where m is the mass per unit area. A radiation loss factor η_{rad} can be defined thus:

$$\bar{P} = \eta_{rad}\omega\bar{E} = \sigma\rho_0 cS\langle\overline{v_n^2}\rangle. \tag{2.24}$$

For a panel of uniform density ρ_s and thickness h executing such vibration, $\bar{E} = \rho_s hS\langle\overline{v_n^2}\rangle$. Hence, in such cases,

$$\eta_{rad} = (\rho_0/\rho_s)(1/kh)\sigma. \tag{2.25}$$

The radiation loss factor indicates the magnitude of acoustic radiation damping; it rarely exceeds 10^{-3} for engineering structures vibrating in air, but in liquids it can exceed the loss factors associated with frictional mechanisms.

Wallace presents the following approximate expressions for the radiation efficiency of rectangular panels at arbitrary frequencies for which the acoustic wavelength is very much greater than either of the panel trace wavelength components ($\lambda_x = 2a/p \ll 2\pi/k$; $\lambda_y = 2b/q \ll 2\pi/k$):

(a) p and q both odd integers,

$$\sigma_{pq} \simeq \frac{32(ka)(kb)}{p^2q^2\pi^5}\left\{1 - \frac{k^2ab}{12}\left[\left(1 - \frac{8}{(p\pi)^2}\right)\frac{a}{b} + \left(1 - \frac{8}{(q\pi)^2}\right)\frac{b}{a}\right]\right\}. \quad (2.26a)$$

(b) p odd and q even,

$$\sigma_{pq} \simeq \frac{8(ka)(kb)^3}{3p^2q^2\pi^5}\left\{1 - \frac{k^2ab}{20}\left[\left(1 - \frac{8}{(p\pi)^2}\right)\frac{a}{b} + \left(1 - \frac{24}{(q\pi^2)}\right)\frac{b}{a}\right]\right\}. \quad (2.26b)$$

(Interchanging p and q gives the result for p even and q odd.)

(c) p and q both even integers,

$$\sigma_{pq} \simeq \frac{2(ka)^3(kb)^3}{15p^2q^2\pi^5}\left\{1 - \frac{5k^2ab}{64}\left[\left(1 - \frac{24}{(p\pi)^2}\right)\frac{a}{b} + \left(1 - \frac{24}{(q\pi)^2}\right)\frac{b}{a}\right]\right\}. \quad (2.26c)$$

Clearly the odd–odd modes are the most effective radiators when ka, $kb \ll 1$; a good approximation to Eq. (2.26a) is then

$$\sigma_{pq} \simeq \frac{2k^2}{\pi^5 ab}\left(\frac{2a}{p}\right)^2\left(\frac{2b}{q}\right)^2 = \frac{2k^2\lambda_x^2\lambda_z^2}{\pi^5 ab}, \quad (2.27)$$

where λ_x and λ_y are the structural trace wavelengths in the x and y directions.

A radiation efficiency based upon the total panel area, but upon the power radiated by only one cell vibrating in isolation, is obtained by integrating Eq. (2.17) over the surface of a large hemisphere of radius r:

$$\sigma_1 = \frac{32(ka)(kb)}{\pi^5 p^2 q^2} = \frac{2k^2\lambda_x^2\lambda_z^2}{\pi^5 ab}. \quad (2.28)$$

We see that the radiation efficiency of all panel modes at low frequencies is at most equal to, and mostly less than, that corresponding to isolated cell vibration and that, for the most efficient odd–odd modes, with the same wavelengths, it is inversely proportional to the area of the panel.

Consider two rectangular panels of different size but of the same aspect ratio, material, and thickness. Modes of the two having equal structural trace wavelengths will have the same natural frequencies. Hence, according to Eq. (2.27) the radiation efficiencies of these two panels *at this frequency* will be different, in inverse proportion to their areas; the total power radiated by vibration of the same level in both is the same. We may ask why it is that the panel of larger area does not radiate more power. One answer lies in the far-field form of the Rayleigh integral [Eq. (2.4)] in which R in the denominator may be taken outside the integral:

$$p(\mathbf{r}, t) \simeq \frac{j\omega\rho_0 e^{j\omega t}}{2\pi R_0}\int_S \tilde{v}_n(\mathbf{r}_s)e^{-jkR}\, dS, \quad (2.29)$$

where R_0 is the average distance from the observation point to the panel. For a mode of order p, q the maximum distance between the centres of adjacent regions of volume velocity of opposite sign is $l = a/p$ or b/q, whichever is the greater. At the observation point the contributions from these regions differ in phase by $\delta\phi = kl \sin\theta$, where θ is measured from the normal to the baffle plane; the maximum value of $\delta\phi$ is kl at positions lying on the plane. If $kl \ll \pi$ or $l \ll \lambda/2$, the phase difference due to distance is negligible compared with the difference of vibrational phase between adjacent cells: hence the contributions to the field from adjacent cells almost cancel.

We may visualise the cancellation process by reference to Fig. 30, which illustrates the one-dimensional case. Since the radiation fields must be either phase-symmetric or phase-antisymmetric about the panel centre, for odd and even modes, respectively, cancellation must involve half-cell pairs, and not whole-cell pairs; it is seen that the half cells at the edges of the panel remain. If, in addition to kl being small, ka is also small, the out-of-phase end half cells of the even mode will also largely cancel, whereas those of the odd mode will add. This reasoning can be extended to the real two-dimensional case to provide a physical understanding of the form of the low-frequency radiation expressions in Eqs. (2.26). In particular, we can see that the four uncancelled corner quarter cells may add their contributions (p,q odd; $ka,kb \ll 1$) or almost cancel (p,q even; $ka,kb \ll 1$). As frequency increases, the reinforcement or cancellation decreases in effect.

An alternative physical picture of the behaviour can be obtained by utilising the principle of acoustic reciprocity for point monopole sources. Under

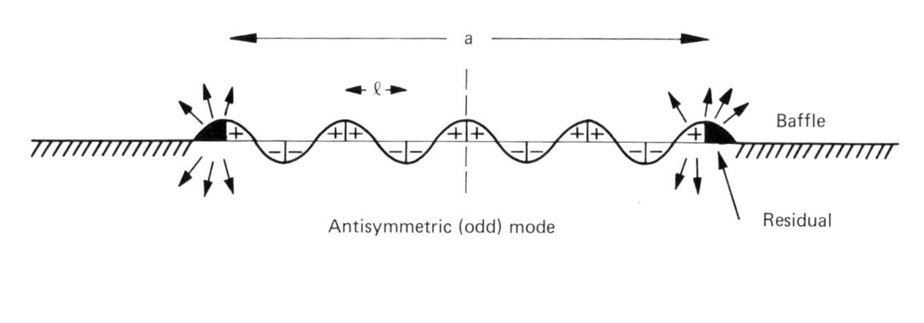

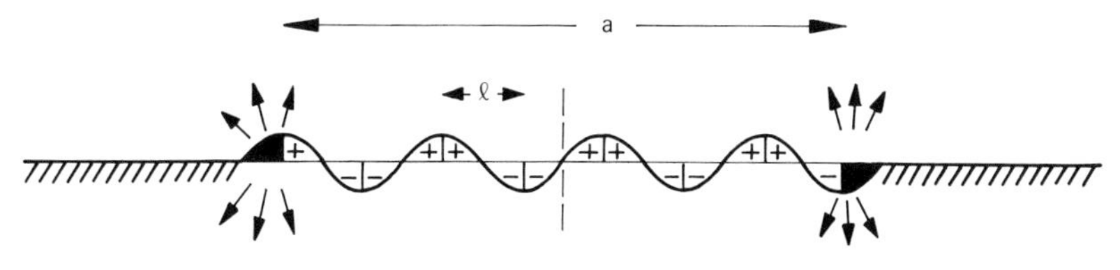

Fig. 30. Intercell cancellation on vibrating plates: $kl \ll 1$.

all time-invariant states of a non-flowing fluid and of the dynamic properties of its boundaries, the positions of a point monopole source and an observer can be reversed without altering the observed sound pressure (Belousov *et al.*, 1975). That this is true for a point source on the surface of an infinite rigid plane can be deduced from the expression for the field of a point source in Eq. (2.2), by considering a point source at any arbitrary position, together with its image in an infinite rigid plane: the introduction of the image allows the plane to be removed. The total pressure at the position of the plane is clearly equal to twice that in the absence of the image; similarly the pressure produced by an elemental volume source operating at the surface of a rigid plane is double that produced by the same source in the absence of the plane.

Now, as we have seen, each element $dx\,dy$ of the integrand of Eq. (2.7) behaves like a point source on a rigid plane. Hence the variation of the exponential term $\exp(-jkR)$ in the integrand with variation of source point can as well be envisaged as the variation with surface position, of the pressure phase on the surface of the panel produced by a point source operating at the observation point. When the observation point is very far from the panel surface, the curvature of the wave fronts produced by such a source at their intersection with the panel surface is negligible, and they can be considered to be plane (Fig. 31). The degree of matching of the vibration velocity and the exponential term can thus be visualised. The reason for the directional

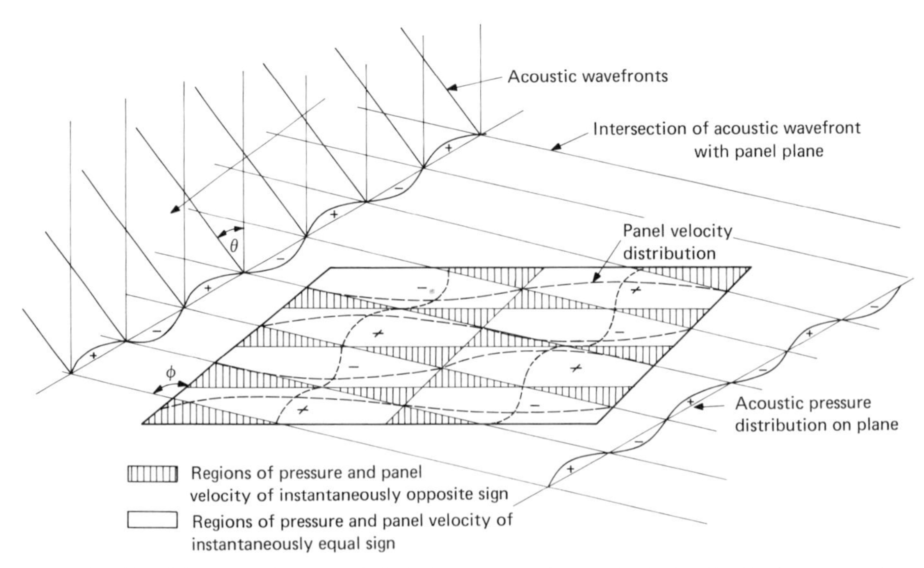

Fig. 31. Matching of pressure distribution and panel normal velocity distribution on panel plane (panel instantaneously in equilibrium position).

maximum radiation characteristics indicated below Eq. (2.13) can now be seen.

It would appear from the construction of Fig. 31 that matching to a given mode would not be poor only at low frequencies, where the acoustic half-wavelength becomes large compared with nodal line separation distances in both the x and y directions, but also at high frequencies, where the inverse is true. However, it must be remembered that the distance between wavefront–plane intersections corresponding to intervals of half an acoustic wavelength increases with elevation of the observer above the source plane; some elevation angle giving optimum matching can therefore always be found below infinite frequency. This fact explains why high-frequency modal radiation is directional not only in azimuth (ϕ) but also in elevation (θ), the direction of maximum radiation moving toward the normal with increasing frequency.

Graphical presentations of radiation efficiency data for rectangular simply supported panels have been published by Wallace (1972). Gomperts (1977) has extended the analysis to other plate boundary conditions. The radiation efficiency of all modes becomes asymptotic to unity at high frequencies or, to be more specific, when the ratio γ of acoustic wavenumber to structural wavenumber exceeds unity: for a simply supported rectangular panel the structural wavenumber is given by $k_b^2 = (p\pi/a)^2 + (q\pi/b)^2$ [see Eqs. (1.40) and (1.41)]. In general, for $\gamma < 1$, the radiation efficiencies of the modes for which both p and q are odd are the highest, that of the fundamental mode (1, 1) being dominant. (Why?) Those of the even–even modes are the lowest, as we have already seen. For a given mode order (p, q) and γ, modes having square internodal areas radiate less well, and modes having internodal areas of aspect ratio aq/bp very different from unity are the most effective radiators. Some representative results are shown in Fig. 32. An alternative method of estimating the total sound power radiated by a single structural mode, which makes direct use of the Rayleigh integral, is described in Section 7.4.

These results are in no way applicable to unbaffled or perforated panels. In the former case, at low frequencies the surface velocity sources on the "underside" of the panel, especially those near the edge, largely cancel the field produced by the "upper side" surface sources of opposite phase, because the distances to the far field are insignificantly different in terms of a wavelength, especially at positions near the plane of the panel. In the language of idealised acoustic sources, the corresponding regions of opposite phase on the two sides of the panel constitute dipoles that are very much less efficient than pure volume velocity (monopole) sources; hence the low-frequency radiation efficiency is far less than the baffled panel equivalent. This cancellation effect decreases as frequency increases and the acoustic wavenumber approaches the structural wavenumber.

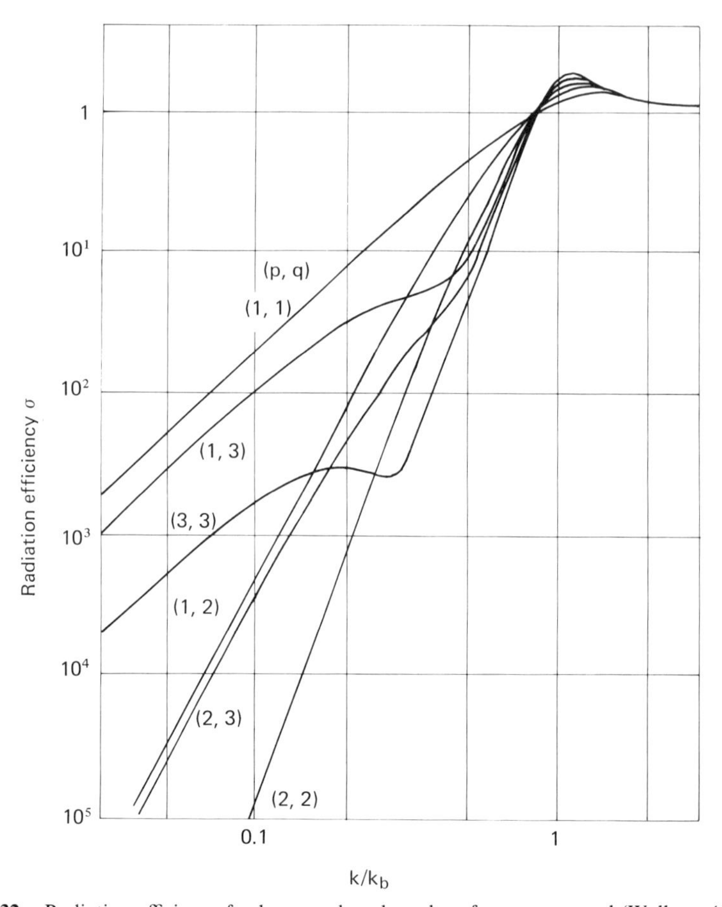

Fig. 32. Radiation efficiency for low-numbered modes of a square panel (Wallace, 1972).

(Try holding a thin panel of plywood, or similar material, about 600 mm × 600 mm in size, parallel to a wall or table top, and tap it while moving it closer to the surface. Can you think of any explanation for the observed effect, perhaps in terms of sources and images?)

In the case of a perforated panel, the effect of pressure differences across the panel, due to opposite phase surface sources, is to drive fluid through the holes, thereby largely cancelling the surface volume velocities and hence the radiation. However, the radiation efficiency of a perforated plate is not zero, and it depends upon the geometry of the perforations and the thickness of the plate. Some measured data are presented in Fig. 33. (Why is volume velocity cancellation not complete?)

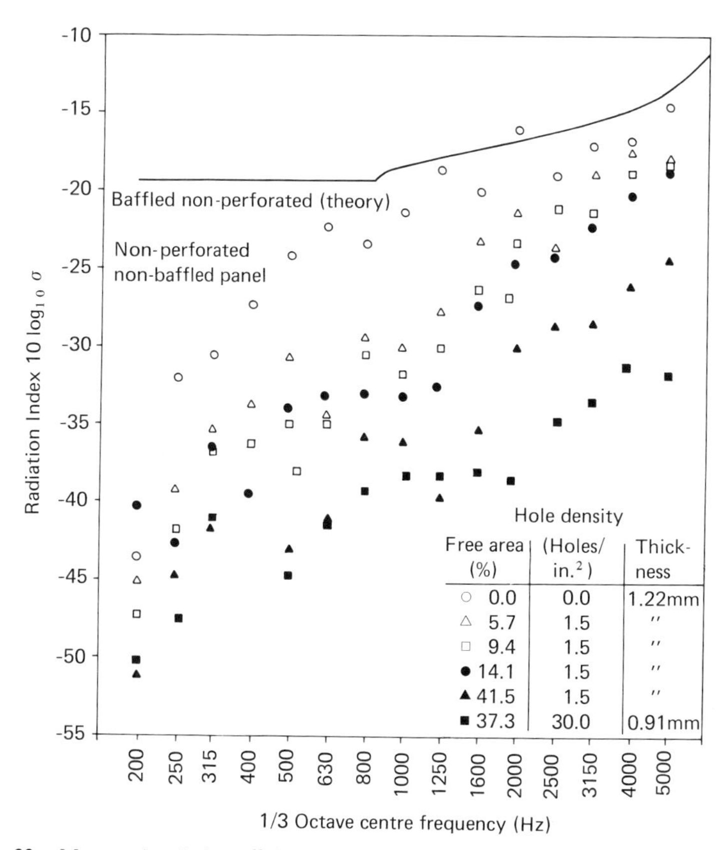

Fig. 33. Measured radiation efficiencies of 1-mm-thick unbaffled perforated panels (Pierri, 1977).

So far we have only briefly mentioned the *natural* frequencies of the baffled rectangular panel studied above. This is because it is possible to analyse radiation from a given mode at *arbitrary* frequencies to produce data in the form presented by Wallace. In most practical cases of vibration of lightly damped structures, the kinetic energy of vibration is associated mainly with resonant modal vibration, which occurs at frequencies close to the *in vacuo* modal natural frequencies. This condition is, however, not sufficient for us to assume that it is only the radiation of a mode at its natural frequency that is important, because the resonance frequency radiation efficiency of thin-panel modes is often rather low and rises with frequency, as Eqs. (2.26) show. Hence there can be a trade-off between reduction of vibration response above the resonance frequency, and increase of efficiency. In most cases of *mechanical* excitation, the reduction in response wins, and only resonant radiation is important. As we shall see in Chapter 4, this is usually not so in the case of *airborne* sound transmission through panels.

2.5 Sound Radiation by Flexural Waves in Plates

Before proceeding to the question of frequency distributions of natural frequencies of plates and thereby to their frequency-average resonant radiation characteristics, we pause to reconsider radiation by panel modes from a different viewpoint. We have seen in Chapter 1 that modal vibration of lightly damped structures arises from the constructive interference between travelling waves, which are multiply reflected from structural boundaries or discontinuities, to form standing patterns of vibration, or modes. It is therefore natural to enquire into the nature of sound radiation by transverse waves travelling along a structure–fluid interface.

We start with the two-dimensional problem illustrated in Fig. 34. An infinite, uniform plate is in contact with a fluid that exists in the semi-infinite space $y > 0$. A plane transverse wave of *arbitrary* frequency ω and wavenumber κ is forced to travel in the plate; the transverse plate displacement is

$$\eta(x, t) = \tilde{\eta} \exp[j(\omega t - \kappa x)]. \quad (2.30)$$

The acoustic field in the fluid is found by using the solution of the two-dimensional wave equation (1.7), together with the fluid momentum equations (1.5) at the plate–fluid interface. The x-wise variation of the acoustic field variable must follow that of the plate wave, i.e., k_x in the fluid equals κ. From Eq. (1.8) we find that

$$k_y = \pm(k^2 - \kappa^2)^{1/2}, \quad (2.31)$$

where $k = \omega/c$.

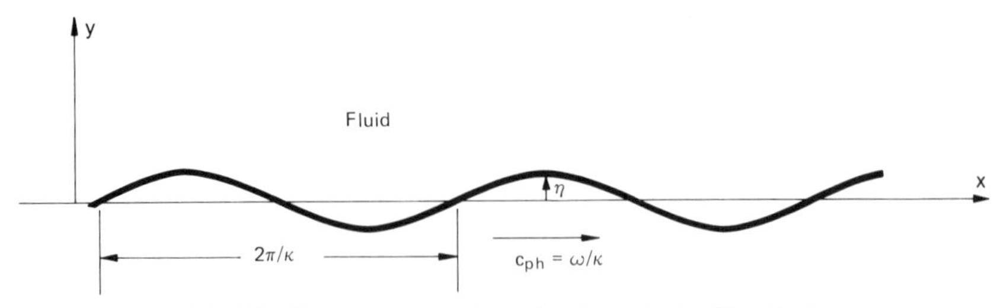

Fig. 34. Transverse wave in a plate in contact with a fluid.

The appropriate sign of the square root is determined by the physics of the model. Three distinct conditions are possible:

(a) Plate wave phase speed is greater than the speed of sound in the fluid: $\kappa < k$. Plane sound waves travel away from the surface of the plate at an angle to the normal given by $\cos\phi = k_y/k = [1 - (\kappa/k)^2]^{1/2}$, as seen from Eq. (1.7): no wave can propagate toward the plate surface, and therefore the negative sign is disallowed.

(b) Plate wave phase speed is less than the speed of sound in the fluid: $\kappa > k$. When $\kappa > k$, k_y is imaginary; the disturbance of the fluid decays exponentially with distance normal to the plate and only a surface wave exists in the fluid. In this case the negative sign of the square root must be selected so that

$$k_y = -j(\kappa^2 - k^2)^{1/2}, \tag{2.32}$$

$$\tilde{p}(x, y) = \tilde{p}\exp(-j\kappa x)\exp\{-k[(\kappa/k)^2 - 1]^{1/2}y\}. \tag{2.33}$$

(c) Plate wave phase speed equals the speed of sound in the fluid: $\kappa = k$. In this case $k_y = 0$: however, the boundary conditions cannot be satisfied in practice, because finite vibration produces infinite pressure, and this condition cannot occur physically.

The amplitude $\tilde{p}$ of the pressure field is determined by the application of the condition of compatibility of normal velocities, or displacements, at the plate–fluid interface. Equation (1.9), which expresses the fluid impedance at the interface, is

$$(\tilde{p}/\tilde{v})_{y=0} = \omega\rho_0/k_y = \pm\omega\rho_0/(k^2 - \kappa^2)^{1/2}. \tag{2.34}$$

Note that this is a particular form of fluid impedance, which can be termed a *wave impedance* because of the wavelike form of the excitation.

In case (a), $\kappa < k$, and

$$(\tilde{p}/\tilde{v})_{y=0} = \rho_0 c/[1 - (\kappa/k)^2]^{1/2}. \tag{2.35}$$

The fluid wave impedance is real and positive; according to Eq. (1.10) the plate does work upon the fluid, and radiation damping results.

In case (b), $\kappa > k$, and

$$(\tilde{p}/\tilde{v})_{y=0} = j\omega\rho_0/(\kappa^2 - k^2)^{1/2} = j\rho_0 c/[(\kappa/k)^2 - 1]^{1/2}. \tag{2.36}$$

The fluid wave impedance is imaginary and positive; no work is done upon the fluid, but inertial fluid loading is experienced, as discussed more fully in Chapter 3.

In case (c), $\kappa = k$, and

$$(\tilde{p}/\tilde{v})_{y=0} \to \infty. \tag{2.37}$$

The fluid loading is infinitely great; in practice, plate wave components satisfying this condition cannot exist.

This analysis shows that only plate wave components with phase speeds greater than the speed of sound create disturbances in the far field and radiate energy into the fluid. In the case of an infinitely extended vibrating plate, the far-field condition necessarily implies an infinitely great distance normal to the plate, whereas in the case of a vibrating plate of finite area, the far-field condition may be satisfied at great distances from the centre of the plate even by points on or near the plane of the plate. Hence we can associate sound energy from any planar structure (strictly, in a baffle) with the presence in its normal vibration velocity distribution of supersonic wave-number components k_s satisfying the condition $k_s < k$ at the frequency concerned. This is the vital conclusion from the aforegoing analysis.

Returning now to the problem of sound radiation from the modes of a simply supported panel, we observe that a mode corresponds to a particular pattern of interference between travelling waves. This suggests that the radiation from the component travelling waves could be estimated individually and then summed to give the total field. However, unlike the waves travelling in an infinite plate, those in a panel exist only within a finite interval of space, between panel boundaries. The corresponding cases in time-dependent signals are those of the pure sinewave of single frequency and the transient signal consisting of a finite duration sample of that signal. The frequency spectrum of the latter is obtained by applying the Fourier integral transform (Randall, 1977)

$$\tilde{F}(\omega) = \int_{-\infty}^{\infty} f(t) \exp(-j\omega t)\, dt. \tag{2.38}$$

The spatial equivalent of Eq. (2.38) is the wavenumber transform

$$\tilde{F}(k) = \int_{-\infty}^{\infty} f(x) \exp(-jkx)\, dx. \tag{2.39}$$

The field $f(x)$ can be considered to be constructed from an infinity of *infinitely extended* pure sinusoidal travelling waves of the form

$$f_k(x) = \tilde{F}(k) \exp(jkx), \tag{2.40}$$

just as the transient time signal can be considered to be made up of an infinity of pure tones,

$$f_\omega(t) = \tilde{F}(\omega) \exp(j\omega t). \tag{2.41}$$

The formal expression of the synthesis of $f(x)$ from $\tilde{F}(k)$ is

$$f(x) = \frac{1}{2\pi} \int_{-\infty}^{\infty} \tilde{F}(k) \exp(jkx)\, dk. \tag{2.42}$$

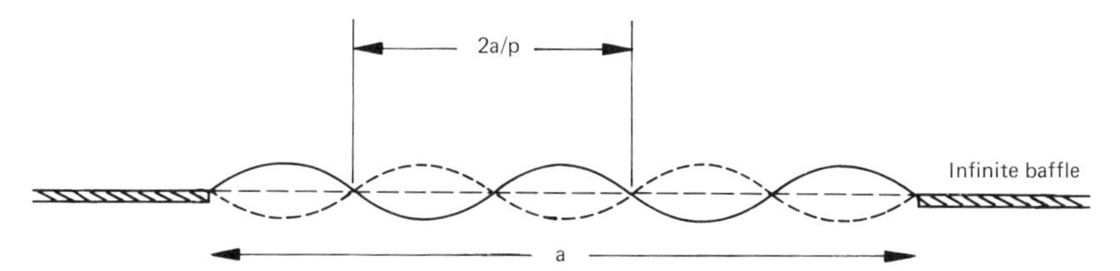

Fig. 35. Modal vibration of an infinitely long strip in a baffle.

The interval of integration is seen to extend from $-\infty$ to $+\infty$, so that waves travelling in both directions are included.

The analytical and conceptual advantage of decomposing a modal pattern into infinitely extended travelling waves is that the plane-wave impedance relationship (2.34) can be applied exactly to each component to allow a radiated pressure field to be related to the surface normal velocity distribution. This process is first illustrated by application to a one-dimensional surface velocity distribution.

Consider a simply supported panel of width a and of infinite length in an infinite rigid baffle (Fig. 35). The velocity distribution is

$$v_n(x, t) = \begin{cases} \tilde{v}_p \sin(p\pi x/a) \exp(j\omega t); & 0 < x < a, \\ 0; & 0 > x > l. \end{cases} \tag{2.43}$$

The wavenumber transform of v_n is

$$\tilde{V}(k_x) = \tilde{v}_p \int_0^a \sin(p\pi x/a) \exp(-jk_x x)\, dx, \tag{2.44}$$

where k_x is used to distinguish it from the acoustic wavenumber k. The solution of Eq. (2.44) is

$$\tilde{V}(k_x) = \tilde{v}_p \frac{(p\pi/a)[(-1)^p \exp(-jk_x a) - 1]}{[k_x^2 - (p\pi/a)^2]}. \tag{2.45}$$

Hence

$$v_n(x, t) = \frac{\exp(j\omega t)}{2\pi} \int_{-\infty}^{\infty} \tilde{V}(k_x) \exp(jk_x x)\, dk_x. \tag{2.46}$$

Associated with each wavenumber component $\tilde{V}(k_x)$ is a surface pressure field wavenumber component given by Eq. (2.34) as

$$[\tilde{P}(k_x)]_{y=0} = \frac{\pm \omega\rho_0}{(k^2 - k_x^2)^{1/2}} \tilde{V}(k_x), \tag{2.47}$$

so that the surface pressure field can be expressed as

$$[p(x, t)]_{y=0} = \frac{\exp(j\omega t)}{2\pi} \int_{-\infty}^{\infty} [\tilde{P}(k_x)]_{y=0} \exp(jk_x x)\, dk_x. \tag{2.48}$$

The power radiated per unit length of the plate is found by evaluating the expression

$$\bar{P} = \frac{1}{T}\int_0^T \int_0^l [p(x,t)]_{y=0}\, v_n(x,t)\, dx\, dt,$$
$$= \frac{1}{2}\,\mathrm{Re}\left\{\int_0^l \tilde{p}(x)\tilde{v}_n^*(x)\, dx\right\}. \tag{2.49}$$

Substitution for v_n and p from Eqs. (2.46) and (2.48) gives

$$\bar{P} = \frac{1}{8\pi^2}\,\mathrm{Re}\left\{\int_{-\infty}^{\infty}\left[\int_{-\infty}^{\infty}[\tilde{P}(k_x)]_{y=0} \times \exp(jk_x x)\, dk_x \int_{-\infty}^{\infty}\tilde{V}^*(k'_x)\exp(-jk'_x x)\, dk'_x\right] dx\right\}, \tag{2.50}$$

where k'_x has been used solely to distinguish between the integrations over k_x associated with $\tilde{P}$ and $\tilde{V}$. Notice that the range of integration over x has changed from 0 to l to $-\infty$ to $+\infty$, because the form of $\tilde{V}(k_x)$ ensures that v is actually zero outside the range $0 \leqslant x \leqslant l$. We can replace $\tilde{P}$ in Eq. (2.50) by substituting from Eq. (2.47) to give

$$\bar{P} = \frac{1}{8\pi^2}\,\mathrm{Re}\left\{\int_{-\infty}^{\infty}\left[\int_{-\infty}^{\infty}\frac{\pm\omega\rho_0}{(k^2-k_x^2)^{1/2}}\tilde{V}(k_x) \times \exp(jk_x x)\, dk_x \int_{-\infty}^{\infty}\tilde{V}^*(k'_x)\exp(-jk'_x x)\, dk'_x\right] dx\right\}. \tag{2.51}$$

Integration is first performed over x: the only functions of x are $\exp(jk_x x)$ and $\exp(-jk'_x x)$, which together form the integrand $\exp[j(k_x - k'_x)x]$. The integral of this term over the doubly infinite range is zero if $k_x \neq k'_x$, and is clearly infinite if $k_x = k'_x$: in fact, it is equivalent to the Dirac delta function $2\pi\delta(k_x - k'_x)$. As explained in Section 1.13, the nature of this function is such that the subsequent integration over k'_x sets k'_x equal to k_x. Hence the integral of Eq. (2.51) becomes

$$\bar{P} = \frac{1}{4\pi}\,\mathrm{Re}\left\{\int_{-\infty}^{\infty}\frac{\pm\omega\rho_0|\tilde{V}(k_x)|^2\, dk_x}{(k^2-k_x^2)^{1/2}}\right\}. \tag{2.52}$$

Now, only wavenumber components satisfying the condition $k_x \leqslant k$ contribute to sound power radiation and to the real part of the integral. Hence the range of integration can be reduced to $-k \leqslant k_x \leqslant k$ to give

$$\bar{P} = \frac{\rho_0 ck}{4\pi}\int_{-k}^{k}\frac{|\tilde{V}(k_x)|^2}{(k^2-k_x^2)^{1/2}}\, dk_x. \tag{2.53}$$

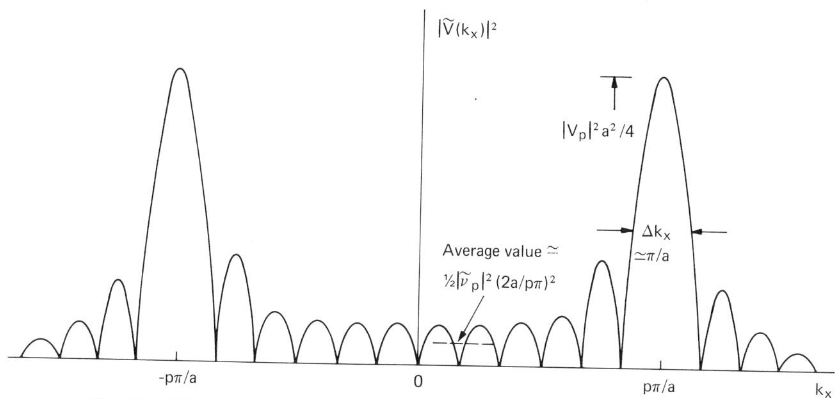

Fig. 36. Wavenumber modulus spectrum of plate velocity (diagrammatic).

Therefore, in order to evaluate the radiated power we need to consider the form of the spectrum of the modulus squared of $\tilde{V}(k_x)$, which is also known as the energy spectrum of $\tilde{V}(k_x)$ (Randall, 1977). Evaluation of $\tilde{V}(k_x)$ gives

$$|\tilde{V}(k_x)|^2 = |\tilde{v}_p|^2 \left[\frac{2\pi p/a}{k_x^2 - (p\pi/a)^2}\right]^2 \sin^2\left(\frac{k_x a - p\pi}{2}\right), \tag{2.54}$$

which is plotted as a spectrum in Fig. 36. As expected, the spectrum peaks when $k_x = p\pi/a = 2\pi/\lambda$, where $\lambda = 2a/p$ is the structural "wavelength": the value of the peak is $|\tilde{v}_p|^2 a^2/4$, independent of p. The exception to this form is the modulus spectrum of the fundamental mode ($p = 1$), which has its maximum value at $k_x = 0$ (the dc value). The "bandwidth" of the major peak in the modulus-squared spectrum is related to the panel width a by $\Delta k_x \simeq \pi/a$, which is independent of p. For values of $k_x \ll p\pi/a$, the average value of the spectrum has an approximately constant value of $\frac{1}{2}|\tilde{v}_p|^2(2a/p\pi)^2$, which is proportional to the square of the structural wavelength. For the condition $ka \ll p\pi$, Cremer *et al.* (1973) evaluates Eq. (2.53) as

$$\bar{P} = \frac{\rho_0 ck|\tilde{v}_p|^2}{2}\left(\frac{a}{p\pi}\right)^2. \tag{2.55}$$

(Try it yourself.) The corresponding radiation efficiency is

$$\sigma = 2kl/p^2\pi^2. \tag{2.56}$$

for $ka \gg p\pi$, the radiation efficiency is unity.

The significance of the shape of the modulus-squared spectrum relates to the condition necessary for sound energy radiation from traveling surface velocity waves, namely, $|k_x| < k$ [Eq. (2.35)]. At any particular frequency,

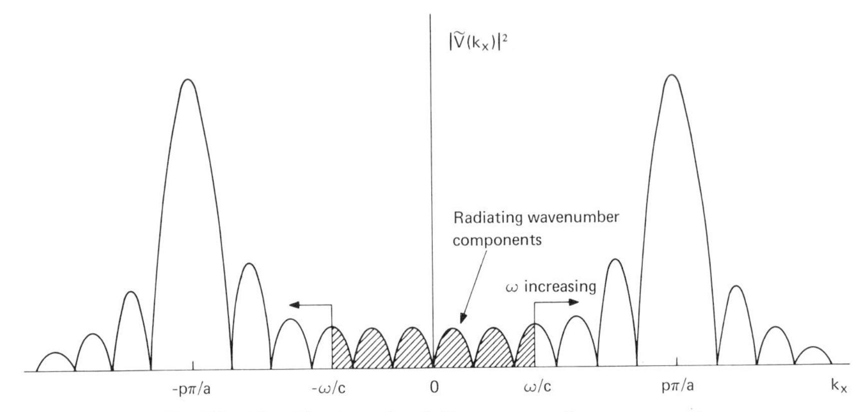

Fig. 37. Identification of radiating wavenumber components.

$k = \omega/c$ can be marked on the wavenumber axis in Fig. 37; only those wavenumber components satisfying the above condition can radiate sound energy; the others simply create nearfield disturbance of the fluid close to the plate surface. As the frequency of vibration in a *given mode shape* is increased, the proportion of radiating wavenumbers increases, as shown in Fig. 37, in which the negative wavenumber spectrum is included. Clearly the radiation increases rapidly as the acoustic wave number approaches the structural wavenumber. The fundamental mode is an exception shown in Fig. 38.

Constructions of the form shown in Figs. 36–38 can help one to form a qualitative picture of the influence of plate parameters on sound energy radiation. For instance in Fig. 39 two panels of the same material and thick-

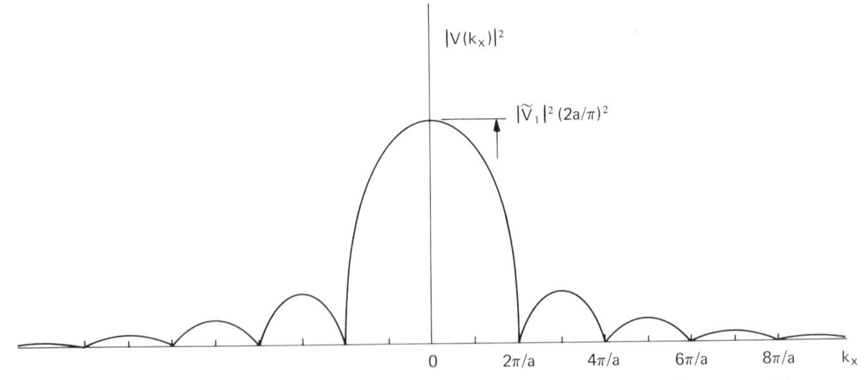

Fig. 38. Wavenumber modulus spectrum of fundamental mode (diagrammatic).

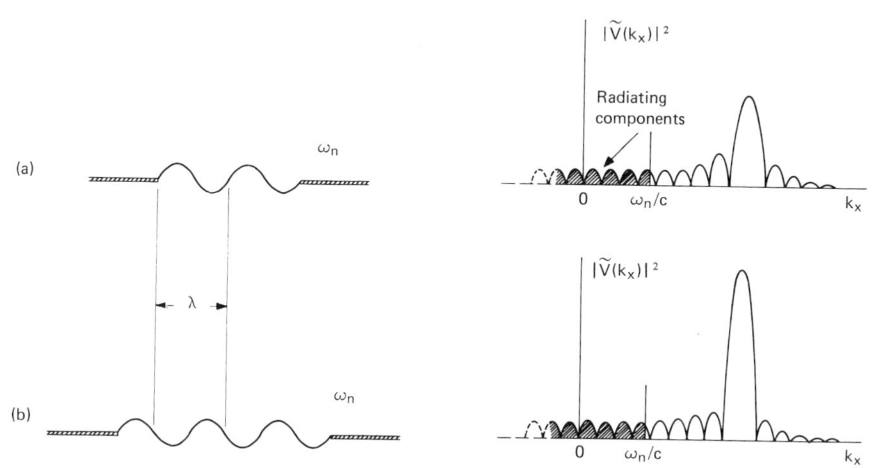

Fig. 39. Modal wavenumber spectra of two plates of the same material but different widths.

ness, but of different width, are shown vibrating in modes of the same structural wavelength λ, which have the same natural frequency ω_n; it is assumed that $2\pi/\lambda \gg k_n$, where $k_n = \omega_n/c$. Although the spectra are different in the region of the main peaks, because of the different widths of the panels, the heights of the peaks at low wavenumbers are almost the same, being approximately equal to $(\lambda/\pi)^2$. Hence it may be concluded qualitatively that for the same vibration velocity amplitude, panel (a) mode and panel (b) mode radiate the same amounts of sound energy at their natural frequencies; therefore, the panel (a) mode has a higher radiation efficiency than the panel (b) mode. The conclusion is in agreement with that based upon the Rayleigh integral approach.

Alternatively, consider two panels of the same width and material, but of different thickness. The panel (a) mode will have a lower natural frequency than the panel (b) mode; however, they have the *same* wavenumber spectra. Figure 40 indicates that the panel (b) mode will have a higher radiation efficiency than the panel (a) mode, because a greater proportion of its wavenumber components are capable of radiating energy.

For modes having natural frequencies ω_n such that their wavenumber spectral peaks occur at wavenumbers less than the acoustic wavenumber corresponding to the natural frequency, i.e., $(k_x)_{\text{peak}} = 2\pi/\lambda = p\pi/a < \omega_n/c$, the great majority of their wavenumber components are supersonic, and their resonant radiation efficiencies are close to unity. One might enquire about the radiation from the wavenumber component that satisfies $k_x = k$, the impedance of which Eq. (2.37) suggests is infinite. Because the impedance is infinite, such a structural wavenumber component cannot physically exist,

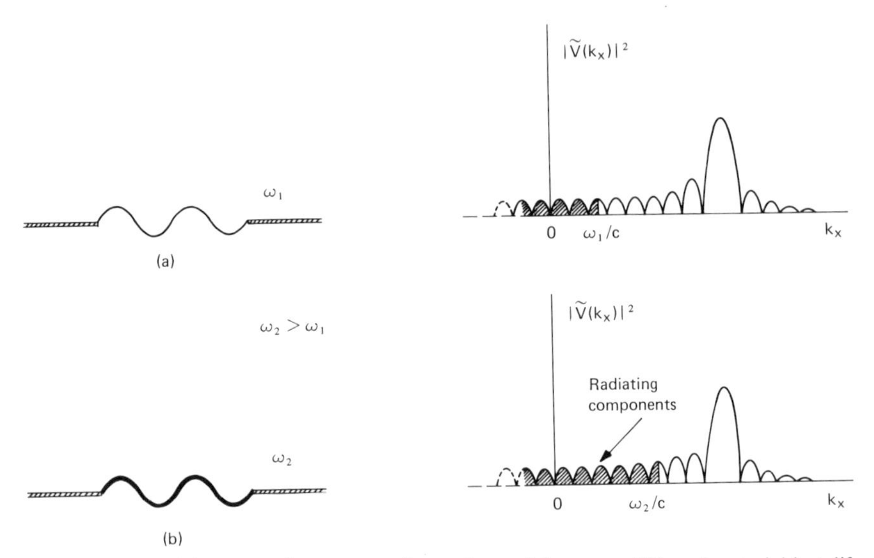

Fig. 40. Modal wavenumber spectra of two plates of the same width and material but different thickness.

and those immediately below it are strongly damped by radiation [see Eq. (2.35) with κ/k slightly less than unity].

The mode shapes assumed in the analysis so far are highly idealised in their sinusoidal form: this means that adjacent internodal cells have volume velocities of equal magnitude and opposite sign, so that they largely cancel each other's ability to radiate sound of a wavelength that is substantially greater than the structural wavelength.

As with time-dependent signals, deviation from a sinusoidal form spreads the frequency spectrum. In the case of modal vibration at frequencies where the structural wavelength is much less than the acoustic wavelength, any distortion from the sinusoidal form that enhances magnitudes of the low-wavenumber (energy-radiating) components increases the radiation efficiency. Consider the idealised case illustrated in Fig. 41. The attachment of supports to the panel allows modes to occur that have a non-zero component at zero wavenumber.

Consequently the presence of the internal supports or constraints increases the radiation efficiency. The radiation efficiency of large, thin beam-

Fig. 41. Two of the lowest band modes of a periodically supported panel.

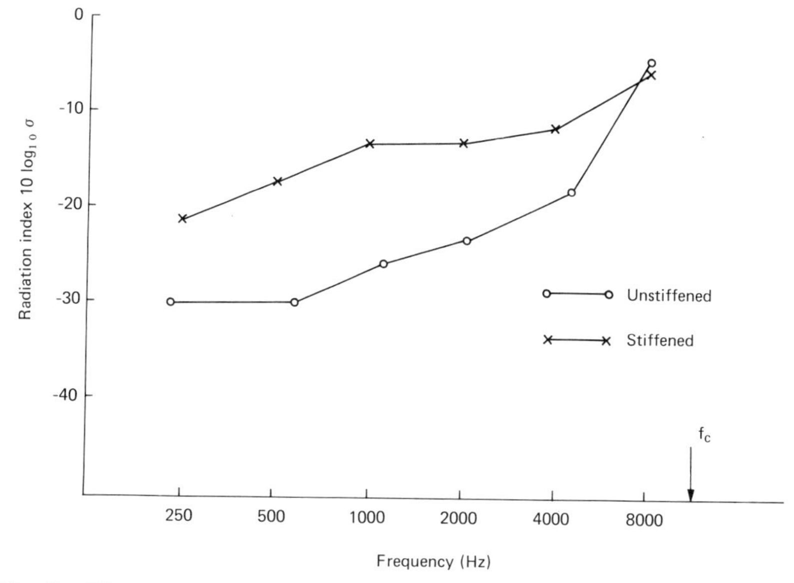

Fig. 42. Effect of stiffening on the radiation efficiency of a model boiler casing structure.

stiffened panels is consequently higher than that of large, thin unstiffened panels (Maidanik, 1962). This is not to say that stiffening a large panel will necessarily increase the total sound power radiated due to the action of a particular form of *excitation*, because the vibration level $\langle \overline{v_n^2} \rangle$ *may* be decreased by stiffening, to a degree sufficient to offset the increase in radiation efficiency σ: remember, $\bar{P} = \sigma \rho_0 c S \langle \overline{v_n^2} \rangle$. As an example, Fig. 42 shows a comparison of the radiation efficiencies of a stiffened and unstiffened panel of a model of a gas-cooled nuclear reactor boiler casing.

Stiffening can also increase airborne sound transmission through plates, as we shall see in Chapter 4. In practice, adverse effects of stiffening can be minimised by increasing the damping of the structure. Section 2.7 discusses the effects of localised constraints on plate radiation in more detail.

2.6 The Frequency-Average Radiation Efficiency of Plates

Thus far we have largely concentrated upon modal radiation at *arbitrary* frequency. In practice, the vibrational response of structures to exciting forces tends to be dominated by the resonant responses of modes having natural frequencies within the frequency range of the excitation. Therefore, we must concentrate our attention on the radiation behaviour of modes at their *natural* frequencies. Again we employ the model of a simply supported, baffled panel for reasons of analytical simplicity; however, the radiation characteristics of all plate structures conform to a similar pattern.

In the application of the wavenumber spectral approach to the qualitative understanding of the influence of plate parameters on radiation, we restricted our attention to one-dimensional panel modes. Of course, most panel modes are two-dimensional in form, and we must therefore consider what effect this has on radiation. In Section 1.7 we saw that the two-dimensional thin-plate equation yielded free-wave solutions of the form

$$\eta(x, y, t) = \tilde{\eta} \exp[j(\omega t - k_x x - k_z z)], \tag{2.57}$$

in which

$$k_x^2 + k_z^2 = k_b^2 = (\omega^2 m/D)^{1/2}, \tag{2.58}$$

and k_b is the free structural wavenumber at frequency ω. We have also seen that the modes of a simply supported rectangular plate take the form

$$v_n(x, z) = \tilde{v}_{pq} \sin(p\pi x/a) \sin(q\pi z/b), \qquad \begin{cases} 0 \leqslant x \leqslant a, \\ 0 \leqslant z \leqslant b. \end{cases} \tag{2.59}$$

This standing wave can be considered to be made up of component travelling waves in the form

$$\begin{aligned} v_n(x, z) = & -(j\tilde{v}_{pq}/2)\{\exp[j(p\pi x/a)] - \exp[-j(p\pi x/a)]\} \\ & \times \{\exp[j(q\pi z/b)] - \exp[-j(q\pi z/b)]\}, \end{aligned} \tag{2.60}$$

which consists of the sum of four travelling waves, existing only within the region $0 \leqslant x \leqslant a$ and $0 \leqslant z \leqslant b$. The corresponding *primary* wavenumber components are

$$k_x = \pm p\pi/a, \tag{2.61a}$$

$$k_z = \pm q\pi/b. \tag{2.61b}$$

It must not be forgotten, however, that the finite extent of the wave-bearing region introduces other wavenumber components, as in the case of the one-dimensional mode. Substitution of these expressions into Eq. (2.58) yields the natural frequency equation

$$\omega_{pq} = (D/m)^{1/2}[(p\pi/a)^2 + (q\pi/b)^2]. \tag{2.62}$$

In considering radiation from modes at their natural frequencies we have to consider the relative magnitudes of the primary structural wavenumber com-

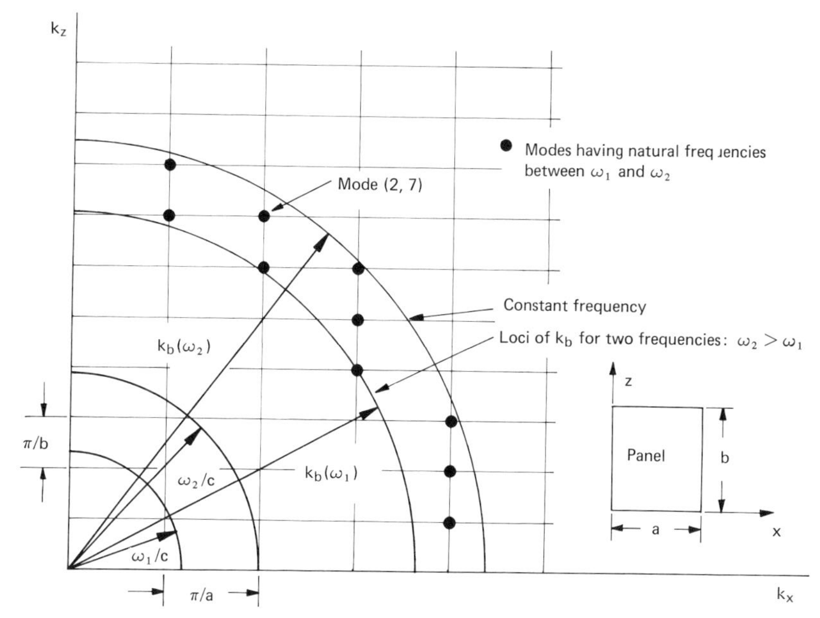

Fig. 43. Constant-frequency loci of acoustic and bending wavenumbers on a wavevector diagram.

ponents and the acoustic wavenumber k; this is most conveniently done by means of a wave vector diagram or grid. In Fig. 43 the constant-frequency locus of k_b, given by Eq. (2.58), is displayed. A grid of lines corresponding to the allowed primary wavenumber components given in Eqs. (2.61) is superimposed. The intersections of the grid lines represent panel modes. We can also superimpose the locus of the acoustic wavenumber $k = \omega/c$ for any arbitrary frequency. The relative radii of the two loci of k and k_b can be determined from Eq. (2.58):

$$k_b^2 = (k^2c^2m/\omega^2D)^{1/2}$$

or

$$(k_b/k) = c(m/D)^{1/4}(1/\omega)^{1/2}. \tag{2.63}$$

This ratio is unity when

$$\omega = c^2(m/D)^{1/2} = \omega_c. \tag{2.64}$$

Hence Eq. (2.63) can be written

$$k_b/k = (\omega_c/\omega)^{1/2}. \tag{2.65}$$

Frequency ω_c is known as the critical frequency of the panel because it separates the two regions (see Fig. 16 in Chapter 1):

$$\text{(a)} \quad \omega < \omega_c, \qquad k_b > k; \tag{2.66a}$$

$$\text{(b)} \quad \omega > \omega_c, \qquad k_b < k. \tag{2.66b}$$

Now, if waves of wavenumber k_b existed in an *infinite* plate, regime (a) would correspond to zero energy radiation, and regime (b) to energy radiation. Unfortunately, the finite extent of the plate makes the situation far more complicated, as we have already seen in the case of one-dimensional modes. Below the critical frequency, modes may satisfy one of the following three conditions:

$$\text{(1)} \quad k > p\pi/a, \qquad k < q\pi/b,$$
$$\text{(2)} \quad k < p\pi/a, \qquad k > q\pi/b,$$
$$\text{(3)} \quad k < p\pi/a, \qquad k < q\pi/b.$$

Figures 44a–c illustrate the wavenumber spectra of the modal velocity distribution along the x and z axes in relation to the acoustic wavenumber.

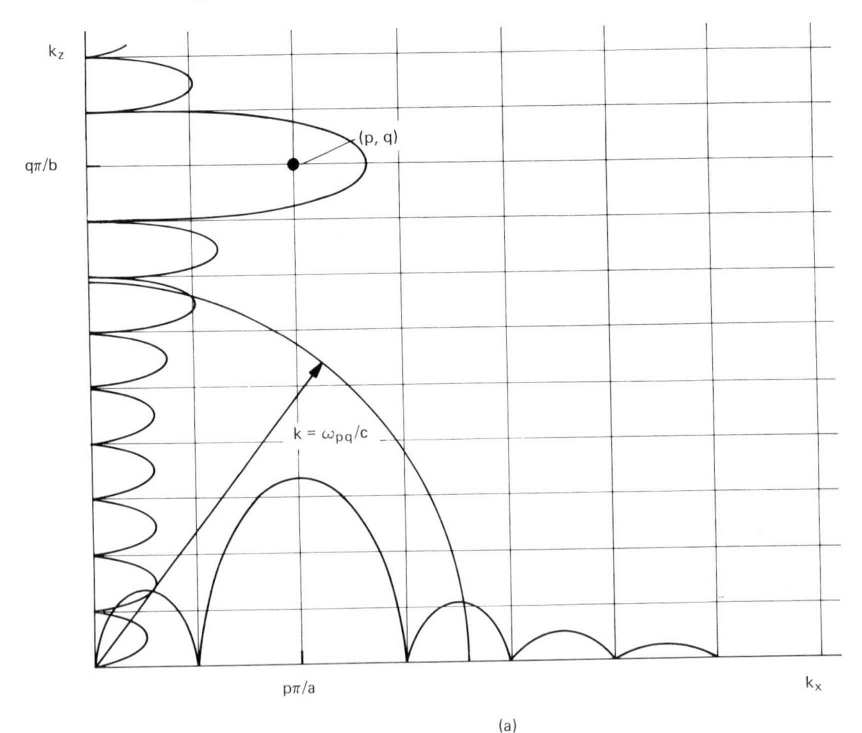

Fig. 44a. Case (1): $k > p\pi/a$; $k < q\pi/b$.

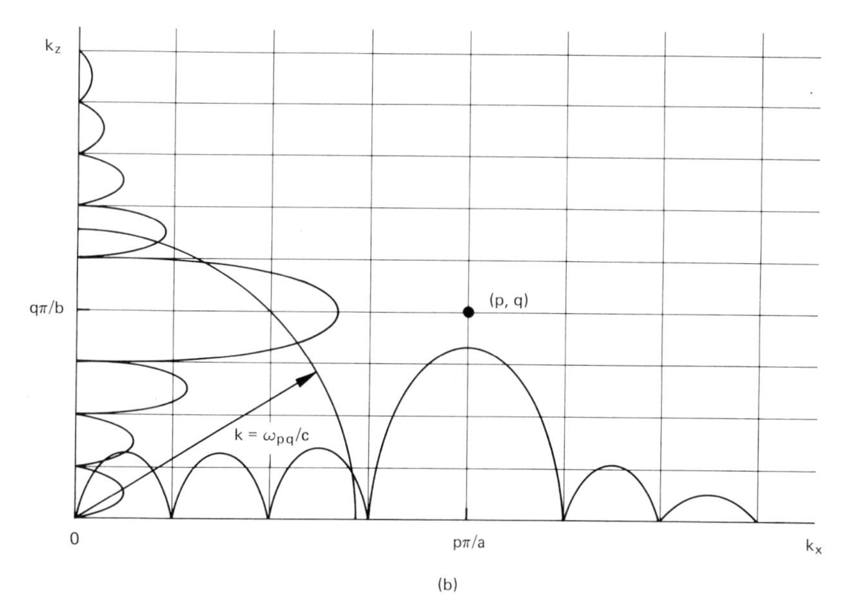

Fig. 44b. Case (2): $k < p\pi/a$; $k > q\pi/b$.

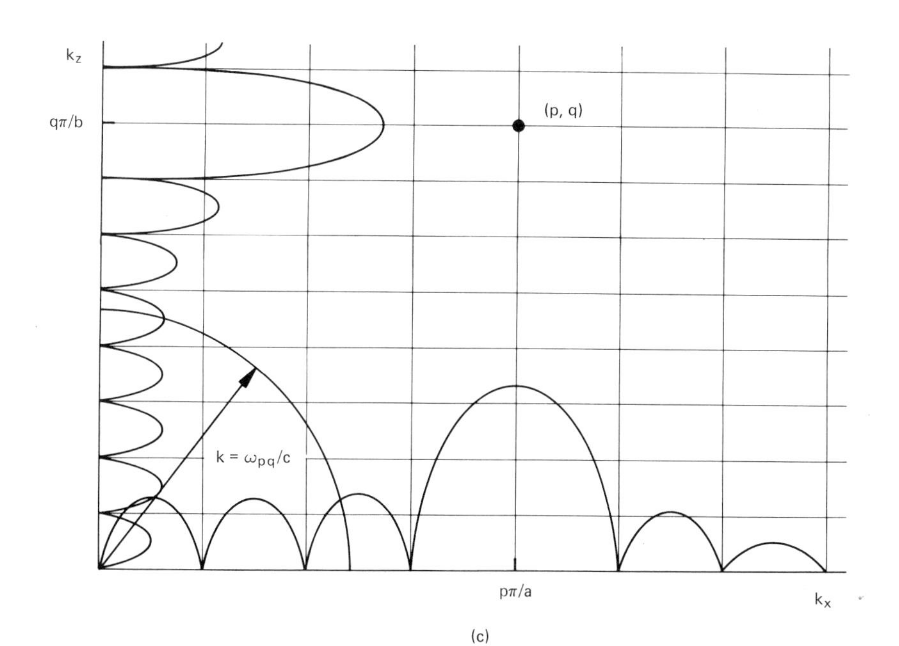

Fig. 44c. Case (3): $k < p\pi/a$; $k < q\pi/b$.

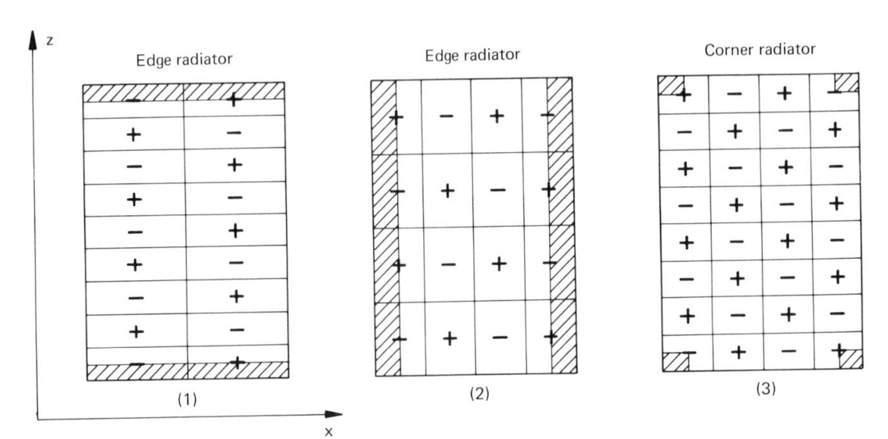

Fig. 45. Regions of uncancelled volume velocity at subcritical frequencies.

Any combination of k_x and k_z components that satisfies the condition

$$k_x^2 + k_z^2 < k^2 \tag{2.67}$$

is seen to produce a component that radiates well. Figure 44 indicates that conditions (1) and (2) produce more effective radiation than condition (3). Modes satisfying the latter condition are sometimes termed *corner modes* for a reason that we have already met in the previous section—only the corner quarter-cells contribute significantly to the radiation. By analogy, modes satisfying the first two conditions are sometimes termed *edge modes*, because strips of half-cell width along the edges normal to the axis for which the primary wavenumber is less than the acoustic wavenumber remain largely uncancelled. Figure 45 illustrates the uncancelled areas in the three cases.

It is not normally possible in practice to find a simple analytical expression for the radiation efficiency corresponding to arbitrary single-frequency excitation, because a number of modes will respond simultaneously, each vibrating with a different amplitude and phase, depending upon its damping and the proximity of its natural frequency to the excitation frequency. Therefore, it is more usual to try to estimate the average radiation efficiency of the modes having natural frequencies in a frequency band; this is called the frequency-average, or modal-average, radiation efficiency. For this purpose it is necessary to assume a distribution of vibration amplitude, or energy, among the modes: a common assumption is that of equipartition of modal energy (since we do not usually know any better!). On this basis, Maidanik (see Vér and Holmer, 1971) has produced a model-average radiation curve shown in Fig. 46a. Some specific applications are presented in Figures 46b–d (Bijl, 1977).

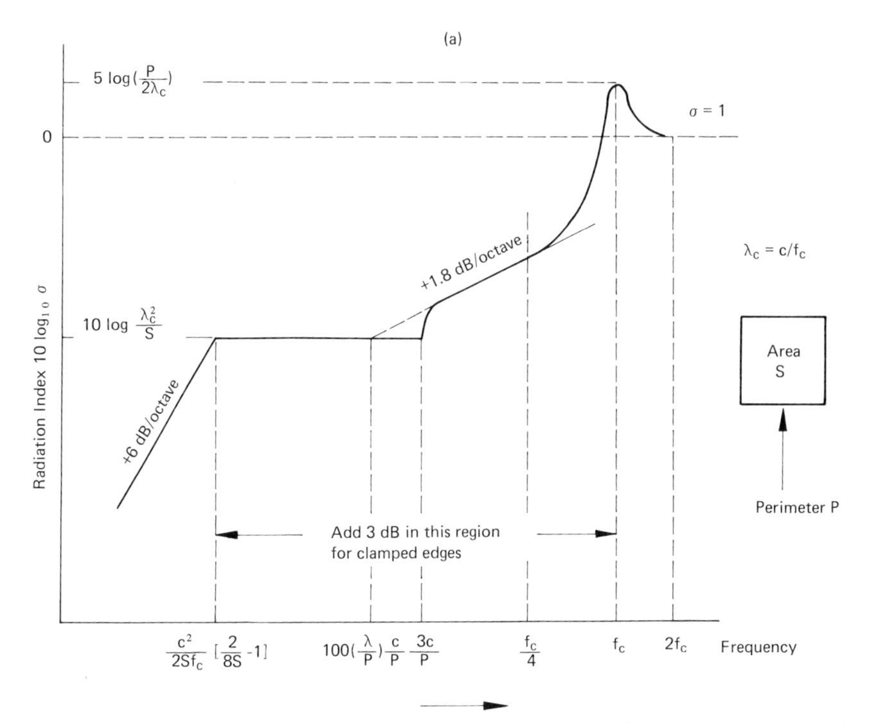

Fig. 46a. Theoretical modal average radiation efficiency of a baffled rectangular panel (Vér and Holmer, 1971).

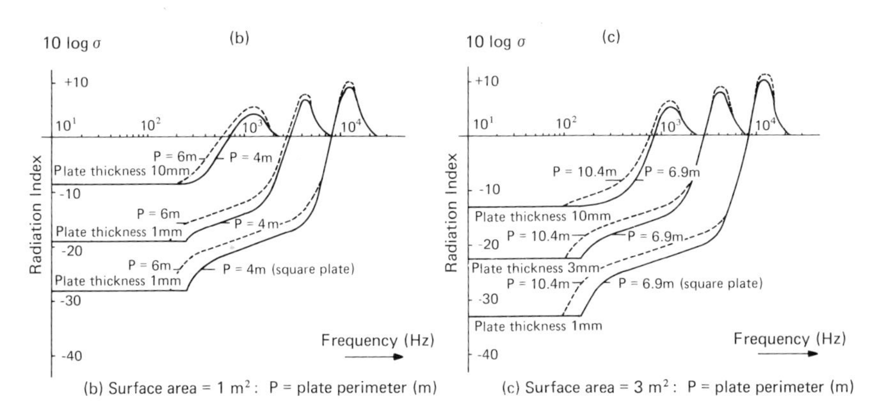

Fig. 46b,c. Modal average radiation efficiencies of baffled rectangular steel or aluminium plates in air (Bijl, 1977).

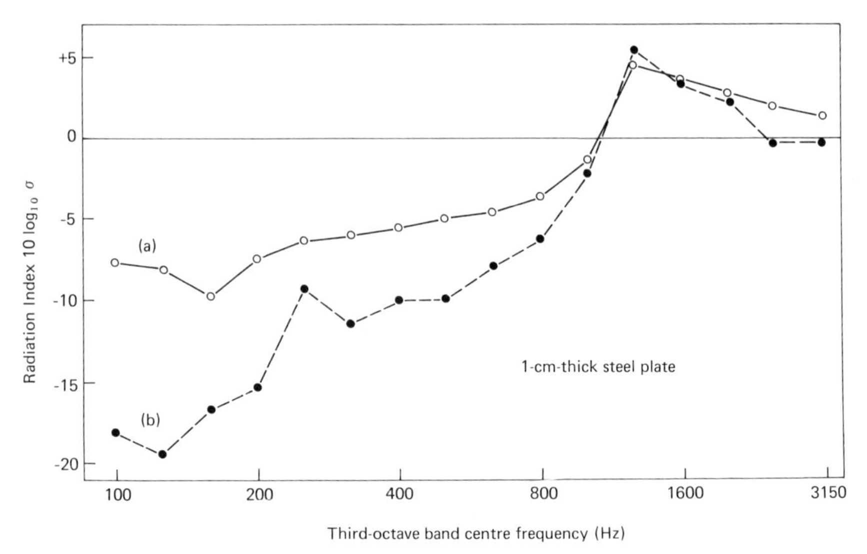

Fig. 46d. Comparison of radiation efficiencies of a plate under (a) airborne excitation and (b) mechanical excitation (Macadam, 1976).

The main parameters of these curves are the ratio of the centre frequency of the band to the critical frequency, and the ratio of the wavelength at the critical frequency to the length of the plate perimeter. This latter parameter is a reflection of the effect of panel size on the radiation efficiency of modes below the critical frequency, as discussed earlier. It should be carefully noticed that the radiation efficiency curves are relevant to *multimode resonant* vibration in response to broad-band mechanical excitation forces: they are not appropriate to acoustically excited structures in which forced wave motion is dominant. (see Figure 46d) Maidanik suggests that the effect of the attachment of stiffeners to a panel is effectively to produce many smaller panels, and therefore to increase its radiation efficiency below the critical frequency by a factor equal to $1 + 2P/B$, where P is the total length of the stiffeners and B the length of the panel perimeter. This is only an approximate correction factor, which should not be allowed to raise the radiation efficiency above unity, but it does indicate that the index of a densely stiffened panel is likely to be close to zero even below the critical frequency. Figure 47 shows the measured radiation efficiency of a model of a stiffened ship deck structure, in which the effect of stiffening can clearly be seen (von Venzke *et. al.*, 1973).

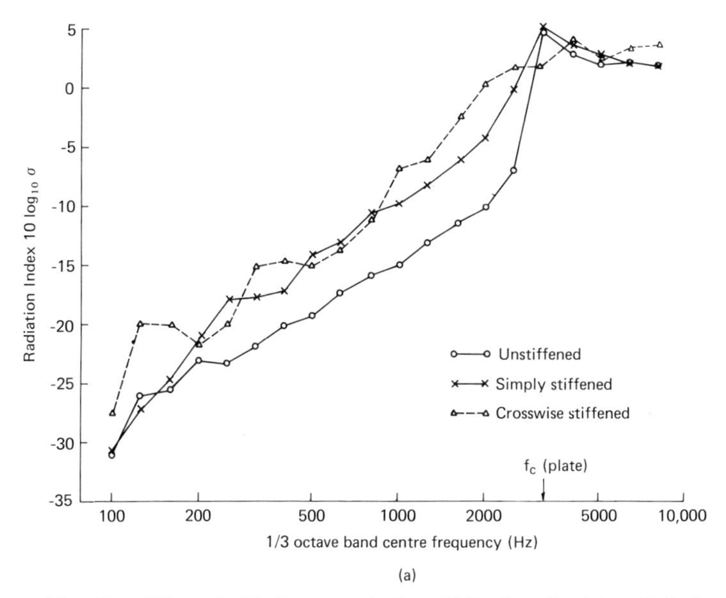

Fig. 47a. Effect of stiffening on radiation efficiencies of point-excited plates (von Venzke *et al.*, 1973).

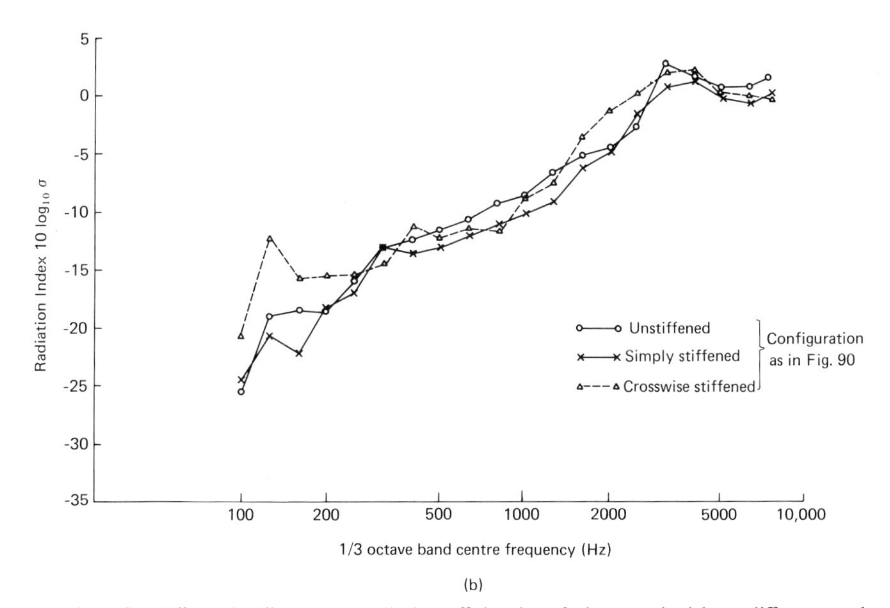

Fig. 47b. Effect of stiffening on radiation efficiencies of plates excited by a diffuse sound field (von Venzke *et al.*, 1973).

2.7 Sound Radiation Due to Concentrated Forces and Displacements

Sound radiation from plates has so far been considered without reference to the location or spatial distribution of excitation. The state of equipartition of modal energy, a concept introduced in the previous section, is likely to obtain if all the modes are lightly damped and if the excitation has no particularly marked spatial structure that allows it preferentially to excite certain classes of modes. One example of a highly structured excitation is an incident plane wave, which we shall meet in Chapter 4.

Another form of excitation that, in theory, excites some modes in preference to others is the point force. In practice, the higher-order mode shapes of structures are generally not so well known that the better excited modes can be distinguished from those less well excited. However, it is known that the mean square vibration level at the location of the excitation point generally exceeds the spatial average mean-square vibration level on which the radiation efficiency is based [Eq. (2.22)]; the reason is as follows.

The velocity response at point $\mathbf{r}$ of a plate structure to single-frequency force excitation at a point $(\mathbf{r}_0)$ may be written

$$v(\mathbf{r}, t) = \exp(j\omega t) \sum_n \frac{\tilde{F}\psi_n(\mathbf{r}_0)\psi_n(\mathbf{r})}{\tilde{Z}_n}, \tag{2.68}$$

where ψ_n is the non-dimensional modal velocity distribution (mode shape) of mode n, and $\tilde{F}$ is the force amplitude. $\tilde{Z}_n$ is the modal impedance, given by

$$\tilde{Z}_n = jm_n(\omega - \omega_n^2/\omega) + \eta_n\omega_n^2/\omega, \tag{2.69}$$

where m_n is the generalised modal mass given by

$$m_n = \int_s m(\mathbf{r})\psi^2(\mathbf{r})\, dS, \tag{2.70}$$

and η_n is the modal loss factor. This result may be found in textbooks on vibration of structures, e.g., Warburton (1976). Readers not familiar with Eq. (2.68) will find further discussion in Sections 6.2 and 6.3. The mean-square response at the driving point is

$$\begin{aligned}\overline{v^2(\mathbf{r}_0)} &= \frac{1}{2}\,\mathrm{Re}\left\{\sum_n \frac{\tilde{F}\psi_n^2(\mathbf{r}_0)}{\tilde{Z}_n} \sum_m \frac{\tilde{F}^*\psi_m^2(\mathbf{r}_0)}{\tilde{Z}_m^*}\right\} \\ &= \frac{|\tilde{F}|^2}{2}\,\mathrm{Re}\left\{\sum_m\sum_n \frac{\psi_n^2(\mathbf{r}_0)\psi_m^2(\mathbf{r}_0)}{\tilde{Z}_n\tilde{Z}_m^*}\right\},\end{aligned} \tag{2.71}$$

and the mean-square response at any other point is

$$\overline{v^2(\mathbf{r})} = \frac{1}{2}\,\mathrm{Re}\left\{\sum_n \frac{\tilde{F}\psi_n(\mathbf{r}_0)\psi_n(\mathbf{r})}{\tilde{Z}_n} \sum_m \frac{\tilde{F}^*\psi_m(\mathbf{r}_0)\psi_m(\mathbf{r})}{\tilde{Z}_m^*}\right\}. \tag{2.72}$$

The spatial-average mean-square response is

$$\langle \bar{v}^2 \rangle = \frac{1}{S}\int_S \overline{v^2(\mathbf{r})}\,dS = \frac{|\tilde{F}|^2}{2S}\sum_n \frac{\psi_n^2(\mathbf{r}_0)A_n}{|\tilde{Z}_n|^2}, \tag{2.73}$$

where $A_n = \int \psi_n^2(\mathbf{r})\,dS$, and the cross-terms disappear because the modes are assumed to be the normal (orthogonal) modes of the structure. The driving-point response exceeds the spatial-average response by virtue of the cross-terms ($m \neq n$) in Eq. (2.71).

An alternative way of understanding this result is to consider Eq. (2.68) at, and remote from, the driving point. At the driving point, the amplitudes of the modal responses contributing to the summation are determined by the square of the modal amplitudes at $\mathbf{r}_0$ and the proximity of the driving frequency ω to the natural frequency ω_n through $|\tilde{Z}_n|$; the phases of $\tilde{Z}_n^{-1}$ lie in the range $\pi/2$ to $-\pi/2$ (Fig. 48a). At any other point the phase can, in addition, take values of π or $-\pi$, by virtue of the standing-wave nature of the modes, i.e., $\psi_n(\mathbf{r})$ may be positive or negative relative to $\psi_n(\mathbf{r}_0)$ (Fig. 48b). Hence the sum expressed in Eq. (2.68) evaluated at the driving point is likely to exceed that at any other point. (Why are the two Figs. for different plates?)

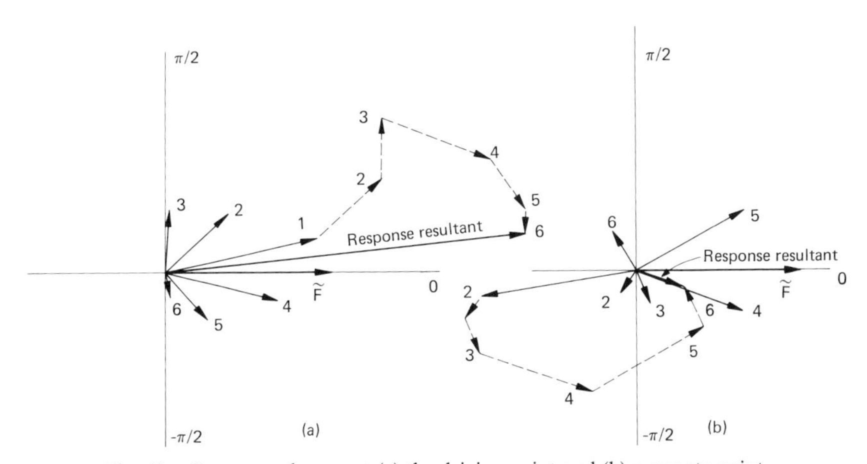

Fig. 48. Response phasors at (a) the driving point and (b) a remote point.

In evaluating a modal average radiation efficiency it is tacitly assumed that the modal vibrations are uncorrelated, so that the modal radiation efficiencies can be mathematically averaged,* the resulting average being multiplied by the spatial-average mean-square velocity in the application of Eq. (2.22). This is not true in the case of point excitation, even if the exciting force varies randomly with time, because $F(t)$ is common to all modal responses. We can view the necessity of including an "extra" point force radiation contribution as being due to the response "concentration" associated with spatially correlated modal motions in the vicinity of the driving point.

The contribution of the point force radiation, relative to that associated with the spatial-average mean-square velocity, increases with the average modal damping. This can be qualitatively understood in terms of Figure 48a, because the rate of change of phase and amplitude of modal response with the ratio of forcing to natural frequency decreases with an increase of damping. Hence, relatively more modes will contribute significantly to the driving point response, whereas the standing-wave cancellation effect seen in Fig. 48b will still operate, and the ratio of driving-point response to spatial-average response will increase. In the limit of high damping, the point force radiation will correspond closely to that produced by a point force acting on an infinite plate. This can be evaluated by using a wavenumber transform technique on the inhomogeneous form of the bending-wave equation corresponding to Eq. (1.39);

$$D\left[\frac{\partial^4 \eta}{\partial x^4} + \frac{2\partial^4 \eta}{\partial x^2 \partial z^2} + \frac{\partial^4 \eta}{\partial z^4}\right] + m\frac{\partial^2 \eta}{\partial t^2} = \tilde{F}\delta(x - x_0)\,\delta(z - z_0)\exp(j\omega t), \quad (2.74)$$

where $\delta(x - x_0)$ and $\delta(z - z_0)$ are Dirac delta functions that concentrate the uniformly distributed force per unit area $\tilde{F}$ onto point (x_0, z_0).

By analogy with the one-dimensional transform expressed in Eq. (2.39), a two-dimensional transform of a two-dimensional force distribution may be defined:

$$\begin{aligned}\tilde{F}(k_x, k_z) &= \int_{-\infty}^{\infty} \tilde{F}\delta(x - x_0)\,\delta(z - z_0)\exp(-jk_x x)\exp(-jk_z z)\,dx\,dz \\ &= \tilde{F}\exp(-jk_x x_0)\exp(-jk_z z_0) \\ &= \tilde{F}, \qquad \text{for} \quad x_0 = 0, \quad z_0 = 0. \end{aligned} \quad (2.75)$$

Hence application of the transform to Eq. (2.74) yields

$$[D(k_x^2 + k_z^2)^2 - \omega^2 m]\tilde{\eta}(k_x, k_z) = \tilde{F}. \quad (2.76)$$

* Or weighted appropriately, if $\langle \bar{v}^2 \rangle$ does not contain equal contributions from all the resonant modes.

Now we can replace the inertia term in Eq. (2.76) by substituting from Eq. (1.41) thus:

$$D[(k_x^2 + k_z^2)^2 - k_b^4]\tilde{\eta}(k_x, k_z) = \tilde{F}. \tag{2.77}$$

This equation expresses the response of the plate to a plane traveling-wave excitation having wavenumber components k_x and k_z. If required, damping can be included by making D complex [see Eq. (1.83)].

The fluid wave impedance associated with a plane traveling wave having wavenumber components k_x and k_z is, of course, the same as that given by Eq. (2.47), with k_x replaced by $(k_x^2 + k_z^2)^{1/2}$:

$$[\tilde{P}(k_x, k_z)]_{y=0} = \frac{\pm\omega\rho_0}{(k^2 - k_x^2 - k_z^2)^{1/2}} \tilde{V}(k_x, k_z). \tag{2.78}$$

Now the surface velocity transform is related to the surface displacement transform by

$$\tilde{V}(k_x, k_z) = j\omega\tilde{\eta}(k_x, k_z). \tag{2.79}$$

Hence the surface pressure transform can be written as

$$[\tilde{P}(k_x, k_z)]_{y=0} = \frac{\pm j\omega^2\rho_0}{(k^2 - k_x^2 - k_z^2)^{1/2}} \frac{\tilde{F}(k_x, k_z)}{D[(k_x^2 + k_z^2)^2 - k_b^4]}. \tag{2.80}$$

The power radiated by the total plate can now be expressed in terms of the product of the surface pressure transform and complex conjugate of the surface velocity transform as in Eqs. (2.50) and (2.51). However, in this case integration over infinite ranges of x, z and k_x, k_z are necessary. Following the arguments applied to the one-dimensional integral [Eq. (2.51)] we can reduce the integral to

$$\bar{P} = \frac{\rho_0 c\omega^2|\tilde{F}|^2}{8\pi^2 D^2} \int_{-k}^{k}\int_{-k}^{k} \frac{k\, dk_x\, dk_z}{[(k_x^2 + k_z^2)^2 - k_b^4]^2(k^2 - k_x^2 - k_z^2)^{1/2}}. \tag{2.81}$$

If we restrict our attention to frequencies well below the critical frequency of the plate, $k \ll k_b$, and since $(k_x^2 + k_z^2)^{1/2} \leqslant k$ in the range of integration, we can neglect $(k_x^2 + k_z^2)^2$ compared with k_b^4 in the denominator of the integral, to give

$$\bar{P} \simeq \frac{\rho_0 c\omega^2 k|\tilde{F}|^2}{8\pi^2 D^2 k_b^8} \int_{-k}^{k}\int_{-k}^{k} \frac{dk_x dk_z}{(k^2 - k_x^2 - k_z^2)^{1/2}}. \tag{2.82}$$

This integral is most readily solved by transforming to cylindrical coordinates so that $k_r^2 = k_x^2 + k_z^2$, $k_x = k_r \cos\phi$, and $k_z = k_r \sin\phi$. The element $\delta k_x\, \delta k_z$ transforms to the element $(k_r\, \delta\phi)\, \delta k_r$, and ϕ ranges from 0 to 2π. The integration over ϕ is straightforward and the integral reduces to

$$\bar{P} = \frac{\rho_0 c\omega^2 k|\tilde{F}|^2}{4\pi D^2 k_b^8} \int_0^k \frac{k_r\, dk_r}{(k^2 - k_r^2)^{1/2}} = \frac{\rho_0 c\omega^2 k^2|\tilde{F}|^2}{4\pi D^2 k_b^8}. \tag{2.83}$$

We replace k_b^8 by $(m\omega^2/D)^2$ to give

$$\bar{P} = \frac{\rho_0 ck^2|\tilde{F}|^2}{4\pi m^2\omega^2} = \frac{\rho_0|\tilde{F}|^2}{4\pi cm^2}. \tag{2.84}$$

This remarkably simple result indicates that the power generated by the excitation of an infinite or very highly damped bounded plate, below the critical frequency, is independent of frequency and plate stiffness, and inversely proportional to the mass per unit area. Above the critical frequency the whole plate surface radiates and all the input power is theoretically converted into sound in the absence of internal damping.

If we assume that this expression holds reasonably well for finite, damped plates, we can obtain a rough estimate of the proportion of sound power generated by this driving-point region and by the distributed vibration field at frequencies below the critical frequency. The power injected into a finite plate by a point force is, on a frequency- or modal-average basis, given by Eqs. (1.55) and (1.89) as

$$\bar{P}_{\text{in}} = \tfrac{1}{2}|\tilde{F}|^2\, \text{Re}(\tilde{Z}^{-1}) = \tfrac{1}{2}|\tilde{F}|^2/8(mD)^{1/2}. \tag{2.85}$$

The total energy of vibration is equal to the sum of the modal energies and, for a uniform plate, is

$$\bar{E} = mS\langle\bar{v}^2\rangle. \tag{2.86}$$

According to the definition of loss factor, the equilibrium energy corresponds to a balance between power in and power dissipated:

$$\tfrac{1}{2}|\tilde{F}|^2/8(mD)^{1/2} = \eta\omega mS\langle\bar{v}^2\rangle$$

or

$$\langle\bar{v}^2\rangle = \tfrac{1}{2}|\tilde{F}|^2/8(mD)^{1/2} m\eta\omega S. \tag{2.87}$$

The sound power radiated by multi-modal vibration is, by definition,

$$\bar{P}_{\text{s}} = \rho_0 cS\sigma\langle\bar{v}^2\rangle = \tfrac{1}{2}\rho_0 c|\tilde{F}|^2\sigma/8(mD)^{1/2} m\eta\omega. \tag{2.88}$$

Hence the ratio of powers radiated from the driving point and the plate surface is

$$\frac{\bar{P}}{\bar{P}_{\text{s}}} = \frac{4}{\pi}\left(\frac{\omega}{\omega_{\text{c}}}\right)\left(\frac{\eta}{\sigma}\right). \tag{2.89}$$

The ratio is seen to increase with the ratio of plate loss factor to radiation efficiency. The apparent proportionality to the frequency ratio ω/ω_c is misleading, since $\sigma \sim (\omega/\omega_c)^{1/2}$ over much of the subcritical frequency range. For many practical structures the ratio η/σ will substantially exceed unity, especially at the lower frequencies of practical interest, and the ratio $\bar{P}/\bar{P}_s$ may well take a value of about unity. Of course, the point force contribution is proportionally more important for artificially damped structures, and Eq. (2.84) represents a lower limit for radiation by a point-excited plate, however well damped it is.

Another case of concentrated excitation that is perhaps of even greater importance than that of the point force is that of the line force. Excitation of vibrations in plate structures by motion of the plate boundaries, perhaps by much stiffer frame structures, is a common practical situation. As opposed to a point force, a line force generates a one-dimensional plane wave in a plate, and hence only a one-dimensional wave-number transform is necessary. We can proceed directly to the plate velocity transform by applying Eq. (1.63) to a plate, thus:

$$\eta(x,t) = \begin{cases} \dfrac{-j\tilde{F}' \exp(j\omega t)}{4Dk_b^3} [\exp(-jk_b x) - j\exp(-k_b x)], & x \geqslant 0, \\ \dfrac{-j\tilde{F}' \exp(j\omega t)}{4Dk_b^3} [\exp(jk_b x) - j\exp(k_b x)], & x \leqslant 0. \end{cases} \tag{2.90}$$

Now

$$\tilde{V}(k_x) = j\omega\tilde{\eta}(k_x) = \frac{\omega\tilde{F}'}{4Dk_b^3}\left\{\int_{-\infty}^{0} [\exp(jk_b x) - j\exp(k_b x)]\exp(-jk_x x)\,dx \right.$$
$$\left. + \int_0^{\infty} [\exp(-jk_b x) - j\exp(-k_b x)]\exp(-jk_x x)\,dx\right\}. \tag{2.91}$$

An alternative and more straightforward approach to the evaluation of this velocity transform is to transform the equation of motion of the plate directly, in which the applied force is represented as a line delta function $F'\delta(0)$: this procedure is the one-dimensional equivalent of the analysis leading to Eq. (2.77). The result is

$$|\tilde{V}(k_x)|^2 = \frac{\omega^2|\tilde{F}'|^2}{D^2(k_x^4 - k_b^4)^2}. \tag{2.92}$$

Applying Eq. (2.53) the power radiated per unit length is

$$\bar{P} = \frac{\rho_0 ck\omega^2|F'|^2}{4\pi D^2}\int_{-k}^{k} \frac{dk_x}{(k_x^4 - k_b^4)^2(k^2 - k_x^2)^{1/2}}. \tag{2.93}$$

[Evaluate the contribution to radiated power of the bending near-field alone and explain the shortfall.]

At frequencies below ω_c we can neglect k_x^4 in comparison with k_b^4 in the denominator of the integrand to give

$$\bar{P} = \frac{\rho_0 c k \omega^2 |\tilde{F}'|^2}{4D^2 k_b^8} = \frac{\rho_0 |\tilde{F}'|^2}{4\omega m^2}. \tag{2.94}$$

This expression is very similar to that in Eq. (2.84) for a point force, the main difference being the presence of the frequency in the denominator, which suggests that the line force is more effective at lower frequencies. As with the point force we can investigate the relative magnitudes of radiation from the excitation region and from the multi-mode vibration of the whole surface.

The power input from a line force to an infinite plate is given by Eqs. (1.55) and (1.88) as

$$\bar{P}_{in} = \tfrac{1}{2}|\tilde{F}'|^2 l/4D^{1/4}\omega^{1/2}m^{3/4}, \tag{2.95}$$

where l is the length of the panel on which it acts. We assume this is distributed among the modes of the plate to give a mean-square velocity

$$\langle \bar{v}^2 \rangle = \tfrac{1}{2}|\tilde{F}'|^2 l/4D^{1/4}\omega^{3/2}m^{7/4}\eta S. \tag{2.96}$$

The sound power radiated by multi-mode vibration is therefore

$$\bar{P}_s = \tfrac{1}{2}\rho_0 c\sigma|\tilde{F}'|^2 l/4D^{1/4}\omega^{3/2}m^{7/4}\eta, \tag{2.97}$$

and the ratio of the powers radiated by the excitation region and the whole panel is

$$\frac{\bar{P}}{\bar{P}_s} = 2\left(\frac{\omega}{\omega_c}\right)^{1/2}\left(\frac{\eta}{\sigma}\right). \tag{2.98}$$

The parameters are the same as those appropriate to point force excitation, but the frequency dependence is different. Again the variation of subcritical radiation efficiency as $(\omega/\omega_c)^{1/2}$ must be noted.

It might be argued that the so-called line force excitation region at a panel boundary cannot be distinguished from the boundary field of the panel modes, particularly if the panel is clamped to a moving boundary. Since the line force is assumed to be of uniform phase over the whole length of the boundary, it may be thought of as producing "edge mode" radiation from that boundary at all frequencies, not only near the resonance frequencies of the panel edge modes appropriate to that boundary; hence it may definitely be considered to enhance the radiation.

The foregoing expressions for force-generated radiation may also be expressed in terms of the plate velocity v_0 at the excitation point, by replacing the force by the product of velocity and appropriate impedance. In the case of point force excitation, Eq. (2.84) becomes

$$\bar{P} = \frac{16\rho_0|\tilde{v}_0|^2 D}{\pi c m} = \frac{16\rho_0|\tilde{v}_0|^2 c^3}{\pi\omega_c^2}. \tag{2.99}$$

Hence the power increases linearly with bending stiffness and inversely as panel mass; again a stiff, light structure, like a honeycomb sandwich panel, is seen to be an effective converter of vibration into sound. The equivalent expression for line velocity excitation is

$$\bar{P} = \frac{2\rho_0 D^{1/2}|\tilde{v}_0|^2}{m^{1/2}} = \frac{2\rho_0 c^2|\tilde{v}_0^2|}{\omega_c}. \tag{2.100}$$

A useful application of the various expressions for force-generated radiation is to the estimation of the effect of locally constraining a plate structure on its sound radiation characteristics. Imagine a bending wave of velocity $\tilde{v}_i$ in a plate to be normally incident upon a line constraint of impedance $\tilde{Z}_c$ as shown in Fig. 49. The reaction force of the constraint is given by

$$\tilde{F} = -\tilde{Z}_c\tilde{v}_0, \tag{2.101}$$

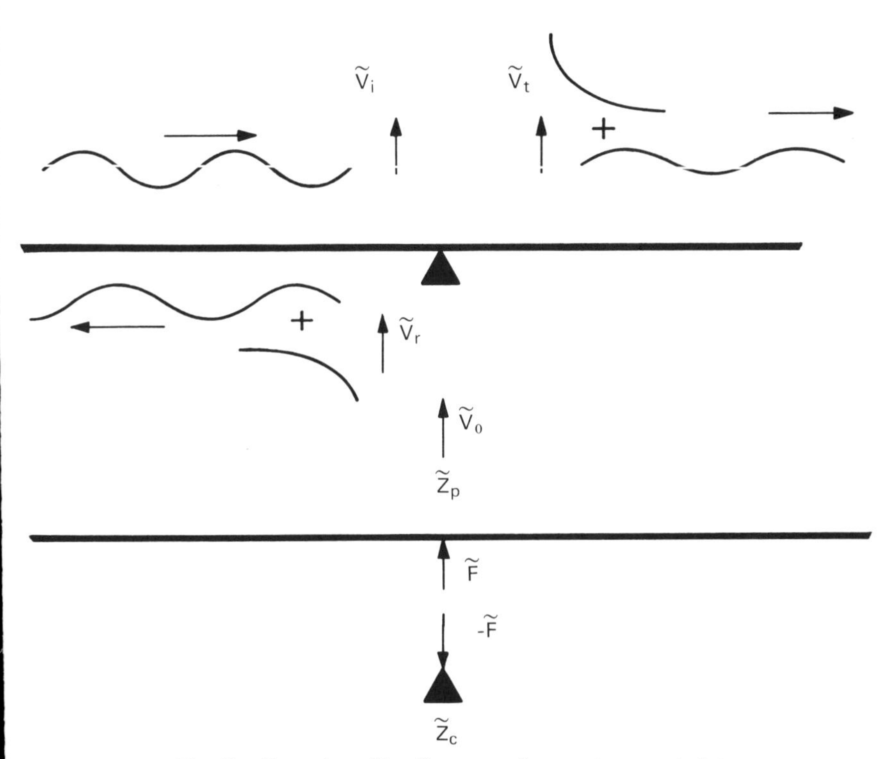

Fig. 49. Scattering of bending waves from a plate constraint.

and the response of the plate to this force is

$$\tilde{v} = \tilde{F}/\tilde{Z}_p, \tag{2.102}$$

where $\tilde{Z}_p$ is the line force impedance of the plate. Scattered waves $\tilde{v}_r$ and $\tilde{v}_t$ are reflected and transmitted:

$$\tilde{v}_0 = \tilde{v}_i + \tilde{v}_r = \tilde{v}_t, \tag{2.103}$$

where $\tilde{v}_r = \tilde{v}$.

Combining Eqs. (2.101)–(2.103) gives

$$\tilde{F} = -\tilde{v}_i[\tilde{Z}_p\tilde{Z}_c/(\tilde{Z}_p + \tilde{Z}_c)] \tag{2.104}$$

and

$$\tilde{v}_0 = \tilde{v}_i[\tilde{Z}_p/(\tilde{Z}_p + Z_c)]. \tag{2.105}$$

If $|\tilde{Z}_p/\tilde{Z}_c| \gg 1$, the incident wave is very weakly scattered by the constraint $\tilde{v}_0 \simeq \tilde{v}_i$ and $\tilde{F} \simeq -\tilde{v}_i\tilde{Z}_c$. On the other hand, if $|\tilde{Z}_p/\tilde{Z}_c| \ll 1$, the incident wave is strongly scattered; $|\tilde{v}_0| \ll |\tilde{v}_i|$ and $\tilde{F} \simeq -\tilde{v}_i\tilde{Z}_p$. The reaction force can be considered in exactly the same way as an applied force and the relevant equations applied to estimate the additional sound radiation due to the presence of the constraint. Nikiforov (1981) uses this type of analysis to evaluate the radiation efficiency of plates.

The practical conclusion to be drawn from this section is that the generation of inhomogeneous, bending-wave nearfields by applied concentrated forces, or by the action of constraints on otherwise freely vibrating plates, gives rise to the radiation of sound power in addition to that produced by the modal vibration of the plate surface. In the case of mechanically excited structures, the sound power radiated by the bending nearfields at sub-critical frequencies forms a significant proportion of the whole, and sets a limit to the reduction in radiation that can be effected by damping a plate. Above the critical frequency the influence of force distribution and constraints is far less important because the whole surface radiates effectively, and the reduction of resonant modal vibration levels by damping is far more effective.

2.8 Sound Radiation from Non-Uniform Plate Structures

The platelike components of many practical structures are not uniform, homogeneous, and isotropic like those so far considered. In order to increase their static stiffness, plates are often corrugated, or of sandwich construction. In comparison with plain plates, the effects are to alter their flexural wave dispersion characteristics, which alters the ratio of structural to acoustic wavenumbers that we have seen is a major parameter in determining radiation efficiency.

The effect of corrugation, which causes the plate to be orthotropic, is greatly to increase the bending stiffness in one direction. The resulting wavenumber diagram is quite different from that of an isotropic plate shown in Fig. 43, the constant frequency contours not being quarter-circles. The equation of flexural wave motion in an orthotropic plate is

$$D_x \frac{\partial^4 \eta}{\partial x^4} + 2(D_x D_z)^{1/2} \frac{\partial^4 \eta}{\partial x^2 \partial z^2} + D_z \frac{\partial^4 \eta}{\partial z^4} + m \frac{\partial^2 \eta}{\partial t^2} = 0, \qquad (2.106)$$

where D_x and D_z are the bending stiffnesses in the two orthogonal directions; compare Eq. (1.39). Substitution of

$$\eta(x, z, t) = \tilde{A} \exp(-jk_x x) \exp(-jk_z z) \exp(j\omega t) \qquad (2.107)$$

yields

$$[(D_x)^{1/2} k_x^2 + (D_z)^{1/2} k_z^2]^2 = \omega^2 m. \qquad (2.108)$$

The resulting wavenumber diagram for a rectangular, orthotropic plate takes the form shown in Fig. 50. It is seen that some plate modes have supersonic wavenumbers below the critical frequency ω_c of the basic plate material and hence radiate well. However, in any one frequency band below ω_c, these modes have a lower density of resonance frequencies in frequency space

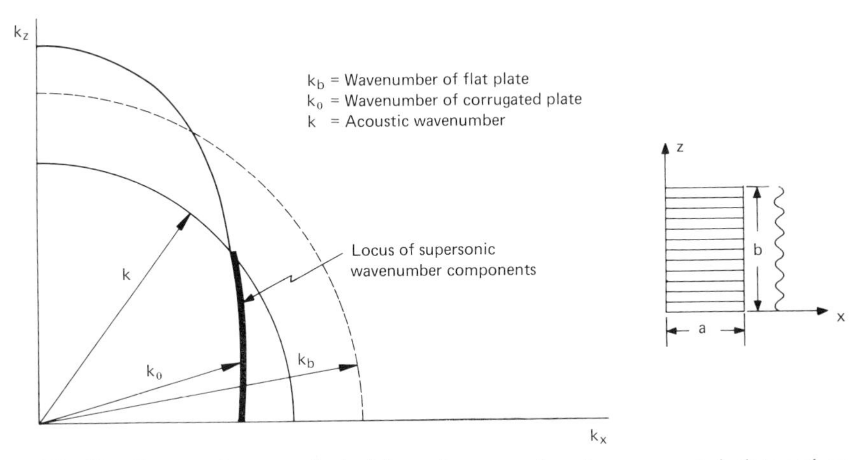

Fig. 50. Constant frequency loci of flexural wavenumbers in a corrugated plate and an equivalent flat plate below the flat-plate critical frequency.

(modal density) than the subsonic modes; hence, if all modes having natural frequencies within a band are assumed to vibrate with equal energy, the modal average radiation efficiency will remain below unity, although it will be higher for a corrugated plate than for an uncorrugated plate of the same material and size (Heckl, 1960).

The dispersion characteristic of an isotropic sandwich panel, which consists of two plates separated by a central layer of rather lower stiffness, is very much affected by the elastic properties and thickness of this layer. The shear stiffness of the layer is normally rather low, in which case transverse wave propagation is controlled by the whole section-bending stiffness at low frequencies, by the central-layer shear stiffness at intermediate frequencies, and by the individual faceplate-bending stiffness at very high frequencies. A typical dispersion curve is sketched in Fig. 51, which shows that the critical frequency is higher than for the equivalent uniform plate, and hence the radiation efficiency may be reduced over part of the frequency range (Cremer, 1968). Unfortunately, elastic layers can produce plate–layer–plate dilatational resonances, which, if not controlled by suitable damping, can greatly increase radiation.

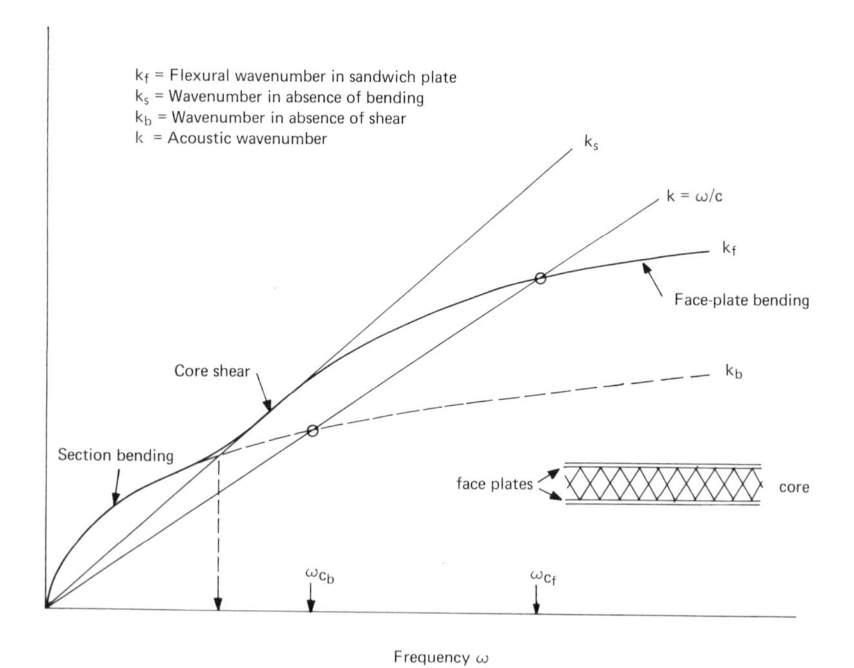

Fig. 51. Typical dispersion curves for a sandwich plate.

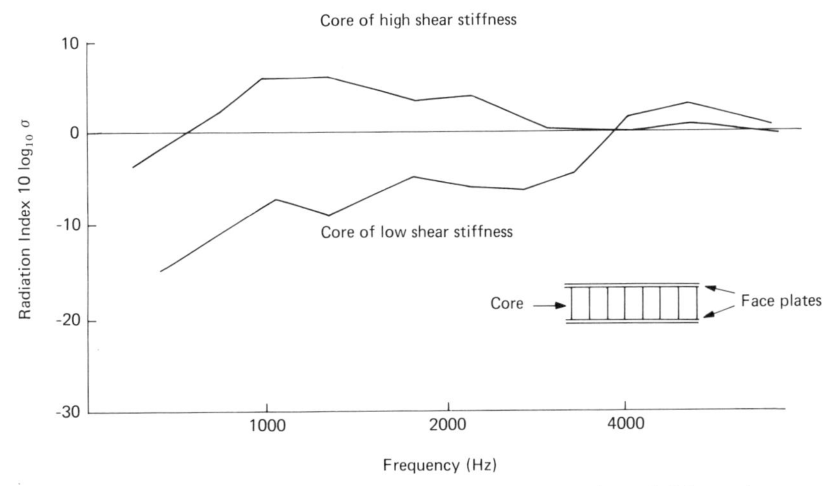

Fig. 52. Typical radiation efficiency curves for honeycomb sandwich panels.

Honeycomb panels, which consist of very thin faceplates separated by cores that are stiff in transverse compression, but weak in shear, are commonly used in aircraft and spacecraft to maximise stiffness-to-weight ratios. The radiation efficiency of such structures is highly dependent upon the shear stiffness of the core, as Fig. 52 shows.

2.9 Sound Radiation from Curved Shells

The influence of shell curvature on sound radiation derives primarily from its effect on the flexural wave dispersion characteristics, particularly at low wavenumbers. Curvature generally increases flexural wave phase velocities through the mechanism of mid-plane strain, with a consequent increase of radiation efficiency. Associated with this increase in wave speed is a reduction in the density of natural frequencies.

The complexity of analysis of curved shells precludes any general analysis from presentation in this book. However one class of curved-shell structures that is of considerable practical importance is that of circular cylindrical shells, which may be used as idealisations of such structures as industrial pipes, aircraft and rocket bodies, and submarine hulls. A brief discussion of cylindrical-shell vibration has been presented in Section 1.6 and an analysis of fluid loading of cylinders is presented in Chapter 3. Here we confine our attention to a comparison of flat-plate and circular cylindrical-shell flexural

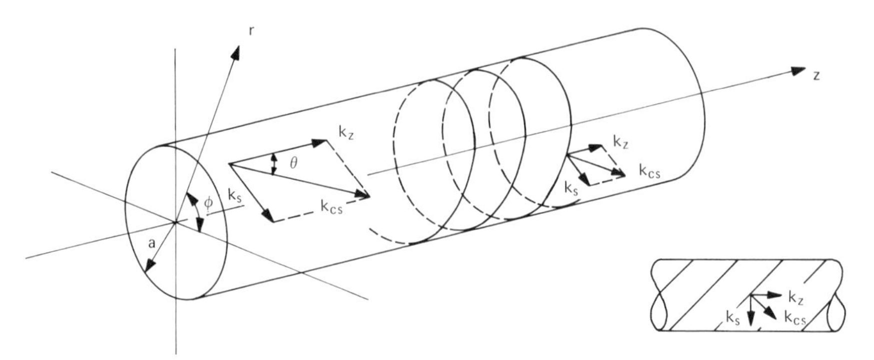

Fig. 53. Cylindrical-shell coordinates, wavenumbers, and helical wavefronts.

wavenumber behaviour, and a qualitative discussion of the consequent difference between radiation behaviours. The surface radial velocity of a wave propagating axially in a cylindrical shell of infinite length may be represented as

$$v_n(z, \phi, t) = \tilde{v}_n \cos n\phi \exp[j(\omega t - k_z z)], \qquad n = 0, 1, 2, \ldots, \qquad (2.109)$$

for the cylindrical coordinate system is shown in Fig. 53. Analysis presented in Chapter 3 shows that sound energy can be radiated as long as the surface axial wavenumber k_z is less than the acoustic wavenumber k, whereas in an infinite plate energy is radiated only if the plate wavenumber $k_b = (k_x^2 + k_z^2)^{1/2}$ is less than k. In a cylinder the wavenumber analogous to k_x is the circumferential wavenumber $k_s = n/a$, where a is the cylinder radius.

Strictly speaking, the cosinusoidal variation with ϕ in Equation (2.109) results from the interference between circumferential wavenumber components travelling in opposite directions around the cylinder; that is to say, Eq. (2.109) represents the interference field of two helical waves of equal and opposite circumferential wavenumber and equal axial wavenumber. It might be thought, therefore, that radiation from a wave represented by Eq. (2.109) would be analogous to that from a baffled flat strip of width $2\pi a$, carrying a sinusoidal standing wave across its width, in which case there exist supersonic wavenumber components in the x direction, as explained earlier in this chapter. However, the edge effect between the strip and the baffle has no counterpart on the cylinder because the surface is continuous, and the equivalent "edges" are coincident.

Under the conditions $k_z < k$, $k_s = n/a > k$, and $k_z^2 + k_s^2 > k^2$, whch are not relevant to the $n = 0$ mode, adjacent zones of positive and negative volume velocity distributed around the circumference, seen in Fig. 54, do not completely cancel; they would in a plane travelling wave on an *infinite* plate below the critical frequency with $k_z < k$ but $k_x > k$, and $k_z^2 + k_x^2 > k^2$.

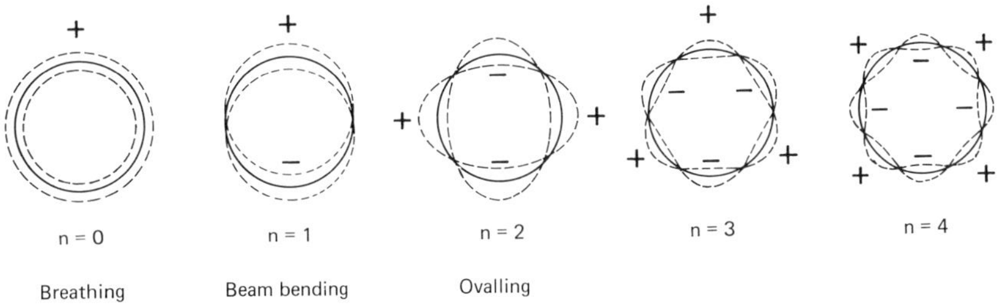

Fig. 54. Cross-sectional radial displacement mode shapes of a circular cylindrical shell.

However, their close proximity, in terms of an acoustic wavelength, makes their radiation very inefficient. The $n = 0$ "breathing" mode radiates as a line monopole; the $n = 1$ "bending" mode radiates as a line dipole; the $n = 2$ "ovalling" mode radiates as a line quadrupole, and so on, the efficiency of radiation at any frequency decreasing with increase in the order of the equivalent source.

The corresponding expressions for power radiated per unit length, for $ka \ll 1$ and $k_z \ll k$, are

$$\bar{P}_{n=0} = \tfrac{1}{2}\pi^2 \rho_0 ca(ka)|\tilde{v}_0|^2, \tag{2.110a}$$

$$\bar{P}_{n=1} = \tfrac{1}{4}\pi^2 \rho_0 ca(ka)^3|\tilde{v}_1|^2, \tag{2.110b}$$

$$\bar{P}_{n=2} = \tfrac{1}{32}\pi^2 \rho_0 ca(ka)^5|\tilde{v}_2|^2. \tag{2.110c}$$

The value of circumferential wavenumber k_s for a given n decreases with increase of cylinder radius. Hence large radius cylinder modes of order n can satisfy the condition

$$k_z^2 + k_s^2 < k^2 \tag{2.111}$$

at frequencies for which the equivalent modes of smaller cylinders, having the same axial wavenumber, give $k_z^2 + k_s^2 > k^2$. In this case, inter-zone cancellation does not occur and the cylinder radiates with a radiation resistance close to $\rho_0 c$ per unit area. This shows that it is not sufficient that the frequency at which a cylinder vibrates in the bending mode ($n = 1$) should exceed the critical frequency based upon the bending-wave phase velocity in order to radiate efficiently; it is also necessary that

$$(k^2 - k_z^2)^{1/2} a > 1. \tag{2.112}$$

For thin-wall cylinders in which the wall thickness is much smaller than the radius, the $n = 1$ bending-mode axial wavenumber is

$$k_z = (2\rho\omega^2/a^2 E)^{1/4}. \tag{2.113}$$

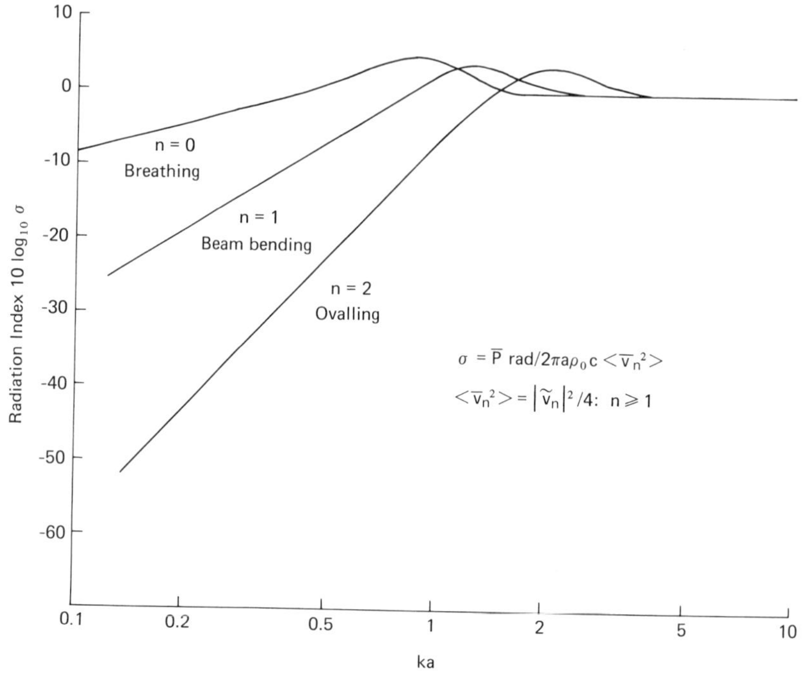

Fig. 55. Radiation efficiencies of uniformly vibrating cylinders.

The transition from inefficient to efficient radiation occurs rather rapidly, as Fig. 55 indicates, so that for most practical purposes sound radiation from transversely vibrating slender bodies can be considered to be negligible if Eq. (2.112) is not satisfied. Similarly, only when Eq. (2.111) is satisfied will radiation from cross-sectional distortion be significant.

As we have seen in the case of flat-plate radiation, wavenumber diagrams can be very useful in helping to identify those forms of vibration most effective in radiating sound. The bending-wave equation for flat plates produces a circular constant-frequency locus, as shown in Fig. 43. The curvature of a cylindrical shell produces coupling between radial, axial, and circumferential motions, and there are consequently three coupled equations of motion and three classes of propagating waves (Leissa, 1973). Although only the radial motion of a shell determines the sound radiation, the form of radial motion, and the associated dispersion characteristics, are significantly affected by mid-plane strains, especially at frequencies well below the ring frequency $f_r = c_l'/2\pi a$. These so-called membrane effects considerably raise the phase velocities of waves whose displacement is predominantly radial, so much so in some cases, that these waves have supersonic phase velocities at frequencies well below the critical frequency based upon the shell wall considered as a flat plate. Above the ring frequency, curvature effects disappear and the shell vibrates like an equivalent flat plate.

An approximate form of dispersion equation is (Heckl, 1962b)

$$\frac{(k_z a\,\beta^{1/2})^4}{[(k_z a\beta^{1/2})^2 + (n\beta^{1/2})^2]^2} + [(k_z a\beta^{1/2})^2 + (n\beta^{1/2})^2]^2 \simeq \Omega^2, \qquad (2.114)$$

where $\beta^2 = h^2/12a^2$, $\Omega = f/f_r$, and h is the wall thickness. This produces a wavenumber diagram of the form shown in Fig. 56a. The membrane effect on wave speed is seen in the bending of frequency loci toward the origin. A strange consequence of this behaviour is that at one frequency two helical waves of the same circumferential wavenumber, but different axial wavenumber, can propagate. Waves of low circumferential wavenumber involve greater membrane strain energy in proportion to flexural strain energy than waves of higher circumferential wavenumber. This leads to a rather unexpected variation of natural frequency with axial and circumferential wavelength.

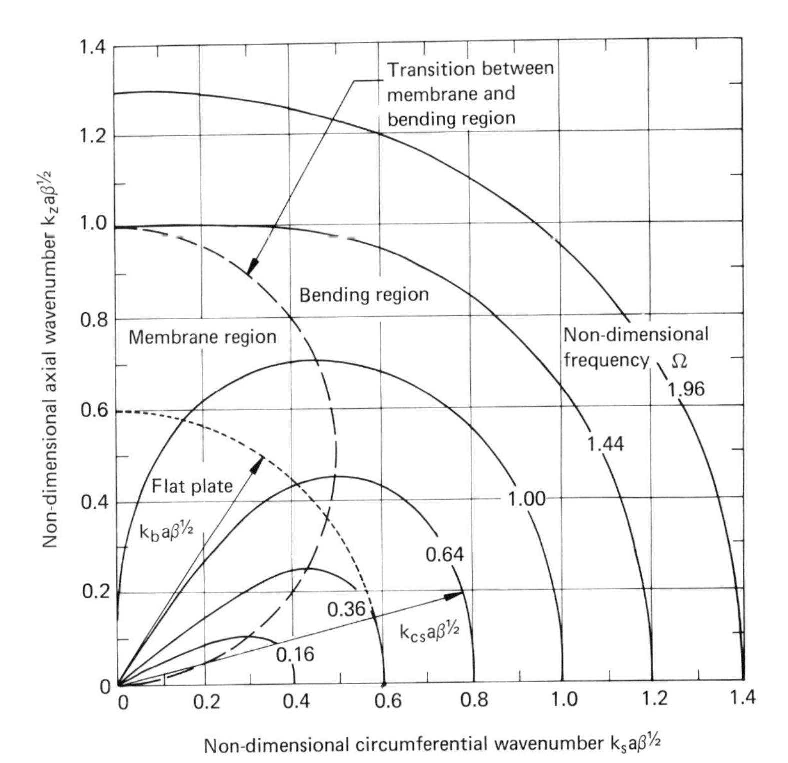

Fig. 56a. Universal constant-frequency loci for flexural waves in thin-walled circular cylindrical shells ($n > 1$).

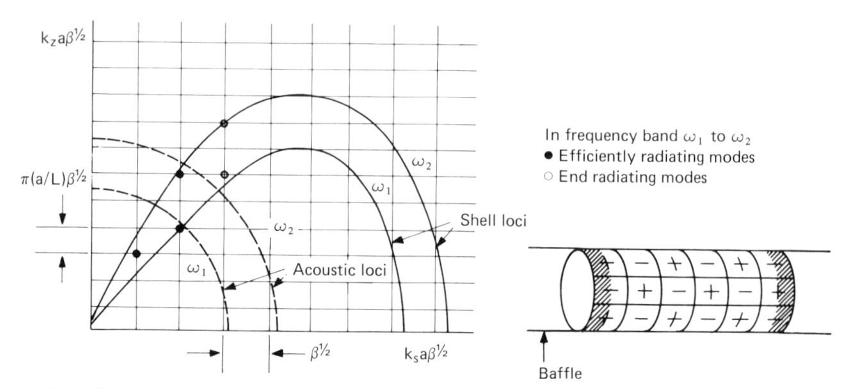

Fig. 56b. End radiating modes of a baffled cylindrical shell and regions of uncancelled volume velocity.

Superimposition of the circular-arc acoustic wavenumber onto the structural wavenumber plots of Fig. 56a indicates the possibility of existence of supersonic structural waves at sub-critical frequencies ($k_z^2 + k_s^2 < k^2$). Such waves clearly dominate low-frequency sound radiation, if they are generated by exciting forces. In a shell of finite length with simply supported edge conditions at the ends, natural modes are indicated by the intersection of the grid lines given by $k_s = n/a$ and the orthogonal set given by $k_z = m\pi/L$, where m is any positive integer and L the cylinder length. Modes for which $k_z^2 + k_s^2 > k^2$, $k_z^2 > k^2$, and $k_s^2 < k^2$ can radiate like plate edge modes: intercell cancellation does not occur at the ends (see Fig. 56b).

A statistical analysis on the basis of equipartition of modal energy in bands, as performed by Maidanik for flat plates, yields a general expression for modal average radiation efficiency. The analysis by Szechenyi (1971), which is appropriate to radiation from cylinders at values of ka much greater than unity, shows that two major frequency parameters influence the radiation efficiency. We have already met one of them, f/f_c, in connection with flat plates; the other is the ratio of critical frequency to ring frequency f_c/f_r. This parameter can also be expressed as

$$f_c/f_r = (c^2/1.8hc_l')/(c_l'/2\pi a) = (1/\beta)(c/c_l')^2. \tag{2.115}$$

Large-diameter, thin-wall shells have values of f_c/f_r greater than unity, whereas industrial pipes have values of the order of or considerably less than unity. The significance of the value of f_c/f_r is that if it is greater than unity, there is a range of frequency between f_r and f_c in which shell curvature effects on flexural wave speed disappear, and the cylinder radiates as a flat plate.

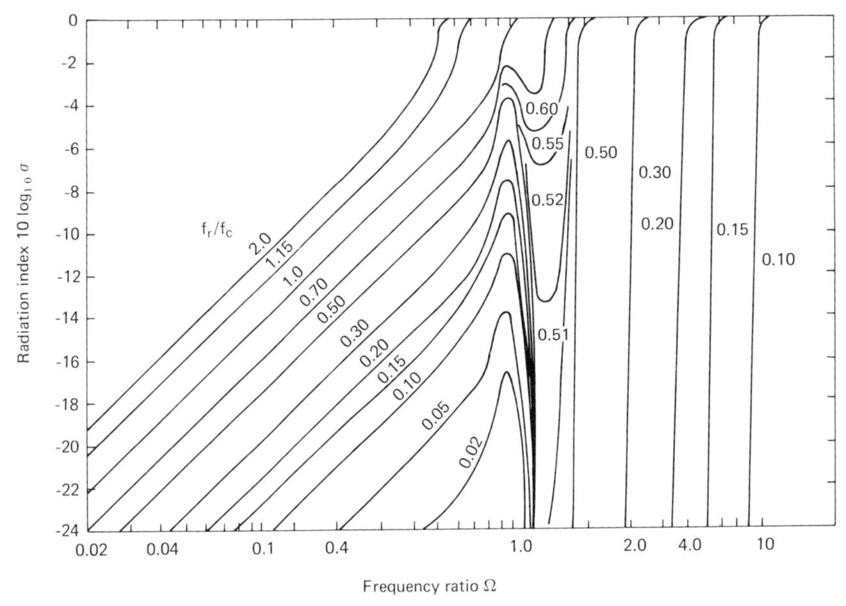

Fig. 57. Modal average radiation efficiency of thin-walled, large-diameter circular cylindrical shells (Szechenyi, 1971).

Hence the radiation efficiency falls above f_r and then rises again as f approaches f_c in the manner associated with flat-plate radiation. Figure 57 shows Szechenyi's radiation efficiency curves. Industrial-type pipes do not exhibit this type of behaviour, the radiation efficiency normally increasing smoothly with frequency and then tending to remain at a value of about unity at high frequencies (Holmer and Heymann, 1980). This behaviour is strongly influenced by the form of internal excitation experienced by pipes (see Chapter 4).

Figure 58 shows radiation efficiencies measured on three straight pipe lengths in a laboratory (Rennison, 1977). In the cases shown the excitation was produced by sound in the contained air. In principle, only stiffness-controlled breathing motion ($n = 0$) can be generated in the pipe wall by the purely plane sound waves that propagate below the cutoff frequency of the first cross mode in the pipe ($\Omega = 0.125$). However, the figure shows that the radiation efficiency generally falls well below that for the $n = 0$ mode shown in Fig. 55. The reason appears to be that higher-order circumferential modes of the pipes are somehow excited by the sound field. The cutoff frequencies of these modes, immediately above which multiple resonances

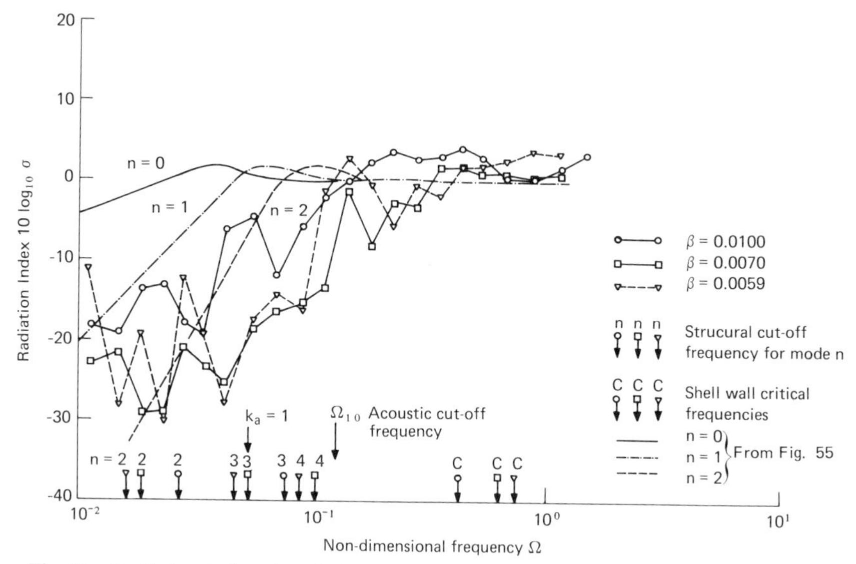

Fig. 58. Radiation index of steel pipes excited by sound in contained air (Rennison, 1977).

in these modes occur, can be calculated from Table V in Chapter 4, and are indicated on Fig. 58. These modes are very inefficient radiators near their cutoff frequencies and, since they dominate the response near their cutoff frequencies, they produce the associated dips in the radiation efficiency curves seen in the figure. The reason for the resonant excitation of such higher-order circumferential modes by an axisymmetric field, which should not couple to them (why not?), is apparently that pipe walls are not uniform in thickness, which produces non-cosinusoidal circumferential distribution of radial displacement, thereby allowing coupling to occur. The actual radiation efficiency curve for any particular pipe at low frequencies is therefore virtually impossible to estimate theoretically. Mechanical excitation of pipes can produce quite different radiation efficiencies from those associated with excitation by fluid flow or sound waves in the internal fluid.

One of the results of the influence of wall curvature on pipe wall wave speed is that the addition of stiffeners and other constraints to cylinders does not increase their radiation efficiencies as much as in the case of a flat plate. Because some resonant shell modes are created from supersonic traveling-wave components, the application of damping to mechanically excited cylindrical shells is likely to reduce radiated power more than it would on a flat plate of the same thickness. However, it should not be assumed that the addition of damping will effectively reduce sound radiated from industrial pipes excited by disturbances in the contained fluids generated by valves, bends, etc.: the primary mechanism of transmission appears to be near coincidence between acoustic waves in the fluid and waves in the pipe wall, and the influence of damping appears to be relatively small. Sound transmission through cylindrical shells is discussed in Chapter 4.

2.10 Sound Radiation from Irregularly Shaped Bodies

Although the analyses of radiation from flat plates and uniform cylinders are useful in revealing the major parameters that control sound radiation, many vibrating structures do not even approximate to such idealised forms, and consequently quantitative evaluation by analytical means is impractical. In such cases it is possible to use numerical methods of analysis, which are introduced in Chapter 7.

Alternatively we may combine experimentally measured "directional Green's functions" for bodies of arbitrary shape, which relate the radiated pressure at distances much greater than both the dimension of the body, and the acoustic wavelength, to unit strength volume velocity sources on the surface of the body, with the actual distribution of surface velocity (Cremer, 1981). These directional Green's functions may be measured by using the principle of reciprocity described in later chapters: the *rigid* body is subjected to the acoustic field of a point source placed in the far field and the acoustic pressures are measured on the body surface. Directional Green's functions for pairs of field and surface points are stored in a computer for subsequent combination with measured surface velocities in a surface integral (or summation) in which the product of the surface distributed volume velocity and the measured Green's functions forms the integrand. Further elucidation of this approach may be found in Sections 5.4 and 5.5.

Problems

2.1 By modeling the human mouth as a monopole source of volume velocity, evaluate the magnitude of the volume velocity produced by a singer of the note middle C if the sound pressure level produced in the open air at a distance of 20 m is 60 dB.

2.2 Evaluate the error in on-axis pressure incurred by replacing the actual radial distance by the average radial distance in the integral of Eq.

(2.4) in the case of a circular piston of diameter 150 mm at average distances of 1, 10, and 50 m and frequencies of 1 and 10 kHz.

2.3 By imagining a spherical-wave field to emanate from any observation point in the far field of a vibrating-plane radiator, and considering the intersection with the plane of wavefronts separated by $\lambda/2$, show how the regions providing contributions of opposite phase to the integral of Eq. (2.4) can be identified.

2.4 Prove the relationship given in Eq. (2.20).

2.5 Evaluate $(\eta_{\text{rad}}/\sigma)$ from Eq. (2.25) for a steel plate of 3-mm thickness in air and water at 500 Hz.

2.6 By assuming that the (1, 1) mode of a square panel radiates as a point volume velocity source, evaluate the radiation loss factor of a 300-mm^2 steel panel of 1-mm thickness vibrating at its fundamental natural frequencies in air and in water. Assume simply supported boundary conditions in a baffle. What is the effect of fluid loading on natural frequency (Chapter 3)?

2.7 Check that the "bandwidth" of the major peak in the wavenumber modulus-squared spectrum of a one-dimensional sinusoidal mode of a panel of width a is given by $\Delta ka \simeq \pi$ and is independent of the modal "wavelength."

2.8 Give a qualitative explanation of the fact that the radiation efficiencies of odd–odd modes of a rectangular panel of side lengths a and b exceed those of other modes when $ka,kb \ll 1$.

2.9 Demonstrate that the modes having equal modal-wavenumber components $k_x = p\pi/a$, $k_y = q\pi/b$ in simply supported rectangular panels of the same thickness and material but of different sizes have the same natural frequencies. What are the necessary conditions on the corresponding side length ratios of the two panels for this to be possible?

2.10 Check the conclusions from Eqs. (2.68)–(2.73) concerning response concentration at the driving point by explicitly evaluating the significant modal response phasors at various positions on a simply supported, 2-mm-thick rectangular steel plate of side lengths 500 and 700 mm when excited at an arbitrary point at a frequency of 500 Hz. Assume a uniform modal loss factor of 10^{-2}.

2.11 Estimate the sound power radiated into air by a 3-mm-thick steel plate at 500 Hz due to the presence of a concentrated mass of 0.1 kg. The incident-wave field can be assumed to produce an acceleration level of 40 dB re.lg rms at 500 Hz. Equations (1.89), (2.99), and (2.105) apply.

2.12 Calculate the approximate critical frequency for bending waves travelling parallel to the corrugations of a 1-mm-thick aluminium alloy plate in which the corrugations have an amplitude of 10 mm and a wavelength of 25 mm. Equation (2.106) applies. Assuming that an orthotropic model applies, plot the bending-wavenumber locus for a frequency of 2 kHz and superimpose the corresponding acoustic-wavenumber locus in air at 20°C. What is the limiting angle of bending-wavenumber propagation for subsonic waves?

2.13 Derive Eq. (2.113).

2.14 Using Eq. (2.115), evaluate the ratio of critical frequency to ring frequency for typical aircraft fuselage structures and industrial pipes.

2.15 By reference to the Kirchhoff–Helmholtz integral equation, or otherwise, explain why the pressure outside but in the plane of an unbaffled plate radiator is zero (see Chapter 3).

3 Fluid Loading of Vibrating Structures

3.1 Practical Aspects of Fluid Loading

In this book the term "fluid loading" refers to the forces a fluid exerts on a vibrating structure with which it is in contact, in reaction to that vibration. There are many cases of practical interest to the engineer in which the interaction between vibrating structures and contiguous fluids has a profound influence upon the magnitude and frequencies of the structural vibration. Examples include dams, chimney stacks, heat exchanger tubes, ships and their propellers, off-shore platforms, aircraft, and electrical transmission cables. In many of these cases the fluid itself is responsible for the vibration, and simultaneously reacts to the vibration. For instance, sea waves may excite the leg of an off-shore platform into vibration, and the reaction of the surrounding water is such as to alter the natural frequencies of the leg from its *in vacuo* values. In this example, the fluid–structure interaction is passive in the sense that the small-amplitude motion of the structure does not significantly alter the exciting forces. In other cases, such as flutter of aircraft wings and galloping of transmission lines, the motion profoundly influences the exciting forces, so that a feedback loop exists and dynamic instability may occur:

these examples belong to the field of aero-, or hydroelasticity, which is not within the scope of this book.

In most of the examples cited above, the fluid reaction forces can be estimated with sufficient accuracy by assuming that the fluid is incompressible, i.e., that its density is unchangeable. Mathematically this assumption leads to the Laplace equation for pressure

$$\frac{\partial^2 p}{\partial x^2} + \frac{\partial^2 p}{\partial y^2} + \frac{\partial^2 p}{\partial z^2} = 0.$$

Comparison with the acoustic wave equation (1.3) shows that this is equivalent to assuming that the speed of sound c is infinite, which is compatible with the concept of an incompressible fluid, since $c^2 = (\partial p/\partial \rho)_0$, the subscript 0 meaning "evaluated at the equilibrium condition."

The reason why incompressible fluid solutions are adequate for the estimation of fluid reactions to structural vibration in many of the cases cited above is that only the pressures on the surface are of interest and the frequency of oscillation is rather small, or more precisely that the non-dimensional parameter $kl = 2\pi l/\lambda \ll 1$, where l is a typical structural dimension: this condition is even more likely to be satisfied in liquids than in gases. A small value of kl generally implies that the vibrating structure is not able efficiently to compress the fluid on its surface and therefore does not produce a strong acoustic field. However, during vibration, the fluid local to the surface must be displaced in an oscillatory fashion, and the surface pressures are then primarily associated with the oscillatory momentum changes of the effectively incompressible fluid. In the case of a loudspeaker, or other structures that vibrate at frequencies for which the acoustic wavelength is not vastly greater than a typical structural dimension, an incompressible-fluid model is generally inadequate. The fluid loading must be estimated by solving the fluid wave equation subject to the boundary conditions imposed by the vibrating surface.

The practical significance of fluid loading on vibrating structures is that it can change both their natural frequencies and dampings—by radiation—and hence can influence the vibration response to excitation forces. Naturally, fluid-loading effects are most strongly exhibited by structures vibrating in dense fluids, because the fluid forces are generally proportional to mean fluid density. However, even air can produce some striking fluid-loading effects, as can be seen by mounting a small loudspeaker at one end of a tube that is terminated by a closed or open end, and then monitoring the loudspeaker displacement as a function of driving frequency. This chapter presents methods of mathematical analysis of fluid loading, and discusses its effects on vibrating plate and shell structures.

3.2 Pressure Fields on Vibrating Surfaces

A large proportion of sound sources of practical importance radiate through the actions of vibrating surfaces on the surrounding fluid. In most cases the geometric forms of the surfaces, and the spatial and temporal distributions of vibration, are very complex. If one wishes to estimate the sound power radiated under free-field conditions, in which the surrounding fluid is essentially unbounded and free of strongly reflective objects, it is possible to employ in the mathematical analysis of the radiated field certain approximations that greatly ease the analytical problems. Because, in the absence of significant sound absorption by the fluid, sound power transmitted through any surface completely surrounding a source is conserved, the far radiation field can be determined at a distance very large compared with a typical dimension of the vibrating object, and the power can be evaluated on that surface (see Chapter 2).

If it is desired to evaluate the fluid loading *on* a vibrating surface the luxury of the far-field approximation is not available. Consequently, there exist far fewer analyses of surface fields than of far fields. It is, however, necessary in a number of cases of practical importance to be able to evaluate acoustic fluid-loading effects on vibrating structures. Among these we may cite underwater transducer design, loudspeaker cone design, analysis of sound propagation in tubes and pipes, analysis of blast resistance of windows, evaluation of heat exchanger tube vibration, fatigue estimation for metal foil heat insulation in reactors, analysis of tympanic musical instruments, noise control studies of paper-handling machinery, and the analysis of microphone membrane dynamics. Clearly, it is not possible within the scope of this book to discuss such diverse problems in detail. Therefore, the basic principles and certain elementary examples are analysed in order to illustrate the general nature of fluid-loading effects.

In principle, the sound field generated by a vibrating surface can be evaluated by solving the wave equation subject to the boundary conditions imposed by the surface and by any other bodies present. It has been shown by Lord Rayleigh (1896), pp. 143 et seq. and in advanced texts on acoustics (e.g., Pierce, 1981) that the equation and boundary conditions can be combined in an integral equation, known as the Kirchhoff–Helmholtz (K–H) integral. For simple harmonic vibration of frequency ω, one form of the equation for acoustic pressure at position vector $\mathbf{r}$ in the fluid produced by vibration of a surface S is

$$\tilde{p}(\mathbf{r})e^{j\omega t} = \frac{1}{4\pi} e^{j\omega t} \int_S \left[\tilde{p}(\mathbf{r}_s) \frac{\partial}{\partial n}\left(\frac{e^{-jkR}}{R} \right) + j\omega\rho_0 \tilde{v}_n(\mathbf{r}_s) \frac{e^{-jkR}}{R} \right] dS, \qquad (3.1)$$

where $p(\mathbf{r}_s)$ is the pressure on the surface at position $\mathbf{r}_s$, v_n the normal velocity of the surface, R the distance $|\mathbf{r} - \mathbf{r}_s|$ between source and field points, and the integral is taken over all surfaces in contact with the fluid. The term $(1/4\pi R)\exp(-jkR)$ is the so-called free-space Green's function, which expresses the variation of amplitude and phase with distance of the pressure in the spherically symmetric field generated by an infinitesimally small acoustic source, which may be idealised as a pulsating sphere (point monopole) operating in free space [see Eq. (2.2) and further discussion in Section 6.2]. Equation (3.1) is only one particular form of the general K–H equation, in which any Green's function that satisfies the wave equation, together with the radiation condition at infinity, may be used.

It would seem from Eq. (3.1) that it would be necessary to specify both the distributions of surface pressure and surface normal velocity; however, these quantities are not independent, and the pressure field is everywhere uniquely determined by a specified distribution of surface velocity on a surface of given geometry. The reason for the presence of $\tilde{p}(\mathbf{r}_s)$ is that each elemental area of surface δS moving with velocity $\tilde{v}_n$ does not radiate like a point monopole source operating in free space because of the presence of the rest of the surface, whether vibrating or not, which acts as a baffle, or reflector. Hence its radiation field is changed from that which a point source of strength $\tilde{Q} = \tilde{v}_n\,\delta S$ would produce in free-field conditions (Cremer, 1981). Numerical approaches to the solution of Eq. (3.1) are discussed in Chapter 7.

Where the vibrating surface is planar and infinitely extended, Eq. (3.1) reduces to the Rayleigh integral (Lord Rayleigh, 1896) already encountered in Chapter 2, Eq. (2.4):

$$\tilde{p}(\mathbf{r})e^{j\omega t} = \frac{j\omega\rho_0}{4\pi}\,e^{j\omega t}\int_S \left[2\tilde{v}_n(\mathbf{r}_s)\,\frac{e^{-jkR}}{R}\right] dS. \tag{3.2}$$

We see that in this case the elemental piston δS radiates like a point source of strength $\tilde{Q} = 2\tilde{v}_n\,\delta S$ in free space, as shown in Fig. 59. Considerations of symmetry show that the presence of the rest of the plane surface S', although subject to acoustic pressures, cannot influence a spherically spreading field because there would be no particle velocity normal to that plane anyway. The Rayleigh integral can strictly be applied only to infinitely extended planar surfaces, but it provides a good estimate for plane vibrating surfaces of dimensions large compared with an acoustic wavelength, except for the source regions close to a free edge, such as on an unbaffled loudspeaker cone.

In order to determine the fluid loading on a vibrating surface it is necessary to evaluate $p(\mathbf{r})$, with $\mathbf{r}$ on the surface. If we consider the pressure produced at a point on the surface by vibration of the immediately surrounding area, i.e., as $kR = k|\mathbf{r} - \mathbf{r}_s|$ goes to zero, the integrand in Eq. (3.2) is seen to

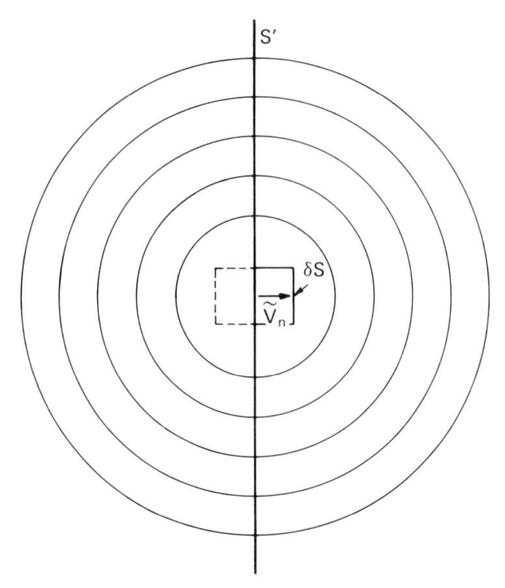

Fig. 59. Equivalence of elemental surface source and free-space source.

be singular, i.e., it goes to infinity, because $e^{-jkR}/R \to 1/R - jk$ as $R \to 0$. However, the integral itself remains finite. This may be seen by assuming $\tilde{v}_n$ to remain constant over the small local region and considering the contribution to pressure of annular strips surrounding the points of interest.

Substitution of the small kR approximation for e^{-jkR}/R into Eq. (3.2) demonstrates that the influence of elemental volume velocity $\tilde{v}_n\,\delta S$ on the pressure at distances very small compared with an acoustic wavelength $(2\pi R/\lambda \ll 1)$ can be separated into two components. One component, which is in phase with $\tilde{v}_n$, is independent of distance, and the other, which is in quadrature with $\tilde{v}_n$, increases as the separation distance decreases [note the j before the integral in Eq. (3.2)]. We may broadly interpret this feature as indicating that on vibrating planar surfaces of small extent compared with an acoustic wavelength, the pressure at any point that is in phase with the surface velocity at that point is determined essentially by the integral of the real components of elemental volume velocities over the surrounding region within the small kR limit, i.e., the net volume velocity of the surface. The quadrature component, on the other hand, is essentially determined by the motion at the point and is much less sensitive to the distribution of volume velocity over the surrounding area. The $1/R$ dependence of this component of the field, which is characteristic of incompressible-fluid pressure fields that satisfy the Laplace equation, justifies its description as the "near-field" component of the surface pressure field.

As an example of this characteristic, consider a rigid circular disc of radius a vibrating in a coplanar rigid baffle. The pressure at the centre due to the motion of an annulus of radius R is given by Eq. (3.2) as

$$\delta\tilde{p}(0) = (j\omega\rho_0/4\pi R)2\tilde{v}_n e^{-jkR} 2\pi R\,\delta R. \tag{3.3}$$

The total pressure at the centre is given by the integral over the limits 0 to a:

$$\tilde{p}(0) = \rho_0 c\tilde{v}_n(1 - e^{-jka}). \tag{3.4}$$

[Plot the real and imaginary components of $\tilde{p}(0)/\rho_0 c\tilde{v}_n$ as a function of ka.] For $ka \ll 1$, Eq. (3.4) becomes

$$\tilde{p}(0) = \rho_0 c\tilde{v}_n[(ka)^2/2 + jka]. \tag{3.5}$$

Even at low ka, the surface pressure component in quadrature with the velocity is not quite uniform over a piston in a baffle, and an integration of the pressure yields an expression for the total fluid reaction force (Pierce, 1981):

$$\tilde{F} = \rho_0 c\tilde{v}_n \pi a^2[(ka)^2/2 + j(8/3\pi)ka]. \tag{3.6}$$

Note that the average resistive component of pressure is equal to that at the centre of the piston. The ratio $\tilde{F}/\tilde{v}_n$ is equivalent to the mechanical impedance discussed in Chapter 1. The ratio of average acoustic pressure $\tilde{F}/\pi a^2$ to the rate of displacement of fluid volume (volume velocity) is known as the *acoustic radiation impedance*, given by

$$\tilde{Z}_{rad} = \tilde{F}/(\pi a^2)^2\tilde{v}_n = (\rho_0 c/\pi a^2)[(ka)^2/2 + j(8/3\pi)ka]. \tag{3.7}$$

The real part of $\tilde{Z}_{rad}$ is termed the acoustic radiation resistance, and the imaginary part is termed the acoustic radiation reactance. Examination of Eq. (3.6) reveals that the reactive reaction of the fluid is inertial in nature, a mass equal to $(8/3)\rho_0 a^3$ apparently being added to that of the piston: physically this is associated with the kinetic energy of fluid motion in the proximity of the piston. The resistive component is associated with energy radiation, which produces radiation damping.

If such a piston is mounted on a damped spring suspension and excited by a force, the *in vacuo* mechanical impedance is

$$\tilde{Z} = j(\omega M - S/\omega) + B, \tag{3.8}$$

where M is the piston mass, and S and B are, respectively, the suspension spring and damping constants. Under the action of fluid loading, the equation of simple harmonic motion of the piston may be written

$$\tilde{Z}\tilde{v}_n = \tilde{F} - (\pi a^2)^2\tilde{Z}_{rad}\tilde{v}_n. \tag{3.9}$$

Equation (3.9) may be physically interpreted as representing the response of a fluid-loaded structure to an applied force, thus;

$$\tilde{v}_n = \tilde{F}/[\tilde{Z} + (\pi a^2)^2 \tilde{Z}_{rad}]. \tag{3.10}$$

Note the different dimensions of mechanical and acoustic radiation impedances. The effective mass of the fluid-loaded structure is $M + (8/3)\rho_0 a^3$ and the effective damping is $B + \frac{1}{2}\rho_0 c\pi a^2 (ka)^2$. Such combination of mechanical and radiation impedances can be made for any structure vibrating in a fluid.

The radiation resistance of a baffled piston at low ka is very small compared with the value of $\rho_0 c/\pi a^2$, which is experienced by a piston at high ka. However, a piston radiating into an anechoically terminated tube at low ka also experiences $\rho_0 c/\pi a^2$; the difference in this case is that fluid cannot escape sideways as it can from a small baffled piston, and is therefore effectively compressed.

Suppose now that the small piston in our example is surrounded by an annular ring that vibrates in opposite phase to the piston (Fig. 60). Let the inner and outer radii of the ring be a_1 and a_2. The pressure at the centre of the piston is

$$\begin{aligned}\tilde{p}(0) &= j\omega\rho_0\tilde{v}_n\left[\int_0^{a_1} e^{-jkR}\,dR - \int_{a_1}^{a_2} e^{-jkR}\,dR\right] \\ &= \rho_0 c\tilde{v}_n[1 - 2\exp(-jka_1) + \exp(jka_2)].\end{aligned} \tag{3.11}$$

For $ka_2 \ll 1$, this expression, correct to second order in ka_2, is

$$\tilde{p}(0) = \rho_0 c\tilde{v}_n[(ka_1)^2 - (ka_2)^2/2 + jk(2a_1 - a_2)]. \tag{3.12}$$

If the areas of the piston and ring are equal, so that the net volume velocity $\int \tilde{v}_n\,dS$ is zero, Eq. (3.12) reduces to

$$\tilde{p}(0) = j\rho_0 c\tilde{v}_n ka_2(\sqrt{2} - 1). \tag{3.13}$$

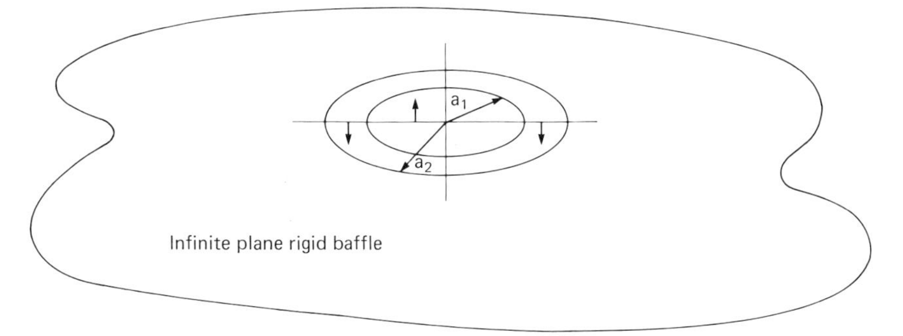

Fig. 60. Pistons vibrating in anti-phase.

This shows that the resistive component of $\tilde{p}(0)$ is negligible. In fact, the radiation resistance of the whole piston–ring system is also negligible for small ka_2, although the radiation reactance remains of the same order as that of a uniformly moving piston of radius a_2. The aforegoing discussion of the Rayleigh integral equation and the example of the vibrating pistons serve to introduce two extremely important characteristics of acoustic fields generated by vibrating surfaces. First, the resistive component of the surface pressure field and indeed the radiated power are functions of the distribution of vibration amplitude and phase over the whole surface, whereas the reactive component tends to be controlled mainly by the *local* motion of the surface, because it depends upon the kinetic energy imparted to the local fluid. Second, at frequencies for which the acoustic wavelength greatly exceeds the surface dimensions, the radiated power is determined essentially by the net volume velocity and is very insensitive to the detail of its spatial distribution.

The radiation impedance of a piston of arbitrary ka is of considerable practical significance in various areas of applied acoustics; for example, the radiation impedance at the exit of a circular-section duct approximates closely to that of a piston, as does that of a baffled loudspeaker cone below the frequency at which it starts to "break up" into complex vibration modes. The details of the analysis, which is based upon Eq. (3.2), may be found in many other textbooks, e.g., Pierce (1981). The result is as follows:

$$\tilde{Z}_{\mathrm{rad}} = R_{\mathrm{rad}} + jX_{\mathrm{rad}}, \tag{3.14}$$

where

$$R_{\mathrm{rad}} = \frac{\rho_0 c}{\pi a^2}\left[1 - \frac{2J_1(2ka)}{2ka}\right], \qquad X_{\mathrm{rad}} = \frac{\rho_0 c}{\pi a^2}\left[\frac{2H_1(2ka)}{2ka}\right],$$

where J_1 is the Bessel function of order one and H_1 the Struve function of order one (Watson, 1966). The impedance components are plotted as functions of $2ka$ in Fig. 61; remember that the acoustic wavelength equals the piston diameter when $2ka = 2\pi$. After an initial variation as $(ka)^2$ at low ka, as indicated by Eq. (3.7), the resistance increases almost linearly with ka and then turns, at about the frequency where the wavelength equals a diameter, to asymptote in an oscillatory fashion to the plane-wave resistance. The reactive component peaks at about half this frequency and then decreases asymptotically to zero at large ka.

In most practical cases, the accurate evaluation of fluid loading using the integral equations is not possible by analytical techniques because the surface geometries and forms of motion are too complex. Therefore, numerical techniques of evaluating the integrals have been developed; these are widely used in the design of underwater transducers, and some of these techniques are discussed more fully in Chapter 7. However, there is a class of surface geo-

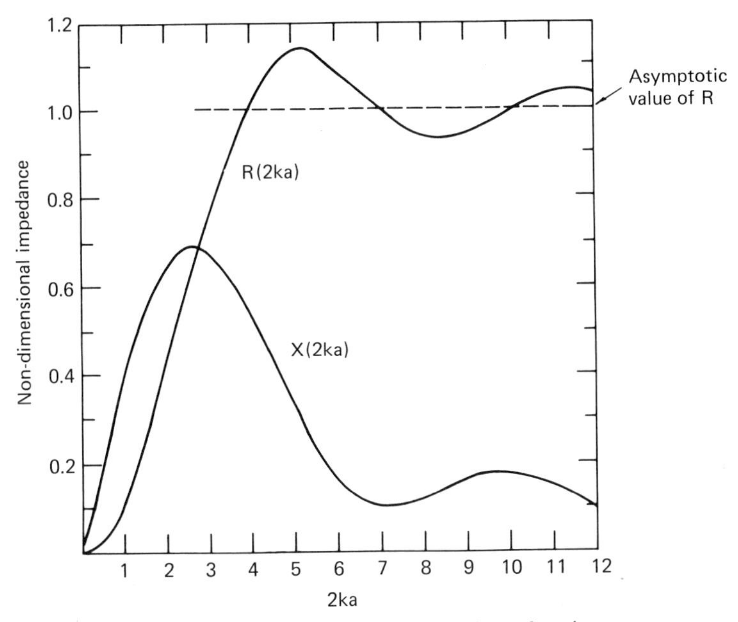

Fig. 61. Baffled circular piston impedance functions.

metries for which analytical solutions are available; these are surfaces that conform to constant coordinate surfaces in coordinate systems in which the acoustic-wave equation is separable, the most common being rectangular, cylindrical, and spherical. The reason why analytical solutions are possible relates to the general form of the K–H integral equation, which shows that if, instead of the point source free-space Green's function $(1/4\pi R)\exp(-jkR)$, a Green's function suitable to the coordinate system and having normal derivative $\partial G/\partial n$ on the surface equal to zero is used, only the surface normal velocity term appears; the surface pressure need not be known, as in the Rayleigh integral for plane surfaces. An equivalent approach more readily understood from a physical point of view is the boundary-matching method, whereby the acoustic particle velocity normal to the structural surface is matched to the normal velocity of the surface.

As an example of considerable practical importance, we consider the fluid loading of an infinitely long circular cylindrical body of which the distribution of normal surface velocity is cosinusoidal in both the axial and circumferential directions. The single-frequency wave equation in appropriate cylindrical coordinates is

$$\frac{\partial^2 p}{\partial r^2} + \frac{1}{r}\frac{\partial p}{\partial r} + \frac{1}{r^2}\frac{\partial^2 p}{\partial \phi^2} + \frac{\partial^2 p}{\partial z^2} + k^2 p = 0. \tag{3.15}$$

The equation has separable solutions $p(r, \phi, z) = p_1(r)p_2(\phi)p_3(z)$. The surface normal velocity distribution is assumed to be

$$v_n(z, \phi, t) = \tilde{v}_n \cos n\phi \cos k_z z \exp(j\omega t), \tag{3.16}$$

in which n/a is the circumferential wavenumber and k_z the axial wavenumber. The radial component of acoustic particle velocity u_r is given by the fluid momentum equation

$$\frac{\partial p}{\partial r} = -\rho_0 \frac{\partial u_r}{\partial t} = -j\omega\rho_0 u_r. \tag{3.17}$$

The acoustic field must be periodic in the axial and azimuthal coordinates with the same periods as the surface vibration. (Why?) Hence $p_3(z) = \cos k_z z$ and $p_2(\phi) = \cos n\phi$.

Equation (3.15) becomes

$$\frac{\partial^2 p_1(r)}{\partial r^2} + \frac{1}{r}\frac{\partial p_1(r)}{\partial r} + \left[k^2 - k_z^2 - \left(\frac{n}{r}\right)^2\right] p_1(r) = 0. \tag{3.18}$$

This is Bessel's equation of which the solutions are linear combinations of Bessel functions of the first and second kind (Watson, 1966):

$$p_1(r) = AJ_n[(k^2 - k_z^2)^{1/2} r] + BY_n[(k^2 - k_z^2)^{1/2} r]. \tag{3.19}$$

The ratio of the constants B/A can be determined from the radiation condition as r tends to ∞. Writing the argument $(k^2 - k_z^2)^{1/2}$ as x,

$$\lim_{x\to\infty} J_n(x) \to (2/\pi x)^{1/2} \cos[x - (2n + 1)\pi/4], \tag{3.20a}$$

$$\lim_{x\to\infty} Y_n(x) \to (2/\pi x)^{1/2} \sin[x - (2n + 1)\pi/4]. \tag{3.20b}$$

At a sufficiently large radial distance from the axis of the cylinder, and with $k_z \neq k$, the radiated field must tend to the form of a plane wave: $p(r) \to \tilde{A} \exp(-jkr) = |\tilde{A}|(\cos kr - j \sin kr)$ as $r \to \infty$. By analogy with Eq. (3.19) for $kr \to \infty$ we see that $B = -jA$, and hence

$$\tilde{p}(r, \phi, z) = \tilde{A}[J_n(k^2 - k_z^2)^{1/2} r - jY_n(k^2 - k_z^2)^{1/2} r] \cos n\phi \cos k_z z. \tag{3.21}$$

The combination $J_n + jY_n$ is the Hankel function H_n of the first kind.

At the surface of a cylinder ($r = a$) Eqs. (3.16), (3.17), and (3.21) link the surface pressure and normal velocity through

$$-j\omega\rho_0\tilde{v}_n = \tilde{A}(k^2 - k_z^2)^{1/2} H_n'[(k^2 - k_z^2)^{1/2} a], \tag{3.22}$$

where the prime denotes differentiation with respect to the argument of the function. Hence,

$$\tilde{A} = \frac{-j\omega\rho_0\tilde{v}_n}{(k^2 - k_z^2)^{1/2} H_n'[(k^2 - k_z^2)^{1/2} a]}. \tag{3.23}$$

The axial variation of surface velocity is manifested in the term $(k^2 - k_z^2)^{1/2}$, which may be real or imaginary, and the circumferential variation determines the order of the Hankel function. If, at the frequency of oscillation, the axial wavenumber k_z exceeds the acoustic wavenumber $k = \omega/c$, the argument of the Hankel function is imaginary and the function can be replaced by a modified Hankel function of real argument

$$K_n(x) = (\pi/2) j^{n+1} H_n(jx). \tag{3.24}$$

The radial variation of pressure in the acoustic field is

$$\tilde{p}_1(r) = \frac{-j\omega\rho_0\tilde{v}_n H_n[(k^2 - k_z^2)^{1/2} r]}{(k^2 - k_z^2)^{1/2} H_n'[(k^2 - k_z^2)^{1/2} a]}, \tag{3.25}$$

and at the surface of the cylinder

$$\frac{\tilde{p}_1(a)}{\tilde{v}_n} = \frac{-j\omega\rho_0 H_n[(k^2 - k_z^2)^{1/2} a]}{(k^2 - k_z^2)^{1/2} H_n'[(k^2 - k_z^2)^{1/2} a]}. \tag{3.26}$$

where $H_n(x)$ is replaced by $K_n(x)$ according to Eq. (3.24) when $k_z > k$.

Equation (3.26) is an expression of cylinder surface *specific* acoustic impedance associated with a particular distribution of surface velocity and describes the fluid loading: it is a form of radiation impedance that we shall denote by $\tilde{z}_{rad}$.

Presentation of the complete details of the variation of the radiation impedance with the parameters involved is beyond the scope of this book: they may be found in a more specialised treatise on the subject of structure–fluid interaction (Junger and Feit, 1972). However, certain approximate expressions presented therein provide an indication of the general physical characteristics of the fluid loading. Where the axial wavenumber exceeds the acoustic wavenumber ($k_z > k$) the radiation impedance is purely imaginary:

$$\begin{aligned} \tilde{z}_{rad} &= -j\omega\rho_0 a \ln[(k_z^2 - k^2)^{1/2} a], \quad && n = 0 \\ &\simeq j\omega\rho_0 a/n, && n \geqslant 1 \end{aligned} \Bigg\} \; (k_z^2 - k^2)a^2 \ll (2n+1) \tag{3.27a}$$

$$\simeq 0, \qquad (k_z^2 - k^2)^{1/2} a \gg n^2 + 1. \tag{3.27b}$$

Both impedances in Eq. (3.27a) are in fact positive (why?) and represent purely inertial, or masslike, fluid reaction. The inertial fluid loading is clearly greatest for the axially symmetric breathing mode ($n = 0$) and decreases with increasing circumferential order n. Hence we would expect the natural frequencies of cylindrical shells of low circumferential order to be most affected by fluid loading, provided their axial distribution of normal surface velocity satisfies the condition $k_z > k$. That this is indeed the case is shown by Fig. 62 (Pallett, 1972).

Where the axial wavenumber is less than the acoustic wavenumber (i.e., the axial wavelength exceeds the acoustic wavelength) the radiation impedance of a cylindrical surface is complex, possessing both real and positive imaginary (inertial) components. Where $(k^2 - k_z^2)^{1/2}a \ll 1$, the resistive component of the breathing mode is dominant and the radiation resistance substantially exceeds $\rho_0 c$. That of the higher circumferential orders peaks at a value of $(k^2 - k_z^2)^{1/2}a \simeq n$ and then tends to $\rho_0 c$ (see Fig. 55).

The evaluation of fluid loading on transversely vibrating slender structures such as wires, pipes, and beams can be important in the analysis of the behaviour of stringed instruments, cross flow heat exchangers, acoustically induced fatigue failure of wire screens, and other systems of practical importance. We have already seen in the case of the piston that where the typical dimension of the region of vibration is substantially smaller than an acoustic wavelength, it is the net volume displacement of fluid and not the details of the distribution of displacement that primarily determines the

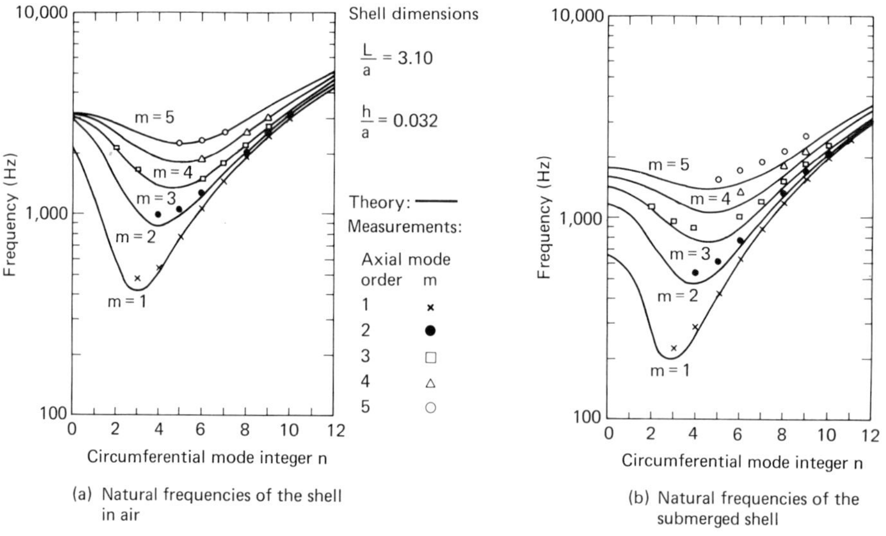

Fig. 62. Effect of water loading on the natural frequencies of an aluminium cylindrical shell with closed ends (Pallett, 1972).

radiated acoustic field. In the case of transverse vibration of a slender body, in the absence of distortion of the cross section normal to the long axis, the net displacement of fluid volume at frequencies where the acoustic wavelength greatly exceeds a typical cross-section dimension is virtually zero, as in the case of the piston and concentric annulus discussed earlier. The fluid has time to "slip around" the circumference toward the opposite side, to avoid being compressed; however it does possess oscillatory kinetic energy.

Although slender bodies are not all circular in cross section, it seems reasonable to analyse the case of a transversely vibrating circular cylindrical rigid body in order to achieve an appreciation of the general nature of fluid loading in such cases: note that $k_z = 0$ and $n = 1$ in this case. In the limit $(k^2 - k_z^2)^{1/2} r \to 0$, we can rewrite Eq. (3.21) in the asymptotic form

$$\tilde{p}(r, \phi, z) = \frac{j2\tilde{A}c}{(k^2 - k_z^2)^{1/2} r} \cos \phi \cos k_z z, \tag{3.28}$$

and from (3.17)

$$-j\omega\rho_0\tilde{v}_1 = \frac{-j2\tilde{A}c}{(k^2 - k_z^2)^{1/2} a^2}. \tag{3.29}$$

Hence

$$\tilde{p}(a, \phi, z) = j\omega\rho_0 a\tilde{v}_1 \cos \phi \cos k_z z. \tag{3.30}$$

This result is equivalent to Eq. (3.27a) with $n = 1$. The transverse fluid reaction force per unit length on the body is given by

$$\tilde{F}(z) = a \int_0^{2\pi} \tilde{p}(a, \phi, z) \cos \phi \, d\phi = j\omega\pi a^2 \rho_0 \tilde{v}_1 \cos k_z z. \tag{3.31}$$

The mechanical impedance per unit length due to fluid loading is

$$\tilde{Z} = \tilde{F}(z)/\tilde{v}_1 \cos k_z z = j\omega\pi a^2 \rho_0 = j\rho_0 c\pi a(ka), \tag{3.32}$$

which is inertial, giving a mass per unit length equal to that of a cylinder of fluid, radius a. A water-filled, thin-walled tube vibrating in water at low frequencies experiences an external inertial loading equal to that produced by the contained water.

This analysis suggests that no acoustic power would be radiated from such a vibrating body. However, a more exact analysis shows that, to third order in ka, the impedance is

$$\tilde{Z} \simeq \rho_0 c[j\pi a(ka) + \tfrac{1}{2}\pi^2 a(ka)^3]. \tag{3.33}$$

The second resistive term is much smaller than the inertial term for $ka \ll 1$, and sound radiation is very inefficient.

3.3 Wave Impedances of Structures and Fluids

In Chapter 1 we considered impedances associated with localised forces and moments applied at fixed points on structures. In this book we are particularly concerned with coupling between waves in structures and in elastic fluids with which they are in contact; such coupling clearly involves distributions of interaction forces over extended surfaces. Hence it is useful to define another form of impedance in which a distribution of forces is implicitly assumed: this is the wave impedance. A discussion of fluid-loading effects on point impedance is delayed until the end of this chapter.

The mathematical justification for the use of wave impedance is that any spatial distribution of a variable, such as force or displacement, can be analysed into and synthesised from a superposition of an infinite number of elementary distributions, each of which is sinusoidal in space and has a particular wavenumber, as described in Section 1.1. When structures and fluids undergo coupled vibration, the spatial distributions of normal displacements at the interfaces are common; hence the wavenumber spectra of the displacements are also common.

A wave impedance for a structure, or a volume of fluid, can be associated with specific wavenumber, or frequency and phase speed combination, $k = \omega/c_{ph}$. In the case of a structure it is evaluated mathematically by applying a force to the structure in the form of a sinusoidal travelling wave and deriving the structural response from the equation(s) of motion. It should be noted carefully that the response to such a force takes the form of a simple travelling wave of wavenumber equal to that of the force only if the structure is both *uniform and infinitely extended;* if not, scattering of the forced wave occurs and the wavenumber spectrum of the response is spread. It should also be remembered that the wavenumber spectrum of a point force is uniformly distributed across all wavenumbers from minus infinity to plus infinity, just as the frequency spectrum of an infinitely narrow temporal pulse is distributed uniformly from minus infinity to plus infinity.

As an example of the derivation of wave impedance we shall consider an infinitely extended, undamped uniform plate subject to a transverse force in the form of a plane travelling wave. The one-dimensional bending equation for an *in vacuo* plate of thickness h, subject to a force $\tilde{f}$ per unit area, is

$$D\frac{\partial^4 \eta}{\partial x^4} + m\frac{\partial^2 \eta}{\partial t^2} = \tilde{f}\exp[j(\omega t - \kappa x)], \tag{3.34}$$

where $D = Eh^3/12(1 - \nu^2)$. Because the plate is uniform and infinite, the solution must take the form

$$\eta(x,t) = \tilde{\eta} \exp[j(\omega t - \kappa x)].$$

Substitution of this form into Eq. (3.34) yields

$$(D\kappa^4 - m\omega^2)\tilde{\eta} = \tilde{f}. \tag{3.35}$$

The wave impedance of the plate is defined as the ratio of the complex amplitude of force per unit area to the complex velocity amplitude:

$$\tilde{z}_{\text{wp}} = \tilde{f}/j\omega\tilde{\eta} = -j(D\kappa^4 - m\omega^2)/\omega. \tag{3.36}$$

In Section 1.7 it was shown that the free-plate bending wavenumber $k_\text{b} = (\omega^2 m/D)^{1/4}$; hence $\tilde{z}_{\text{wp}} = 0$ when $\kappa = k_\text{b}$. Excitation by a traveling force wave having $\kappa = k_\text{b}$ may be likened to excitation of a simple undamped oscillator at its natural frequency; an infinitely small force produces an infinitely large response. Equation (3.36) also reveals that the wave impedance of the plate is springlike for $\kappa > k_\text{b}$, and masslike for $\kappa < k_\text{b}$.

If the plate is damped according to the complex modulus introduced in Chapter 1,

$$\tilde{z}_{\text{wp}} = -j(D\kappa^4 - m\omega^2)/\omega + D\kappa^4\eta/\omega. \tag{3.37}$$

Hence $\tilde{z}_{\text{wp}}$ never equals zero, but when $\kappa = k_\text{b}$ it is purely real and equal to $Dk_\text{b}^4\eta/\omega$. Since η is normally of the order 10^{-2}, the real part of the impedance is dominant only when the excitation force wavenumber is very close to k_b.

We now suppose the plate to be in contact on one side with a semi-infinitely extended, two-dimensional layer of fluid ($y > 0$) of which it forms the boundary. We assume further that a bending wave of arbitrary wavenumber k_x is generated in the plate by a force wave applied to the other side, as illustrated in Fig. 63. The acoustic pressure field in the fluid is found by using the solution of the two-dimensional wave equation (1.7), together with the fluid momentum equations (1.5a,b) at the plate–fluid interface.

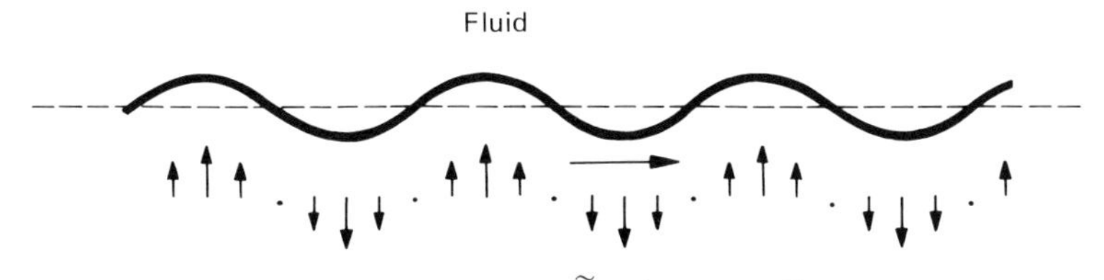

Fig. 63. Excitation of a plate by a traveling force wave.

The x-wise variation of all acoustic variables must follow that of the force and plate displacement. From Eq. (1.8) we find that $k_y = \pm(k^2 - k_x^2)^{1/2}$. The appropriate sign of the square root is determined by the physics of the model. When $k_x < k$, the positive sign corresponds to a plane sound wave travelling away from the plate surface, as seen from Eq. (1.7): no wave can propagate toward the plate surface and therefore the negative sign is disallowed. On the other hand, when $k_x > k$, k_y becomes imaginary, and the disturbance decays exponentially with distance from the plate surface: in this case the negative sign of the square root must be selected so that $k_y = -j(k_x^2 - k^2)^{1/2}$.

Equation (1.9) yields the acoustic pressure at the plate surface ($y = 0$) as

$$p(x, 0, t) = \tilde{p} \exp[j(\omega t - k_x x)] = \frac{\omega \rho_0 \tilde{v}}{\pm(k^2 - k_x^2)^{1/2}} \exp[j(\omega t - k_x x)], \quad (3.38)$$

where $\tilde{v} = j\omega\tilde{\eta}$ is the complex amplitude of the plate velocity. Hence the wave impedance of the fluid space, which is effectively a specific acoustic impedance, is

$$\tilde{z}_{\text{wf}} = (\tilde{p})_{y=0}/\tilde{v} = \pm\omega\rho_0/(k^2 - k_x^2)^{1/2}. \quad (3.39)$$

When $k_x < k$,

$$\tilde{z}_{\text{wf}} = \rho_0 c/[1 - (k_x/k^2)]^{1/2}. \quad (3.40)$$

The impedance is purely real and positive, indicating, according to Eq. (1.10), that the vibrating plate does work on the fluid, the resulting energy being radiated away in the form of acoustic plane waves of which the wave vector makes an angle of $\cos^{-1}(k_x/k)$ to the plate plane; the fluid loading thus acts on the plate as a form of damping. The impedance approaches infinity as k_x approaches k, and it asymptotes to $\rho_0 c$ as k_x approaches zero. The physical interpretation of this result is that the fluid impedance approaches that of a progressive plane wave travelling purely in the y direction as the ratio of phase speed of the forcing field to the speed of sound tends to infinity; or as the ratio of the wavelength of the plate motion to the wavelength of sound tends to infinity. When $k_x > k$,

$$\tilde{z}_{\text{wf}} = j\omega\rho_0/(k_x^2 - k^2)^{1/2} = j\rho_0 c/[(k_x/k)^2 - 1)^{1/2}. \quad (3.41)$$

Equation (3.41) indicates that when the phase speed of the forcing field is less than the speed of sound in the fluid, the fluid impedance is purely reactive, and the fluid loading is inertial in nature, corresponding in a numerical sense to a fluid layer of thickness $(k_x^2 - k^2)^{-1/2}$ moving with a uniform transverse velocity equal to that of the plate. (Check.) Both the inertial and damping loading effects are greatest when $k_x \simeq k$ and are, in principle, infinite when $k_x = k$: this is a product of the idealised mathematical model, which need not worry us in dealing with real, bounded structures.

When considering excitation of a coupled structure–fluid system it is useful to combine the impedances of the two to form the impedance of the fluid-loaded structure, as in the earlier example of the piston. This can be done by reference to the equation of motion of the fluid-loaded structure, which for the plate is

$$D\frac{\partial^4\eta}{\partial x^4} + m\frac{\partial^2\eta}{\partial t^2} = \tilde{f}\exp[j(\omega t - k_x x)] - p(x, 0, t). \tag{3.42}$$

The last term represents the acoustic pressure at the interface which is produced by plate motion. In terms of impedances, Eq. (3.42) can be written $\tilde{v}\tilde{z}_{\mathrm{wp}} = \tilde{f} - \tilde{v}\tilde{z}_{\mathrm{wf}}$ or

$$\tilde{f}/\tilde{v} = \tilde{z}_{\mathrm{wp}} + \tilde{z}_{\mathrm{wf}}. \tag{3.43}$$

Hence the wave impedance of the coupled system equals the sum of the two wave impedances.

We have briefly considered the fluid loading associated with waves generated in a plate by an applied force wave. The reaction of a fluid to vibration of a structure with which it is in contact must clearly also influence the process of *free*-wave propagation in the structure. In an undamped wave-bearing system, the equation that has to be solved for the free propagation wavenumber corresponds to setting the wave impedance equal to zero, i.e., zero input force is equivalent to free vibration. The resulting equation is the dispersion equation. In the case of the fluid-loaded, undamped plate of impedance given by Eq. (3.43), we cannot state a priori that the fluid impedance will be wholly reactive, because its nature depends upon k, which is the quantity sought. The general solution of the problem is beyond the scope of this book. However, reference to Fig. 16 in Section 1.8, shows that below the critical frequency the *in vacuo* bending wavenumber is greater than the acoustic wavenumber in the fluid. We have seen that if the plate wavenumber exceeds that of the sound in the fluid, the reaction is masslike. Recalling that the *in vacuo* free bending wavenumber is given by $k_{\mathrm{b}} = (\omega^2 m/D)^{1/4}$, it is clear that additional mass will increase k_{b} and reduce the phase speed of free waves in the plate. Hence we may obtain an approximate solution for the *low-frequency*, fluid-loaded free wavenumber k_{b}' by assuming that $k_{\mathrm{b}}' \gg k$. From Eq. (3.41),

$$\tilde{z}_{\mathrm{wf}} \simeq j\omega\rho_0/k_{\mathrm{b}}', \tag{3.44}$$

and substitution from Eqs. (3.36) and (3.44) into Eq. (3.43), together with setting the total wave impedance to zero, yields

$$Dk_b'^4 - \omega^2(m + \rho_0/k_b') = 0. \tag{3.45}$$

Although it is not simple to solve Eq. (3.45) for k_b' explicitly, since it is a fifth-order equation in k_b', it is clear that the term ρ_0/k_b' represents an addition to the plate mass per unit area m, just as the term $8\rho_0 a^3/3$ does for a small piston, and that this contribution increases as k_b' falls. Since *in vacuo* $Dk_b^4 = \omega^2 m$ and fluid mass loading will reduce the phase speed and increase the bending wavenumber, we may assume that $Dk_b'^4 \gg \omega^2 m$ and approximate Eq. (3.45) by

$$k_b' \simeq (\omega^2\rho_0/D)^{1/5}. \tag{3.46}$$

This rather surprising result indicates that, in dense fluids, at frequencies very much below the critical frequency, the plate mass has no influence on the free bending wavenumber because the inertial fluid loading is dominant. The corresponding phase speed is given by

$$c_{ph} = (\omega^3 D/\rho_0)^{1/5}. \tag{3.47}$$

Further qualitative observations may be made about fluid-loaded, free flexural waves in infinite, thin plates. It is clearly not possible for a wave to propagate freely with a wavenumber equal to the acoustic wavenumber because the fluid wave impedance is infinite. In fact, detailed mathematical analysis shows that the only physically significant purely real solution for the free flexural wavenumber represents a surface wave that travels subsonically even above the critical frequency (Strawderman *et al.*, 1979), because as k_x approaches k, the inertial loading is seen from Eq. (3.41) to increase indefinitely. Fortunately for engineers concerned only with bounded plates, arguments among mathematicians about the physical significance of other complex roots of Eq. (3.45) are mainly of academic interest.

3.4 Fluid Loading of Vibrating Plates

We have seen that, for infinitely extended plates carrying sinusoidal plane waves, there is a strict division of the nature of the fluid reaction between purely inertial loading for subsonic phase speeds, and purely resistive or damping-like loading for supersonic phase speeds. Real structures are bounded, and we must now turn to the question of fluid loading of bounded plates.

So far in this chapter we have considered the analysis of fluid loading by means of a Green's function approach for a source region of finite extent, and a boundary-matching approach for vibrations of surfaces of infinite extent. The latter approach can be applied to vibrating surfaces of limited extent, provided that they form part of an infinitely extended surface of which the geometry conforms to a coordinate system in which the wave equation is separable. Examples include a flat vibrating panel set in an otherwise infinite rigid planar baffle and a vibrating tube set in coaxial rigid tubular extensions of semi-infinite extent.

The cynic might observe that we seem to be no nearer to reality, because infinitely extended surfaces of regular geometry do not exist, whether vibrating or not. It transpires, however, that in many cases the fluid loading on vibrating structures of finite extent approximates closely to that on the equivalent infinitely extended system; in particular, the inertial component of fluid loading on a region of structure tends to be related to the kinetic energy of the local fluid and is rather insensitive to the state of vibration at remote points.

We have already encountered the concept of spatial frequency (wavenumber) analysis and synthesis of spatially distributed variables. This mathematical technique is now applied to the analysis of fluid loading on the two-dimensional plate–baffle system shown in Fig. 64.

It is assumed that the simply supported plate vibrates in one of its *in vacuo* modes in which the normal velocity distribution is given by

$$v(x,t) = \begin{cases} \tilde{v}_{\mathrm{p}} \sin(m\pi x/l) e^{j\omega t}, & 0 < x < l, \\ 0, & 0 > x > l. \end{cases} \tag{3.48}$$

The wavenumber transform of v is

$$\tilde{V}(k_x) = \tilde{v}_{\mathrm{p}} \int_0^l \sin(m\pi x/l) \exp(-jk_x x)\, dx. \tag{3.49}$$

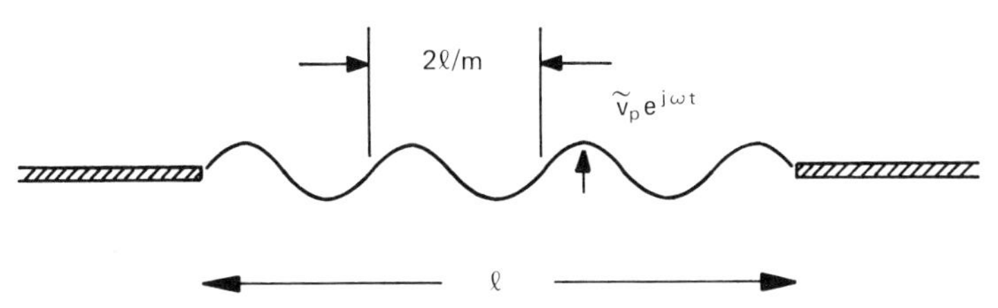

Fig. 64. Modal vibration of a baffled plate.

In this case we can simplify the integration by expressing the standing wave as the sum of two traveling waves. Letting $k_m = m\pi/l$,

$$V(k_x) = -\frac{j}{2}\tilde{v}_p \int_0^l [\exp(jk_m x) - \exp(-jk_m x) \exp(-jk_x x)\,dx$$

$$= -\frac{1}{2}\tilde{v}_p \left[\frac{\exp[j(k_m - k_x)x]}{k_m - k_x} + \frac{\exp[-j(k_m + k_x)x]}{k_m + k_x}\right]_0^l$$

$$= -\frac{1}{2}\tilde{v}_p \left[\frac{\exp[j(k_m - k_x)l]}{k_m - k_x} + \frac{\exp[-j(k_m + k_x)l]}{k_m + k_x} - \frac{2k_m}{k_m^2 - k_x^2}\right]. \quad (3.50)$$

The modulus of the transform is

$$|\tilde{V}(k_x)| = |\tilde{v}_p| \left(\frac{2\pi ml}{(k_x l)^2 - (m\pi)^2}\right) \sin\left(\frac{k_x l - m\pi}{2}\right). \quad (3.51)$$

and is plotted in Fig. 65. The wavenumber spectrum of the surface acoustic field generated by the wavenumber component $\tilde{V}(k)$ is given by Eq. (3.39) as

$$[\tilde{P}(k_x)]_{y=0} = \tilde{V}(k_x)\tilde{z}_{wf}(k_x). \quad (3.52)$$

At this stage we have to specify the frequency of vibration: The analysis of the wavenumber spectrum was solely concerned with the variation of v with x: the result, Eq. (3.50), contains no frequency term. The frequency becomes vital once the question of the type of acoustic field generated by the vibration arises. This is because the form, and indeed associated physical nature, of fluid loading represented by $\tilde{z}_{wf}$ depends upon how the wavenumber spectrum $\tilde{V}(k_x)$ is distributed with respect to the acoustic wavenumber k *at the particular frequency of vibration.* For $|k_x| > k$, the corresponding component

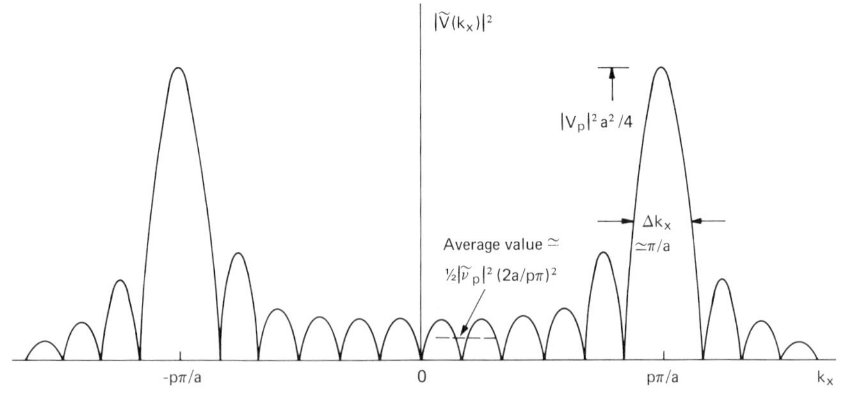

Fig. 65. Wavenumber modulus spectrum of plate velocity (diagrammatic).

of $\tilde{z}_{wf}(k_x)$ is purely imaginary and is given by Eq. (3.41); for $|k_x| < k$ the component is purely real and is given by Eq. (3.40): the modulus of k_x is used because the bounded plate carries waves travelling in both directions. These conditions are illustrated in Fig. 65, which indicates that, because the truncation or windowing of a sinusoid of "wavelength" λ spreads its wavenumber spectrum around the wavenumber $2\pi/\lambda$, the fluid loading generally comprises both reactive and resistive components at all frequencies.

Equation (3.52) represents the wavenumber spectral decomposition of the surface pressure. The actual surface pressure amplitude distribution $\tilde{p}(x)$ is given by the inverse Fourier transformation of $\tilde{P}(k_x)$:

$$\tilde{p}(x, 0) = \frac{1}{2\pi} \int_{-\infty}^{\infty} [\tilde{P}(k_x)]_{y=0} \exp(jk_x x)\, dk_x. \tag{3.53}$$

The integral must be split into three parts, viz.,

$$\begin{aligned} \tilde{p}(x, 0) = {} & \frac{\rho_0 c}{2\pi} \int_{-k}^{k} \tilde{V}(k_x)(1 - k_x^2/k^2)^{-1/2} \exp(jk_x x)\, dk_x \\ & + \frac{j\rho_0 c}{2\pi} \int_{k}^{\infty} \tilde{V}(k_x)(k_x^2/k^2 - 1)^{-1/2} \exp(jk_x x)\, dk_x \\ & + \frac{j\rho_0 c}{2\pi} \int_{-\infty}^{-k} \tilde{V}(k_x)(k_x^2/k^2 - 1)^{-1/2} \exp(jk_x x)\, dk_x. \end{aligned} \tag{3.54}$$

The first integral represents the resistive component of the fluid loading and the other two represent the reactive part. These integrals do not have general closed-form solutions, and anyway a knowledge of the detailed distribution of surface pressure is not always necessary. Of more common interest are the radiated sound power and the effective mass added to the plate by the reactive component of the fluid loading.

The sound power radiated from bounded plates has been treated in Chapter 2 by means of a Rayleigh integral approach (Wallace, 1972). Extension into three dimensions of the two-dimensional wavenumber spectrum approach of Heckl, outlined in Eqs. (2.43)–(2.55), produces very similar results. This form of analysis can be adopted to provide an estimate of the inertial loading on plates vibrating well below the critical frequency. The reactive power generated by a vibrating plate in a fluid is put into near-field kinetic energy. Over alternate quarter cycles of oscillation, fluid kinetic energy is created, and over the other two quarter-cycles the same energy returns to the plate, so that over a half-period, or multiple thereof, zero net work is done by the plate (Fig. 66). The effective mass per unit area associated with the near field can be obtained from the imaginary part of the product of the complex pressure and the associated normal fluid velocity by the

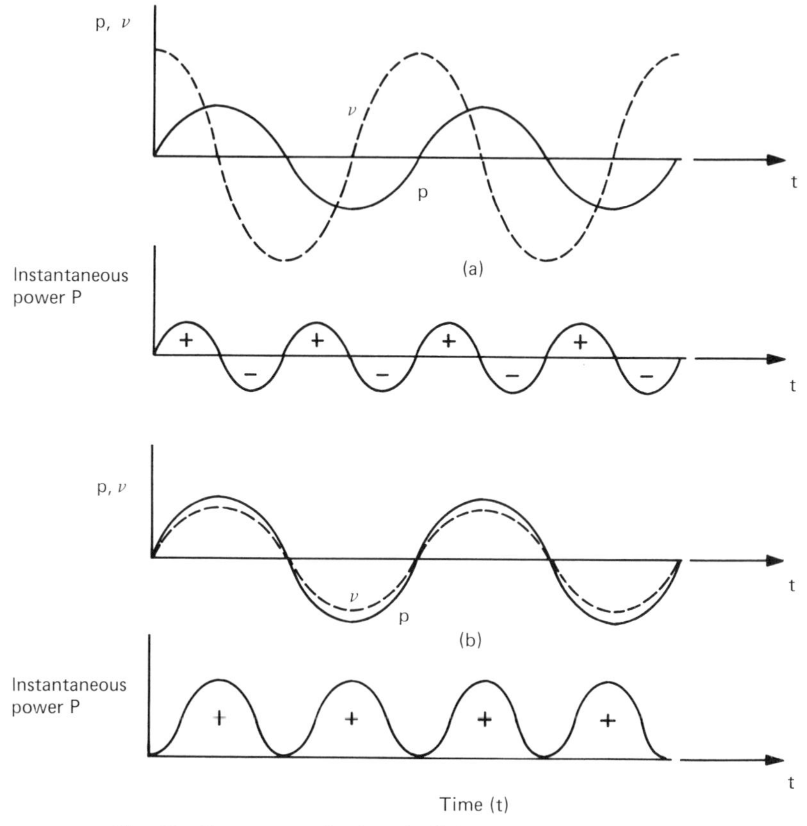

Fig. 66. Power transmitted to the fluid: (a) reactive; (b) resistive.

relationship

$$\frac{1}{4} m_e l v_p^2 = \frac{1}{2\omega} \operatorname{Im} \int_0^l \{\tilde{p}(x,0)\tilde{v}(x)\}\, dx, \tag{3.55}$$

for a one-dimensional vibration field in which $\tilde{v}_p$ is assumed to be purely real. The extra factor of one-half appears on the left-hand side because of the sinusoidal spatial variation of v with x.

Substitution of the expressions for $\tilde{p}(x, 0)$ and $\tilde{v}(x)$ in terms of wavenumber transforms, and integration over the length of the plate, in a manner similar to that presented in Eqs. (2.44)–(2.51), yields the following expression for the effective mass per unit area of the plate:

$$m_e = \frac{2\rho_0}{\pi l v_p^2} \int_k^\infty \frac{|\tilde{V}(k_x)|^2\, dx_x}{(k_x^2 - k^2)^{1/2}}. \tag{3.56}$$

The plot of $|\tilde{V}(k_x)|^2$ in Fig. 65 suggests that the main contribution to the integral is associated with wavenumbers around $k_x = m\pi/l$. Comparison with the contribution in the region $k_x \simeq k$ shows that the former is generally the larger, except when m is 1 or 2, and for all m when $m\pi/l$ approaches k. Hence we may evaluate m_e for values of $m\pi/l \gg k$ by substituting $m\pi/l$ for k_x in the denominator of the integrand, removing it from the integrand, and extending the range of integration from zero to infinity. Substitution into Eq. (3.56) of the expression for $|\tilde{V}(k_x)|$ in Eq. (3.51), and performance of the integration, yields

$$m_e \simeq \rho_0/(m\pi/l) = \rho_0/k_m. \tag{3.57}$$

Note that this expression is equivalent to that given by Eq. (3.44) for a plane flexural wave traveling in an unbounded uniform plate. This equivalence indicates that the near-field inertial loading at subcritical frequencies is not very sensitive to the boundary conditions of the plate, unlike the resistive loading. This result reinforces our previous conclusion that reactive fluid loading is associated essentially with local vibration. For modes that have natural frequencies just below the critical frequency the inertial component of fluid loading of this one-dimensional plate will be somewhat larger than that given by Eq. (3.57), because the peak in the modal wave number spectrum coincides with an asymptotic trend of the denominator of Eq. (3.56) toward zero. The physical interpretation of the trend is that a greater depth of fluid is being displaced in a reactive oscillatory fashion. The inertial loading for plate modes that have natural frequencies above the critical frequency is smaller than that given by Eq. (3.57) because the range of integration (k to ∞) excludes the peak in the modal wavenumber spectrum. The inertial loading of two-dimensional plate modes of order (p,q) at frequencies well below the critical frequency will also be given by Eq. (3.57), with k_m equal to $[(p\pi/a)^2 + (q\pi/b)^2]^{1/2}$. The localised nature of inertial fluid loading leads us to expect that a similar expression would apply to the modes of cylindrical shells, provided the axial wavenumber exceeds the acoustic wavenumber. Indeed this is so, and Eq. (3.27a) with $k < k_z \ll n/a$ confirms our expectations.

The inertial loading on very low order modes of a rectangular plate are not accurately given by Eq. (3.57) because the width of the main peak of the modal wavenumber spectrum extends over a range $\pm 2\pi/l$ about $m\pi/l$, and when $m = 2$ it extends to $k_x = 0$, hence including the range for which $k \simeq k_x$, and the loading is underestimated by Eq. (3.57). The wavenumber spectrum of the fundamental mode ($m = 1$) has its maximum at $k_x \simeq 0$. As well as making this mode far more efficient as a radiator at subcritical frequencies than the other low-order modes, it also produces a rather high inertial loading, which can be estimated fairly accurately by treating the plate as a piston oscillating with an equivalent volume velocity, and using piston radiation impedance expressions. The results of detailed analyses of fluid loading on rectangular plates by Lomas and Hayek (1977) and Kamkar (1981) confirm that Eq. (3.57) provides a resonable estimate.

3.5 Natural Frequencies of Fluid-Loaded Plates and Shells

In cases where it is necessary to include fluid-loading effects in the estimation of natural frequencies of structures, such as ship hulls, sonar transducers, liquid-filled pipes, and liquid-cooled nuclear reactor internals, it is important to distinguish between cases in which the fluid is contained within a region bounded wholly or largely by the structure under consideration, and cases in which the structure is enveloped in the fluid, and the fluid can be considered virtually unbounded. The essential physical difference between these two cases may be described qualitatively as follows. A bounded fluid can store energy in the form of standing waves and possesses natural frequencies proper to its geometry and physical properties; these may be evaluated in the case where the boundaries are considered to be rigid. In this case, for the purpose of evaluating the natural modes and frequencies, the total system of structure plus fluid should, in principle, be analysed as a coupled system: in practice this is normally only necessary for substantial engineering structures when the fluid is a liquid of high density (see Chapter 6). Where the fluid is essentially unbounded by the structure, reactive fluid-loading forces are associated only with the near-field pressures, which usually manifest themselves as inertial loadings, as we have seen in the previous section.

The natural modes of a strongly coupled structure–fluid system of the closed type have natural frequencies different from those of the two components when uncoupled. Even so, they may generally be divided into two sets—those in which the vibrational energy is stored primarily in the fluid, and those where it resides primarily in the structure. Fortunately for the analyst, the corresponding coupled-mode shapes observed in the dominant component often differ relatively little from mode shapes identified in the uncoupled state. This is even more apparent where a fluid is essentially unbounded; the structural mode shapes remain almost unchanged, and the natural frequencies fall below their *in vacuo* values in proportion to the square root of the ratio of the loaded to unloaded modal masses. The analysis of reactive loading on structural waves having wavenumbers much greater than an acoustic wavenumber has shown that the effective added mass per unit area is ρ_0/k_m, where k_m is the primary effective wavenumber component of the vibration. Hence we may approximate the fluid-loaded structure natural frequencies by

$$\omega'_m \simeq \omega_m(1 + \rho_0/mk_m)^{-1/2}, \tag{3.58}$$

where ω_m is the *in vacuo* natural frequency, k_m the primary modal wavenumber component, and m the average structural mass per unit area. The important conclusion from this result is that it is normally only the natural frequencies of the very low order modes of a plate or shell structure that are significantly affected by fluid loading. It must be noted that the natural modes of plates and shells surrounded by essentially unbounded fluids, in which energy can be radiated away to infinity, are not mathematically orthogonal (Davies, 1971; Mkhitarov, 1972; Stepanishen, 1982). The natural frequencies of fluid-loaded plates and shells are treated by Lomas and Hayek (1977) and Warburton (1978). An example of fluid loading effects on cylinder natural frequencies is shown in Fig. 62. The vibration characteristics of structures that enclose fluid volumes are discussed further in Chapter 6.

The free-vibration equations of strongly coupled structure-bounded fluid systems are not usually amenable to analytic solution in closed form, and numerical methods are now commonly applied to such problems, of which further details will be found in Chapter 7. However, it is possible to solve the one-dimensional coupled problem rather simply, and the solution reveals features common to more complex systems. We consider the one-dimensional model of a spring-mounted piston sliding in a tube terminated by a rigid closure (Fig. 67). The analysis is restricted to frequencies below the lowest acoustic cutoff frequency of the tube, $f_{10} = 1.84c/2\pi a$.

The acoustic impedance at the right-hand face of the piston is obtained by considering acoustic waves travelling to the left and right with amplitudes $\tilde{A}$ and $\tilde{B}$, respectively. A force $\tilde{F}e^{j\omega t}$ acts on the piston, and fluid-loading effects on the left-hand face of the piston are ignored. The acoustic impedance of the column of fluid is given by

$$\tilde{Z}_a = \frac{\rho_0 c}{\pi a^2}\left[\frac{\tilde{A} + \tilde{B}}{\tilde{A} - \tilde{B}}\right]. \tag{3.59}$$

Application of the boundary conditions at the closed end of the tube gives

$$\tilde{B} = \tilde{A}\exp(-2jkl). \tag{3.60}$$

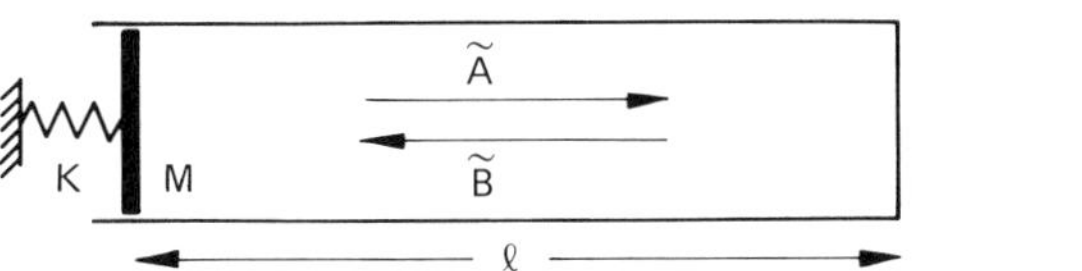

Fig. 67. Spring-mounted piston in a closed tube.

Hence

$$\tilde{Z}_a = -(j\rho_0 c/\pi a^2) \cot kl. \tag{3.61}$$

The corresponding form of impedance of the *in vacuo* piston is

$$\tilde{Z}_p = j(\omega M - K/\omega)/(\pi a^2)^2. \tag{3.62}$$

Note that the acoustic impedance can simulate both inertial and elastic mechanical impedances, depending on the sign of cot kl. These impedances add; hence the equation of simple harmonic motion of the piston may be written in terms of its displacement ξ as

$$\tilde{F} = j\omega\tilde{\xi}(\pi a^2)^2(\tilde{Z}_a + \tilde{Z}_p). \tag{3.63}$$

The natural frequencies ω_n of the coupled system correspond to the condition $\tilde{F} = 0$, i.e., the total impedance vanishes. Hence

$$\cot[(\omega_n/\omega_0)k_0 l] = (M\omega_0/\rho_0 c\pi a^2)(\omega_n/\omega_0)[1 - (\omega_0/\omega_n)^2], \tag{3.64}$$

where $\omega_0^2 = K/M$ and $k_0 = \omega_0/c$. The term $\rho_0 c\pi a^2/M\omega_0$ may be considered to be a form of non-dimensional fluid-loading parameter, since it represents the ratio of the characteristic specific acoustic impedance of the fluid to the magnitude of the impedance of the piston mass per unit area at the *in vacuo* natural frequency; we shall symbolise it by β. The ratio of acoustic wavelength at ω_0 to the tube length is represented by the parameter $k_0 l$.

Solutions to Eq. (3.64) may be obtained numerically or by graphical means. In Fig. 68, cot(kl) and $\beta^{-1}(kl/k_0 l)[1 - (k_0 l/kl)^2]$ are plotted against kl for various values of $k_0 l$ and β. A number of special cases may be identified. Suppose first that the *in vacuo* natural frequency of the mass-spring system is well below that at which $kl = \pi/2$, or $l = \lambda/4$, i.e., the first acoustic resonance frequency of the tube. In this case $\omega_0 \ll \pi c/2l$, and cot kl can be approximated by $(kl)^{-1}$. Then

$$\omega_1^2 \simeq \omega_0^2(1 + \beta/k_0 l), \tag{3.65}$$

and the effective stiffness of the piston system has been increased by a factor $1 + \beta/k_0 l$.

The physical explanation of this result is the fact that at frequencies where the acoustic wavelength is much greater than the tube length, the fluid reaction is produced by almost pure bulk compression. The pressure change produced by a volume displacement $\xi\pi a^2$ is equal to $\rho_0 c^2 \xi\pi a^2/\pi a^2 l = \rho_0 c^2 \xi/l$. Hence the effective fluid stiffness is $\pi a^2 \rho_0 c^2/l$. The term $\beta/k_0 l$ can be written as $\rho_0 c^2 \pi a^2/M\omega_0^2 l = \rho_0 c^2 \pi a^2/Kl$, giving a fractional increase of stiffness as indicated. The effect of the fluid on the natural frequency clearly increases as the piston mass per unit area decreases, as the fluid impedance increases, and as the length of the tube decreases.

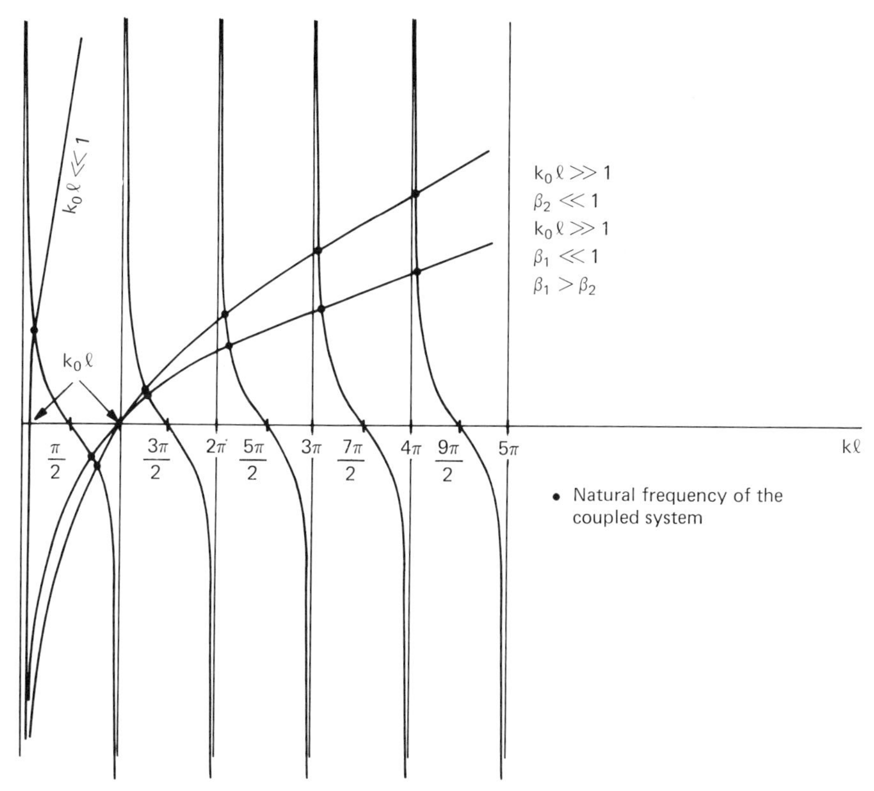

Fig. 68. Plot of cot kl and $\beta^{-1}(kl/k_0l)[1-(k_0l/kl)^2]$ vs. kl.

As another special case we may assume that a natural frequency lies far above ω_0, and that β is considerably less than unity. In this case

$$\tan[(\omega_n/\omega_0)k_0l] \simeq (\omega_0/\omega_n)\beta \ll 1,$$

and

$$(\omega_n/\omega_0)k_0l \to n\pi, \tag{3.66}$$

where n is any integer. Hence

$$\omega_n \simeq n\pi c/l, \tag{3.67}$$

frequencies that correspond to the acoustic natural frequencies of the tube when closed at both ends. This makes physical sense since the impedance of the piston is dominated by the inertial term $j\omega M$, which is large at high frequencies and can combine with the high negative acoustic impedance (stiffness) of the fluid in the tube to produce zero total impedance. The single natural frequency of the *in vacuo* system has been joined by an infinite host of colleagues due to the extended form of the elastic-fluid system.

Intermediate cases have to be evaluated for specific combinations of β and $k_0 l$. However, this example shows that fluid in a bounded space can have a profound influence on the values and *numbers* of natural frequencies of a coupled structural system. Such dramatic fluid stiffness effects as we have encountered here are not very common in practice, but have been shown to influence the resonances of windows coupled to small rooms and are, of course, important in low-frequency cabinet loudspeaker performance. They are also at work in double-leaf partitions, as we shall see in the next chapter. [As an experiment, try measuring the impedance of a loudspeaker placed at one end of a tube that is open, or closed, at the other end.]

If the tube in our example were terminated by an impedance that contained a real (resistive) term the natural vibration of the piston would be damped, even if the piston suspension were not itself damped. Analysis of the power radiated by a piston into an acoustically "lossy" tube shows that it is very much influenced by the resonant behavior of the system. The influence of waves reflected from a tube termination, or impedance discontinuity, can have a profound influence on the acoustic source and is central to the operation of many musical instruments, including pipe organs and wind instruments. In these cases, analysis of the coupled source–tube system is far more complicated than in the case of the piston, because the aerodynamic source characteristics are essentially non-linear.

3.6 Effects of Fluid Loading on Vibration of and Sound Radiation from Point-Excited Plates

It is of considerable practical importance to those concerned with the vibration of plate and shell structures in contact with water to be able to estimate, and to have a qualitative feel for, the effects of the water loading on the generation of structural and acoustic waves by concentrated applied forces.

The detailed mathematical analyses of fluid loading effects involve rather advanced analytical techniques, and in many cases quantitative results can be obtained only by numerical procedures; however, some general qualitative trends can be deduced. Maidanik and Kerwin (1966) show that at frequencies well below the critical frequency the sound power radiated by a thin plate, excited by a simple harmonic point force, into a fluid present only on one side of the plate, is given by

$$\bar{P} = k^2 \beta^2 |\tilde{F}|^2 / 4\pi \rho_0 c, \tag{3.68}$$

when the fluid-loading parameter $\beta = \rho_0 c/m\omega$ is much less than unity: the associated directivity is uniform except very close to the plate where the pressure drops to zero. This condition is one of "light fluid loading" and the result corresponds to Eq. (2.84), in which the effect of fluid loading on the plate vibrations was entirely neglected. When the fluid loading is heavy and $\beta \gg 1$, the radiated power becomes

$$\bar{P} = k^2|\tilde{F}|^2/12\pi\rho_0 c, \tag{3.69}$$

and the directivity becomes that associated with a dipole source having its axis coincident with the applied force. Note that Eq. (3.69) does not contain any plate parameters, the plate mass not affecting the free-plate wavenumber, as seen in Eq. (3.46).

The expressions for power radiated per plate unit length, when excited by a line force $\tilde{F}' \exp(j\omega t)$, are as follows:

$$\bar{P} = \begin{cases} \pi k|\tilde{F}'|^2\beta^2/8\rho_0 c, & \text{for } \beta \ll 1, \quad (3.70) \\ \pi k|\tilde{F}'|^2/16\rho_0 c & \text{for } \beta \gg 1. \quad (3.71) \end{cases}$$

As with the point force, increase of fluid loading causes the directivity to change from that of a line monopole to that of a line dipole.

Fluid-loading effects on point and line force impedances are more difficult to calculate than those on radiation. Table III summarises the main results (Crighton, 1972, 1977).

TABLE III

Low-Frequency Mobility Formulas for Fluid-Loaded Plates[a]

Line force:	$\tilde{Y}_F = \dfrac{\omega}{5Dk_b^3\sigma^3}\left(1 + j\tan\dfrac{\pi}{10}\right)$	$\sigma = (\rho_0/mk_b)^{1/5}$.
	$\tilde{Y}_{F_0} = \dfrac{\omega}{4Dk_b^3}(1 + j)$	
Line moment:	$\tilde{Y}_M = \dfrac{k_b^3}{5m\omega\sigma}\left(1 - j\cot\dfrac{\pi}{5}\right)$	
	$\tilde{Y}_{M_0} = \dfrac{k_b^3}{4m\omega}(1 - j)$.	
Point force:	$\tilde{Y}_F = \left(\dfrac{1}{8(Dm)^{1/2}}\right)\dfrac{4}{5}\left(\dfrac{k}{\nu k_b}\right)^{2/5}\left(1 - j\tan\dfrac{\pi}{10}\right)$	$\nu = \rho_0 k/mk_b^2$
	$\tilde{Y}_{F_0} = \dfrac{1}{8(Dm)^{1/2}}$	

[a] The subscript 0 indicates *in vacuo* values. Frequency condition: $\sigma \gg 1$. *In vacuo* bending wavenumber $k_b = (\omega^2 m/D)^{1/4}$.

Problems

3.1 A baffled loudspeaker cone of diameter 150 mm and mass 5×10^{-3} kg has an *in vacuo* natural frequency of 35 Hz. Estimate the change of natural frequency due to the inertial component of air loading. Equation (3.7) applies.

3.2 Estimate the approximate bending-wave phase velocities at frequencies of 100 and 2 kHz of a 5-mm-thick steel plate when submerged in air and in water. Equation (3.44) applies.

Estimate the difference between the natural frequencies of the (2, 2) and the (10, 10) vibration modes of a simply supported rectangular 5-mm-thick steel plate of dimensions 500 × 700 mm when submerged in air and water.

3.3 The added mass per unit area due to fluid loading on a long, uniformly pulsating cylinder can be shown to approach $m = -\rho_0 a \ln(ka)$ for $ka \ll 1$. Show that, in spite of the fact that this increases without limit as ka tends to zero, fluid loading, even by a liquid, is not likely in practice to introduce an extra ring frequency at a very low value of ka.

3.4 A piston slides freely in a tube that is closed at the other end by a rigid plug pierced by a small hole. By assuming that the fluid velocity through the hole is proportional to the pressure difference across the hole, appropriately modify the equation in Section 3.5 to investigate the influence of such a hole on the impedance characteristics of the piston–tube system. Neglect inertial radiation loading on the fluid in the hole. Under what conditions is such neglect justified?

3.5 Derive an expression for power radiated by a piston vibrating with frequency-independent velocity amplitude into a uniform tube of length l and diameter d that is terminated by a plug of acoustic impedance $\tilde{Z}_l$. Determine, by numerical or graphical means, the lowest frequencies of maximum and minimum power radiation in air when $d = 50$ mm, $l = 500$ mm, and $\tilde{Z}_l = 2 \times 10^5 (5-10\,\mathrm{j})$ kg m^{-4} s^{-1}. Explain your results in physical terms.

4 Transmission of Sound through Partitions

4.1 Practical Aspects of Sound Transmission through Partitions

There are two main methods of inhibiting the transmission of sound energy from one region of fluid to another. In the first, sound energy is absorbed in transit by materials that are specially chosen to accept energy efficiently from waves in the contiguous fluid, and then efficiently to dissipate it into heat: systems that utilise this principle include room wall sound absorbers, absorbent duct liners, and splitter attenuators in ventilation systems. Alternatively, sound in transit may be reflected by means of introducing a large change of acoustic impedance into the transmission path. Examples include internal combustion engine exhaust expansion chambers, in which the changes of cross section are effective; hydraulic line silencers, in which the wave in the oil encounters an acoustically "soft" pipe section surrounded by pressurised gas; and partitions of solid sheets such as room walls and industrial noise control enclosures. Partition of adjacent fluid regions may, of course, not be total, in which case we use the terms "barrier" and "screen."

The design and construction of effective partitions in a central element in the practice of noise control by engineers and architects, and an awareness of the basic physical principles and of good design practice is important to a wider group, including local authority planners, environmental health officers, buildings and works officers, and industrial management. In this chapter the basic principles of the subject are illustrated by analyses of sound transmission through some simple idealisations of uniform single- and double-leaf partition constructions. A review of some of the more important extensions of these analyses to account more completely for features of practical systems is accompanied by a brief presentation of a range of typical experimental data. Finally, the transmission characteristics of some non-uniform and non-plane structures are discussed, with emphasis on circular cylindrical shells.

4.2 Transmission of Normally Incident Plane Waves through an Unbounded Partition

The idealised system is shown in Fig. 69. A uniform, unbounded, non-flexible partition of mass per unit area m is mounted upon a viscously damped, elastic suspension, having stiffness and damping coefficients per unit area of s and r, respectively. This represents an approximation to the fundamental mode of a large panel. The partition separates fluids of different characteristic specific acoustic impedances, $\rho_1 c_1$ and $\rho_2 c_2$. A plane sound wave of frequency ω is incident upon the partition from the region $x < 0$;

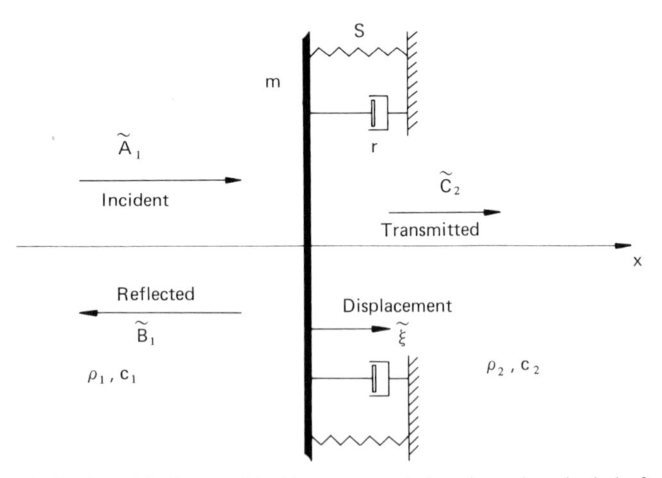

Fig. 69. Idealised model of normal incidence transmission through a single-leaf partition.

the incident pressure field is written as

$$p_i(x, t) = \tilde{A}_1 \exp[j(\omega t - k_1 x)], \tag{4.1}$$

where $k_1 = \omega/c_1$. The pressure field of the wave reflected from the partition is written

$$p_r(x, t) = \tilde{B}_1 \exp[j(\omega t + k_1 x)]. \tag{4.2}$$

The coefficients $\tilde{A}_1$ and $\tilde{B}_1$ are linked by the normal particle velocity at the left-hand surface of the partition, which moves with a normal velocity equal to $j\omega\tilde{\xi}$. Hence

$$\tilde{A}_1 - \tilde{B}_1 = j\omega\rho_1 c_1 \tilde{\xi}. \tag{4.3}$$

As we have seen in Section 3.3, the mechanical impedance of an *in vacuo* structure may be combined with the fluid-loading impedance associated with structural motion to form the total effective impedance of a "fluid-loaded structure" as presented to mechanically applied forces [see Eq. (3.43)]. We may enquire whether such a concept is relevant to acoustically applied forces.

The acoustic pressure field radiated in the negative x direction by a displacement $\tilde{\xi}$, whatever the cause of partition motion, is

$$p_r^-(x, t) = \tilde{C}_1 \exp[j(\omega t + k_1 x)], \tag{4.4}$$

where $\tilde{C}_1 = -j\omega\rho_1 c_1 \tilde{\xi}$. The corresponding wave radiated in the positive x direction is

$$p_r^+(x, t) = \tilde{C}_2 \exp[j(\omega t - k_2 x)], \tag{4.5}$$

where $\tilde{C}_2 = j\omega\rho_2 c_2 \tilde{\xi}$ and $k_2 = \omega/c_2$. These fields may be termed the "radiated fields."

According to Eqs. (4.1)–(4.3), the total pressure field on the left-hand side of the partition is

$$\begin{aligned} p^-(x, t) &= \tilde{A}_1 \exp[j(\omega t - k_1 x)] + (\tilde{A}_1 - j\omega\rho_1 c_1 \tilde{\xi}) \exp[j(\omega t + k_1 x)] \\ &= 2\tilde{A}_1 \cos k_1 x \exp(j\omega t) - j\omega\rho_1 c_1 \tilde{\xi} \exp[j(\omega t + k_1 x)]. \end{aligned} \tag{4.6}$$

Equation (4.6) may be rewritten, using Eq. (4.4), as

$$p^-(x, t) = 2\tilde{A}_1 \cos k_1 x \exp(j\omega t) + \tilde{C}_1 \exp[j(\omega t + k_1 x)]. \tag{4.7}$$

Now, the first term on the right-hand side of Eq. (4.7) represents the standing interference field created by the incidence upon and reflection from a completely immobile partition; we may term this the "blocked pressure field." The second term represents the pressure field generated by partition motion. Hence the total field on the incident side equals the sum of the blocked field and the radiated field: the total field on the right-hand side is simply the radiated field represented by Eq. (4.5).

The equation of motion of the partition is

$$m\ddot{\xi} + r\dot{\xi} + s\xi = p(x = 0^-, t) - p(x = 0^+, t). \tag{4.8}$$

where $x = 0^-$ and $x = 0^+$ refer to the left- and right-hand faces of the partition. Substitution from Eqs. (4.5) and (4.6) gives

$$(-\omega^2 m + j\omega r + s)\tilde{\xi} = 2\tilde{A}_1 - j\omega\rho_1 c_1\tilde{\xi} - j\omega\rho_2 c_2\tilde{\xi}. \tag{4.9}$$

The fluid-loading (radiation) pressure terms on the right-hand side of this equation may be incorporated into the term on the left-hand side, which represents the *in vacuo* partition properties, to give

$$[-\omega^2 m + j\omega(r + \rho_1 c_1 + \rho_2 c_2) + s]\tilde{\xi} = 2\tilde{A}_1. \tag{4.10}$$

The fluid-loading terms represent radiation damping to be added to mechanical damping. If we express the left-hand side in terms of the partition velocity $\tilde{v} = j\omega\tilde{\xi}$, instead of the displacement $\tilde{\xi}$, we can rewrite this equation as

$$[j(\omega m - s/\omega) + (r + \rho_1 c_1 + \rho_2 c_2)]\tilde{v} = 2\tilde{A}_1, \tag{4.11}$$

or

$$(\tilde{z}_p + \tilde{z}_f)\tilde{v} = 2\tilde{A}_1, \tag{4.12}$$

where $\tilde{z}_p$ and $\tilde{z}_f$ are the partition (*in vacuo*) and fluid-loading impedances, respectively. The forcing term $2\tilde{A}_1$ on the right-hand side is, of course, the blocked surface pressure field [compare with Eq. (3.43)]. Equation (4.12) proves that we may treat the problem as one of the response of a fluid-loaded structure to the surface pressure distribution of a blocked incident field. In fact, such a decomposition of the total field, which leads to this concept, is valid for any elastic structure immersed in a fluid; however, in most practical cases the analysis is far more complicated than in this simple one-dimensional idealisation.

Having obtained an expression for the velocity of the partition in terms of the amplitude of the incident pressure wave, we can now write an expression for $\tilde{C}_2$, the transmitted wave amplitude. Using Eqs. (4.5) and (4.12),

$$\tilde{C}_2 = \rho_2 c_2 \tilde{v} = 2\tilde{A}_1 \rho_2 c_2/(\tilde{z}_P + \tilde{z}_f)$$

$$= \frac{2\tilde{A}_1}{j(\omega m - s/\omega)/\rho_2 c_2 + (r/\rho_2 c_2 + \rho_1 c_1/\rho_2 c_2 + 1)}. \tag{4.13}$$

The transmission coefficient τ is defined as the ratio of transmitted to incident intensities:

$$\tau = \frac{|\tilde{C}_2|^2/2\rho_2 c_2}{|\tilde{A}_1|^2/2\rho_1 c_1} = \frac{4n}{[(\omega m - s/\omega)/\rho_2 c_2]^2 + (\omega_0 m\eta/\rho_2 c_2 + n + 1)^2}, \tag{4.14}$$

where $n = \rho_1 c_1/\rho_2 c_2$, and r has been replaced by $\omega_0 m\eta$, where η is the *in vacuo* loss factor. The internationally recognised index of sound transmission is the *sound reduction index* R defined by

$$R = 10 \log_{10}(1/\tau) \quad \text{dB}.$$

The index is also known in some countries as the *sound transmission loss* (TL).

The transmission coefficient clearly has a maximum value at the undamped natural frequency of the partition. Three special cases may be identified:

(i) $\omega \ll \omega_0 = (s/m)^{1/2}$, well below the *in vacuo* natural frequency:

$$\tau \simeq \frac{4n}{(s/\omega\rho_2 c_2)^2 + (s\eta/\omega_0\rho_2 c_2 + n + 1)^2} \simeq \frac{4n}{(s/\omega\rho_2 c_2)^2 + (n+1)^2}, \tag{4.15}$$

because η is normally much less than unity. Now, $s/\omega\rho_2 c_2 = (\omega_0/\omega) \times (\omega_0 m/\rho_2 c_2)$ and $\omega_0 m/\rho_2 c_2$ is normally much greater than unity for typical structures at audio frequencies in gases, but not necessarily in liquids. If the fluid on both sides is air, Eq. (4.15) can, under this frequency condition generally be reduced to

$$\tau \simeq (2\rho_0 c\omega/s)^2. \tag{4.16}$$

The equivalent sound reduction index is

$$R = 20 \log_{10} s - 20 \log_{10} f - 20 \log_{10}(4\pi\rho_0 c) \quad \text{dB}, \tag{4.17}$$

where $f = \omega/2\pi$ Hz. R is seen to be determined primarily by the elastic stiffness of the mounting and is insensitive to mass and damping. It decreases with frequency by 6 dB per octave.

If the fluid impedance ratio n is very large, or if the mass per unit area of the partition is very low (e.g., thin plastic sheet), Eq. (4.17) is not valid. If $n \gg (\omega_0/\omega)(\omega_0 m/\rho_2 c_2)$, which means $\rho_1 c_1 \gg \omega_0^2 m/\omega$, then

$$\tau \to 4/n, \tag{4.18}$$

and is independent of the mechanical properties of the partition. For example, if a partition separates air and water, $\tau \simeq 1.1 \times 10^{-3}$, or $R \simeq 29.5$ dB.

(ii) $\omega \gg \omega_0$, well above the natural frequency:

$$\tau \simeq \frac{4n}{(\omega m/\rho_2 c_2)^2 + (n+1)^2}, \tag{4.19}$$

because $\eta < 1$ [Eq. (4.14)]. If the fluid on both sides is air, then normally $\omega m/\rho_2 c_2 \gg 1$ and

$$\tau \simeq (2\rho_0 c/\omega m)^2. \tag{4.20}$$

Correspondingly,

$$R = 20 \log_{10} m + 20 \log_{10} f - 20 \log_{10}(\rho_0 c/\pi) \quad \text{dB},$$

or

$$R \simeq 20 \log_{10}(mf) - 42 \quad \text{dB}. \tag{4.21}$$

R is seen to be determined primarily by mass per unit area, and is largely independent of damping and stiffness; it increases with frequency at 6 dB per octave and 6 dB per doubling of mass. Equation (4.21) is known as the *normal incidence mass law*.

Very lightweight films at low frequencies may not behave according to Eq. (4.21) because $\omega m/\rho_2 c_2$ may not be much greater than unity. If $n \gg \omega m/\rho_2 c_2$, or $\omega m \ll \rho_1 c_1$, then τ is given by Eq. (4.18).

(iii) $\omega = \omega_0$, the natural frequency:

$$\tau = \frac{4n}{[\eta(\rho_2 c_2/\omega_0 m)^{-1} + (n+1)]^2}. \tag{4.22}$$

If the fluid on both sides of the partition is the same and if $\eta \ll \rho_0 c/\omega m$, then

$$\tau \simeq 1. \tag{4.23}$$

If $\eta \gg \rho_0 c/\omega m$, then

$$\tau \simeq (2\rho_0 c/\eta\omega_0 m)^2. \tag{4.24}$$

The corresponding sound reduction indices are

$$R = 0 \quad \text{dB}, \tag{4.25}$$

and

$$R = 20 \log_{10} f_0 + 20 \log_{10} m + 20 \log_{10} \eta - 20 \log_{10}(\rho_0 c/\pi) \quad \text{dB}. \tag{4.26}$$

The latter value is different from the mass law value at $f = f_0$ by

$$20 \log_{10}(\eta) \quad \text{dB}.$$

Equations (4.23) and (4.25) indicate total transmission at resonance when radiation damping exceeds mechanical damping. Equation (4.26) shows that the mass, stiffness, and damping all influence transmission at resonance, provided that the mechanical damping exceeds the radiation damping. If $n > 1$, then $\eta(\rho_2 c_2/\omega_0 m)^{-1}$ must be comparable with n for mechanical damping to have any effect.

It is tempting to use this model to evaluate the transmission characteristics of bounded flexible panels vibrating in their fundamental modes of vibration, in which the phase of the displacement is uniform over the whole surface. Examples include windows and the panels of enclosures. In the former example, unfortunately, at fundamental natural frequencies typical of glazing panels (10–30 Hz), the acoustic wavelength is so large compared with the typical aperture dimension that the transmission is controlled as much by aperture diffraction as by the window dynamics: the partition acts like a piston of small ka in a baffle and therefore does not radiate (transmit) effectively. The same stricture applies to the transmission characteristics of acoustic louvres at low frequencies; at low audio frequencies a simple hole in the wall has a reasonable transmission loss. In these cases it is the insertion loss (received sound pressure level with and without the insertion of the particular item) that is significant.

The results of the aforegoing analysis suggest that in cases where the characteristic acoustic impedance of one medium is much greater than the other (e.g., air/water) the mechanical properties of a partition have little influence on the transmission, which is controlled simply by the ratio of the impedances. It should also be noted that all the expressions for τ are reciprocal in n, so that τ is independent of the direction of the normally incident plane wave.

4.3 Transmission of Sound through an Unbounded Flexible Partition

Having established the principle of applying the blocked surface pressure as the forcing field on a fluid-loaded structure, we may now apply it to the case of an unbounded, thin, uniform, elastic plate upon which acoustic plane waves of frequency ω are incident at an arbitrary angle ϕ_1: the model is shown in Fig. 70.

The component of the incident wave vector $\mathbf{k}$, which is directed parallel to the partition plane (sometimes called the trace wavenumber), is $k_z = k \sin \phi_1$. Since the partition is uniform and unbounded, no one point is different dynamically from any other: therefore, the flexural wave induced in the partition must have a wavenumber $k_z = k \sin \phi_1$. The blocked pressure at the partition surface is

$$p_{\text{bl}}(x = 0^-, z, t) = 2\tilde{A}_1 \exp[j(\omega t - k \sin \phi_1 z)]. \tag{4.27}$$

The coefficient $2\tilde{A}_1$ is equivalent to the applied force per unit area in Eq. (3.41); hence

$$2\tilde{A}_1 = (\tilde{z}_{\text{wp}} + \tilde{z}_{\text{wf}})\tilde{v}. \tag{4.28}$$

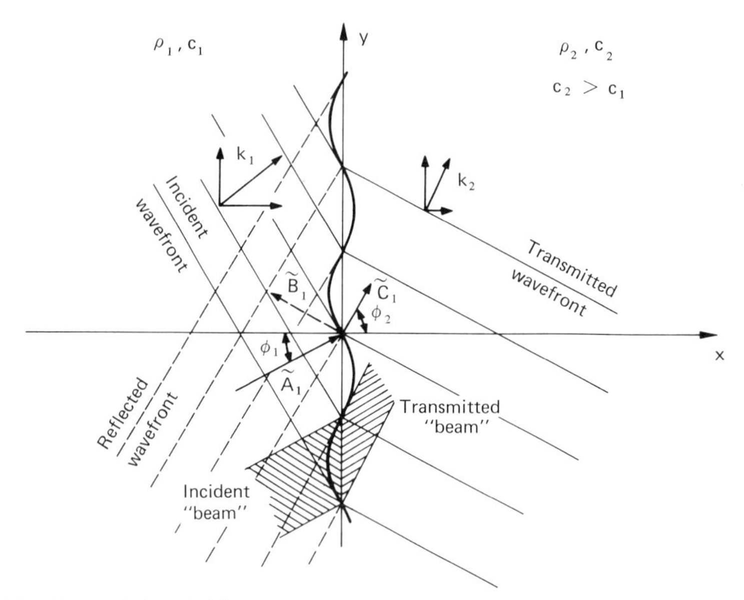

Fig. 70. Transmission of obliquely incident sound through an unbounded flexible partition.

The fluid wave impedance is given by Eq. (3.38) as

$$\begin{aligned}\tilde{z}_{\mathrm{wf}} &= \rho_1 c_1 (1 - \sin^2\phi_1)^{-1/2} + \rho_2 c_2[1 - (k_1 \sin\phi_1/k_2)^2]^{-1/2} \\ &= \tilde{z}_{\mathrm{wf}_1} + \tilde{z}_{\mathrm{wf}_2},\end{aligned} \tag{4.29}$$

and the partition wave impedance is given by Eq. (3.35) as

$$\tilde{z}_{\mathrm{wp}} = -(j/\omega)(Dk_1^4 \sin^4\phi_1 - m\omega^2) + Dk_1^4 \sin^4\phi_1 \eta/\omega. \tag{4.30}$$

Note that the component of wave impedance $\tilde{z}_{\mathrm{wf}_2}$ produced by the fluid to the right of the partition is only real if

$$\sin\phi_1 < k_2/k_1 = c_1/c_2. \tag{4.31}$$

Hence energy transmission is limited to a range of ϕ_1 satisfying this condition. For instance, irrespective of the properties of a partition, plane-wave energy cannot be transmitted by a uniform partition from air into water at angles of incidence greater than 13.6°. The transmitted pressure coefficient $\tilde{C}_2$ is related to the partition normal velocity $\tilde{v}$ by

$$\tilde{C}_2 = \tilde{z}_{\mathrm{wf}_2}\tilde{v}. \tag{4.32}$$

Equations (4.28) and (4.32) yield

$$\tilde{C}_2 = \frac{2\tilde{A}_1 \tilde{z}_{\mathrm{wf}_2}}{\tilde{z}_{\mathrm{wf}_1} + \tilde{z}_{\mathrm{wf}_2} + \tilde{z}_{\mathrm{wp}}}. \tag{4.33}$$

The intensity transmission coefficient τ is given by

$$\tau = \frac{|\tilde{C}_2|^2/2\rho_2 c_2}{|\tilde{A}_1|^2/2\rho_1 c_1}. \tag{4.34}$$

However, this is not generally the ratio of sound power transmitted per unit area of partition to sound power incident per unit area of panel because of refraction when $c_1 \neq c_2$. Reference to Fig. 70 will reveal that the widths of corresponding "beams" on the two sides are in the ratio

$$\begin{aligned}\cos\phi_1/\cos\phi_2 &= (1 - \sin^2\phi_1)^{1/2}/(1 - \sin^2\phi_2)^{1/2}\\ &= (1 - \sin^2\phi_1)^{1/2}/[1 - (c_2 \sin\phi_1/c_1)^2]^{1/2}.\end{aligned} \tag{4.35}$$

The sound *power* transmission coefficient is therefore given by

$$\tau_{\mathrm{P}} = \frac{4|\tilde{z}_{\mathrm{wf}_2}|^2}{|\tilde{z}_{\mathrm{wf}_1} + \tilde{z}_{\mathrm{wf}_2} + \tilde{z}_{\mathrm{wp}}|^2}\left[\frac{\rho_1 c_1}{\rho_2 c_2}\right]\left[\frac{1 - (c_2 \sin\phi_1/c_1)^2}{1 - \sin^2\phi_1}\right]^{1/2}. \tag{4.36}$$

This rather complicated expression reduces to a much simpler form when the fluids on the two sides are the same. Then $\phi_1 = \phi_2 = \phi$, and

$$\tau_{\mathrm{P}} = \tau = \left|\frac{\tilde{z}_{\mathrm{wf}}}{\tilde{z}_{\mathrm{wf}} + \tilde{z}_{\mathrm{wp}}}\right|^2, \tag{4.37}$$

where $\tilde{z}_{\mathrm{wf}} = \tilde{z}_{\mathrm{wf}_1} + \tilde{z}_{\mathrm{wf}_2} = 2\tilde{z}_{\mathrm{wf}_1} = 2\tilde{z}_{\mathrm{wf}_2}$. The explicit form of Eq. (4.37) is

$$\tau = \frac{(2\rho_0 c \sec\phi)^2}{[2\rho_0 c \sec\phi + (D/\omega)\eta k^4 \sin^4\phi]^2 + [\omega m - (D/\omega)k^4 \sin^4\phi]^2}. \tag{4.38a}$$

In order to investigate the relative influences of partition mass, stiffness, and damping, it is helpful to consider the conditions under which the incident wave is coincident with the flexural wave in the partition. The wavenumber of the wave induced in the partition by the incident field is, as we have seen, equal to the trace wavenumber $k_z = k \sin\phi$. The expression for the free flexural wavenumber in a plate is $k_{\mathrm{b}}^4 = \omega^2 m/D$. Hence Eq. (4.38a) may be rewritten as

$$\tau = \frac{(2\rho_0 c/\omega m)^2 \sec^2\phi}{[(2\rho_0 c/\omega m)\sec\phi + (k/k_{\mathrm{b}})^4\eta \sin^4\phi]^2 + [1 - (k/k_{\mathrm{b}})^4 \sin^4\phi]^2}. \tag{4.38b}$$

The *coincidence* condition is

$$k \sin\phi = k_{\mathrm{b}} = (\omega^2 m/D)^{1/4}, \tag{4.39}$$

which corresponds to the disappearance of the reactive contribution to the denominator of Eqs. (4.38). Rewriting Eq. (4.39) as

$$\omega_{\mathrm{co}} = (m/D)^{1/2}(c/\sin\phi)^2 \tag{4.40}$$

shows that for a given angle of incidence ϕ there is a unique coincidence frequency ω_{co}, and vice versa. However, since $\sin\phi$ cannot exceed unity, there is a lower limiting frequency for the coincidence phenomenon given by

$$\omega_c = c^2(m/D)^{1/2}, \tag{4.41}$$

where ω_c is known as the *critical frequency*, or lowest coincidence frequency. Equation (4.40) can therefore be rewritten as

$$\omega_{co} = \omega_c/\sin^2\phi, \tag{4.42a}$$

or

$$\sin\phi_{co} = (\omega_c/\omega)^{1/2}, \tag{4.42b}$$

where ϕ_{co} is the coincidence angle for frequency ω. These relationships are illustrated graphically in Fig. 71. The nature of coincidence is further illustrated in Fig. 72.

It is clear from Eq. (4.41) that lightweight, stiff partitions, such as honeycomb sandwiches, tend to exhibit lower critical frequencies than homogeneous partitions of similar weight but of lower stiffness. Critical frequencies of homogeneous partitions can be raised by making a series of parallel grooves in the material, but this is not usually acceptable because of static stiffness reduction.

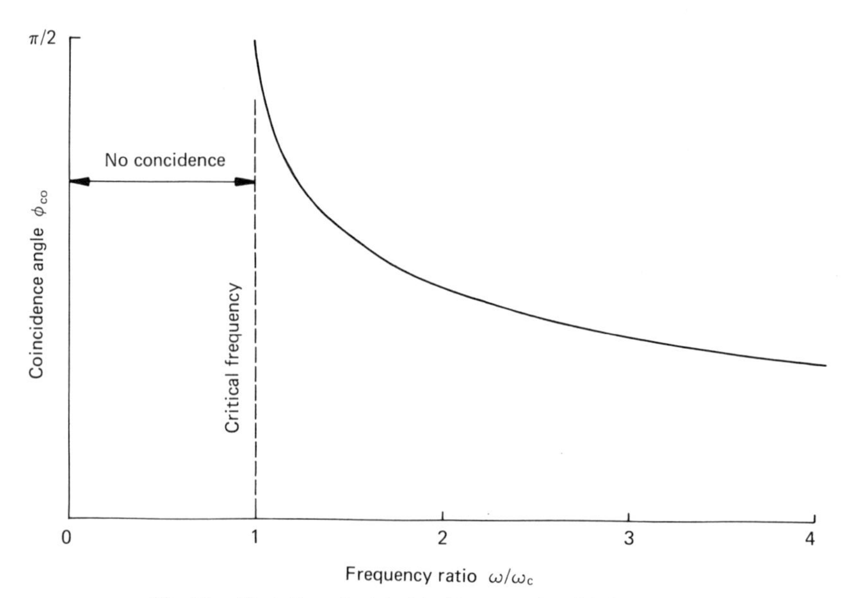

Fig. 71. Variation of critical incidence angle with frequency.

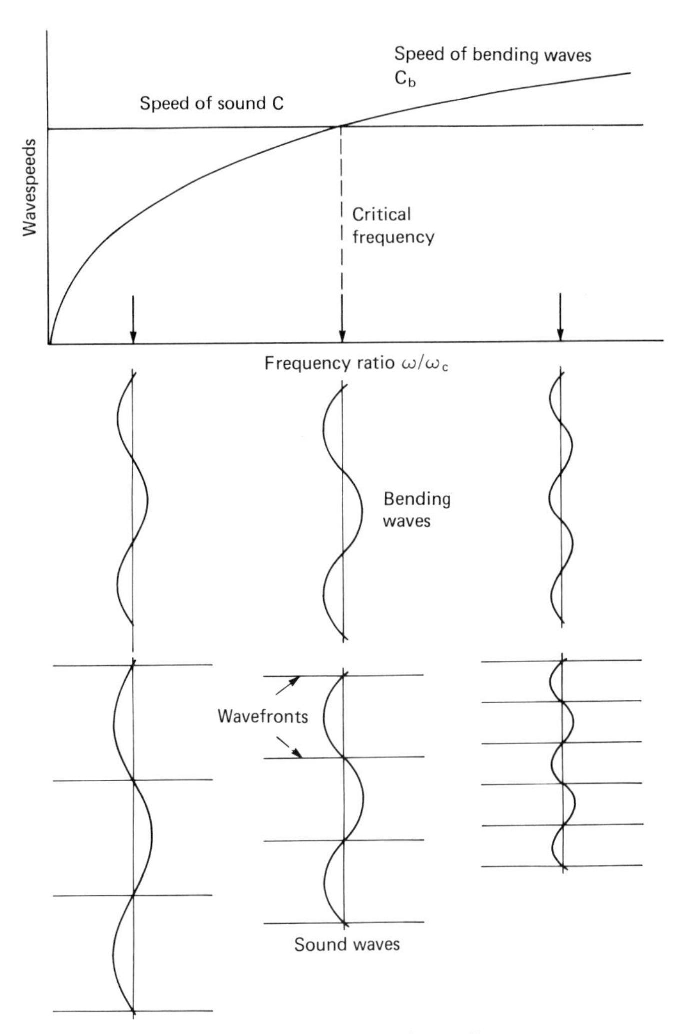

Fig. 72. Coincidence diagram. (Trace the sound wavefronts onto a transparent sheet and superimpose them on the corresponding bending waves at various angles to illustrate coincidence.)

In the case of uniform homogeneous flat plates of material density ρ_s, Eq. (4.41) can be written as

$$\omega_c = c^2(\rho_s h)^{1/2}[(Eh^3/12(1 - v^2)]^{-1/2},$$

or

$$f_c = c^2/1.8hc_{l'} \quad \text{Hz}, \tag{4.43}$$

TABLE IV

Product of Plate Thickness and Critical Frequency in Air (20°C)[a]

Material	hf_c (m sec^{-1})
Steel	12.4
Aluminium	12.0
Brass	17.8
Copper	16.3
Glass	12.7
Perspex	27.7
Chipboard	23[b]
Plywood	20[b]
Asbestos cement	17[b]
Concrete	
dense	19[b]
porous	33[b]
light	34[b]

[a] To obtain values in water, multiply by 18.9.
[b] Variations of up to ±10% possible.

where h is the plate thickness and $c_{l'}$ the phase speed of longitudinal waves in the plate. Thus the product hf_c is a function only of the material properties of the fluid and solid media. This product is tabulated for a range of common materials in air at 20°C in Table IV. As an example, the critical frequency of 6-mm-thick steel plate in air is 2060 Hz. In water, the values would be greater by a factor of approximately 19 than those for the same plate in air. Hence, in marine applications, frequencies greater than f_c are rarely of practical importance.

Returning to the transmission coefficient Eq. (4.38), it is now clear that the influence of the coincidence phenomenon, which corresponds to the disappearance of the reactive term in the denominator, will affect the value of τ at all frequencies in the range $\omega_c \leqslant \omega \leqslant \infty$. It is instructive to examine the variation of τ with angle of incidence for a fixed frequency.

Consider first the range of frequency below the critical frequency of the partition. The ratio of the trace wavenumber of the exciting field to the free flexural wavenumber is given by

$$\frac{k_z}{k_b} = \frac{k \sin \phi}{(\omega^2 m/D)^{1/4}}, \tag{4.44a}$$

which from Eq. (4.40) may be written as

$$k_z/k_b = (\omega/\omega_c)^{1/2} \sin\phi. \tag{4.44b}$$

The physical interpretation of the fact that for $\omega < \omega_c$ this ratio is necessarily less than unity is that the phase speed of *free* bending waves is *less* than the trace wave speed of the incident field at all angles of incidence. The influence of this condition on transmission is seen in the dominance of the inertia term ωm over the stiffness term $(D/\omega)k^4 \sin^4\phi$ in the denominator of Eq. (4.38a). Clearly the mechanical damping term, which is η times the stiffness term, is also negligible compared with the inertia term. Hence the transmission coefficient at frequencies well below the critical frequency is to a good approximation

$$\tau(\phi) = 1/[1 + (\omega m \cos\phi/2\rho_0 c)^2]. \tag{4.45}$$

Provided that $\omega m \cos\phi \gg 2\rho_0 c$, which is normally true except for $\phi \simeq \pi/2$, the corresponding sound reduction index is given by

$$R(\phi) = 20 \log_{10}(\omega m \cos\phi/2\rho_0 c) \quad \text{dB}. \tag{4.46}$$

Comparison with the normal incidence mass law [Eq. (4.21)] shows that

$$R(0) - R(\phi) \simeq 20 \log_{10}(\cos\phi) \quad \text{dB}, \tag{4.47}$$

and hence the difference increases as the angle of incidence approaches $\pi/2$ (grazing).

Now the condition $k_z/k_b < 1$, although always true when $\omega < \omega_c$, is not restricted to this frequency range. Reference to Eq. (4.42a) shows that Eq. (4.44b) may be written as

$$k_z/k_b = (\omega/\omega_{co})^{1/2}.$$

Thus the conclusions drawn above concerning the dominance of the inertia term apply, for a *given angle of incidence*, not just for $\omega \ll \omega_c$ but for $\omega \ll \omega_c/\sin^2\phi$. As ω approaches ω_{co}, the magnitude of the stiffness term in the transmission expression approaches that of the inertia term, a maximum in the transmission coefficient occurs at $\omega = \omega_{co}$, and

$$\tau = 1/(1 + \eta\omega_{co} m \cos\phi/2\rho_0 c)^2. \tag{4.48}$$

Comparison of this expression with that for purely mass controlled transmission at the same frequency [Eq. (4.45)] shows that the difference between the corresponding sound reduction indices is at least $20 \log_{10} \eta$ dB. If $\eta > 2\rho_0 c/\omega_{co} m \cos\phi$, the transmission of sound energy in the vicinity of coincidence is controlled by mechanical damping.

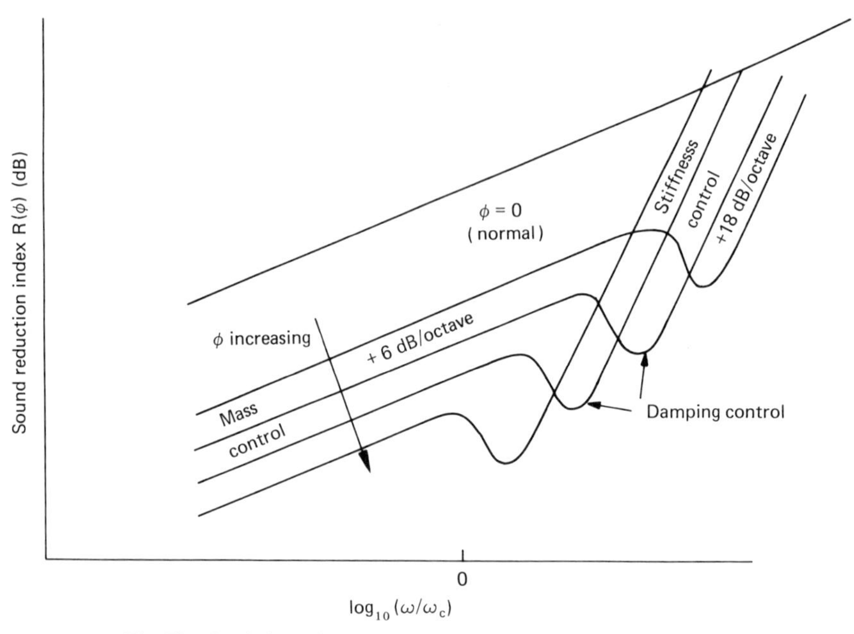

Fig. 73. Variation of R with frequency for single angle of incidence.

At frequencies above ω_{co}, the stiffness term dominates in the transmission expression and

$$\tau \simeq 1/[1 + (Dk^4 \sin^4\phi \cos\phi/2\rho_0 c\omega)^2]. \tag{4.49}$$

In most cases of sound transmission in air, the stiffness term greatly exceeds unity and hence the sound reduction index for a given ϕ increases at approximately 18 dB per doubling of frequency; the damping exerts no influence in this range. The form of variation of R with frequency for constant ϕ is shown in Fig. 73.

An alternative view of this rather complicated behaviour is obtained by considering transmission over the whole range of angle of incidence at *fixed frequency*. Below the critical frequency, transmission at all angles is, of course, mass controlled. At any frequency above the critical frequency, Eq. (4.42b) determines the coincidence angle. If $\sin\phi$ is less than $\sin\phi_{co}$, i.e., $\phi < \phi_{co}$, Eq. (4.38) shows that the inertia term dominates: if $\phi > \phi_{co}$ then stiffness dominates. At $\phi = \phi_{co}$ damping is in control providing it is sufficiently large to exceed acoustic radiation damping. This behaviour is illustrated in Fig. 74.

In practice, sound waves are usually incident upon a partition from many angles simultaneously, e.g., the wall of a room or a window exposed to traffic

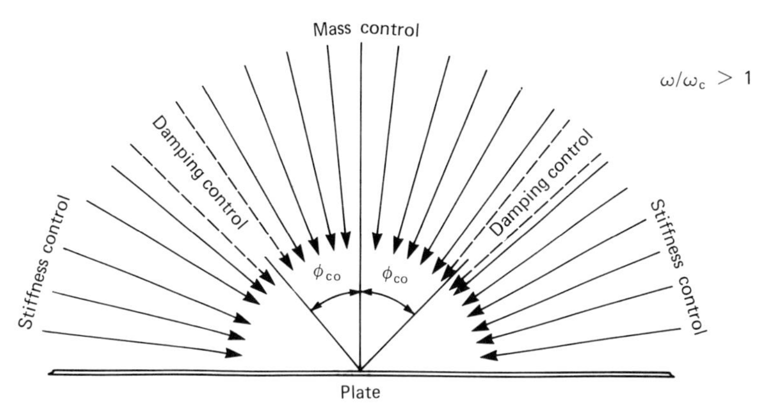

Fig. 74. Regions of incidence angle associated with mass, stiffness, and damping control of transmission at a single supercritical frequency.

noise. The appropriate transmission coefficient can in principle be derived from Eq. (4.38) by weighting according to the directional distribution of incident intensity and integrating over angle of incidence. In practice the directional distribution of incident intensity is rarely known, and therefore an idealised diffuse field model is usually assumed, in which plane waves are incident from all directions with equal probability and with random phase.

It may be shown (Pierce, 1981) that the appropriate weighting leads to the following expression for the diffuse field transmission coefficient:

$$\tau_d = \frac{\int_0^{\pi/2} \tau(\phi) \sin \phi \cos \phi \, d\phi}{\int_0^{\pi/2} \sin \phi \cos \phi \, d\phi} = \int_0^{\pi/2} \tau(\phi) \sin 2\phi \, d\phi. \qquad (4.50)$$

The cos ϕ term arises from the variation with ϕ of the plane-wave intensity component normal to the partition, and the sin ϕ term relates the total acoustic power carried by the incident waves to their angle of incidence. The general expression for τ is not amenable to analytic integration, but the restricted expression in Eq. (4.45) may be evaluated for frequencies well below the critical frequency. The result, in terms of sound reduction index, is

$$R_d = R(0) - 10 \log_{10}[0.23R(0)] \quad \text{dB}. \qquad (4.51)$$

It is generally found that experimental results do not agree very well with Eq. (4.51), tending to higher values more in accord with an empirical expression

$$R_f = R(0) - 5 \quad \text{dB} \qquad \text{or} \qquad R_f \simeq 20 \log_{10}(mf) - 47 \quad \text{dB}, \qquad (4.52)$$

which is called the *field incidence mass law*. This formula is closely approximated by using Eq. (4.45) in the integrand of Eq. (4.50) and performing the

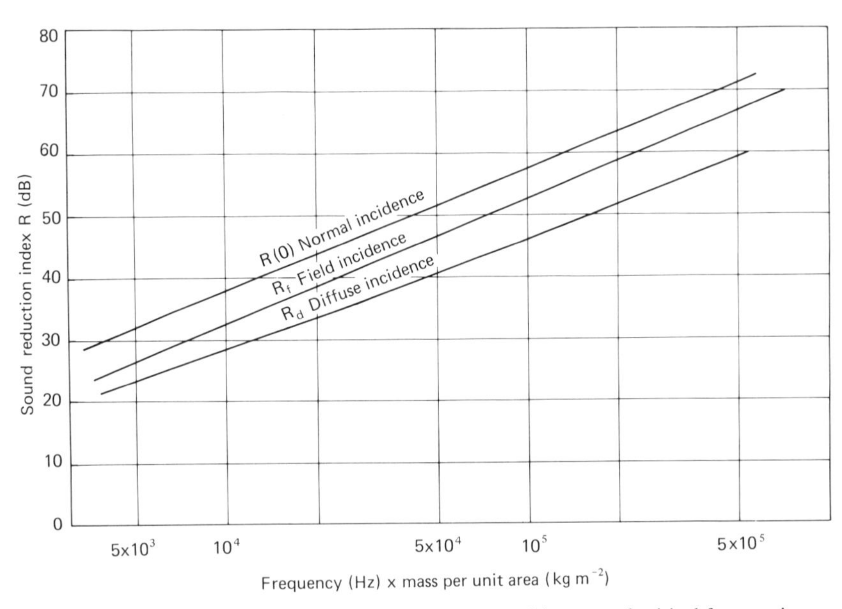

Fig. 75. Sound reduction indices for unbounded partitions at subcritical frequencies.

integration from 0 to 78°. Theories of sound transmission through panels of finite area, discussed in the following sections, provide evidence to support the omission of waves at close to grazing incidence in the case of a bounded panel. Curves of $R(0)$, R_d, and R_f for subcritical frequencies are compared in Fig. 75.

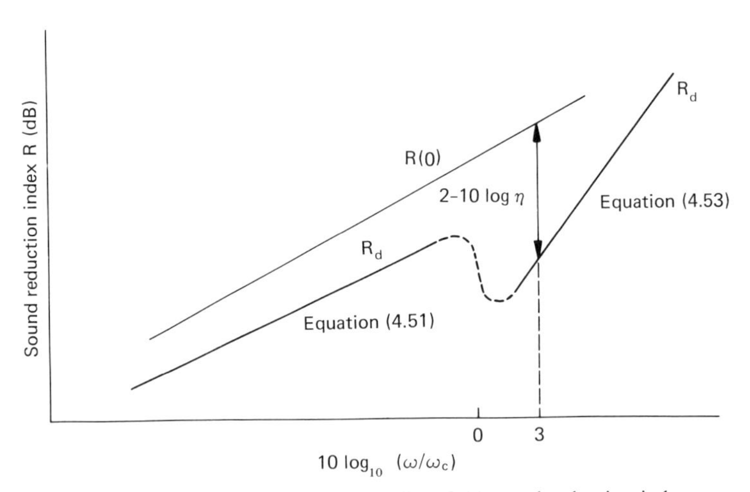

Fig. 76. General form of the theoretical diffuse field sound reduction index curve.

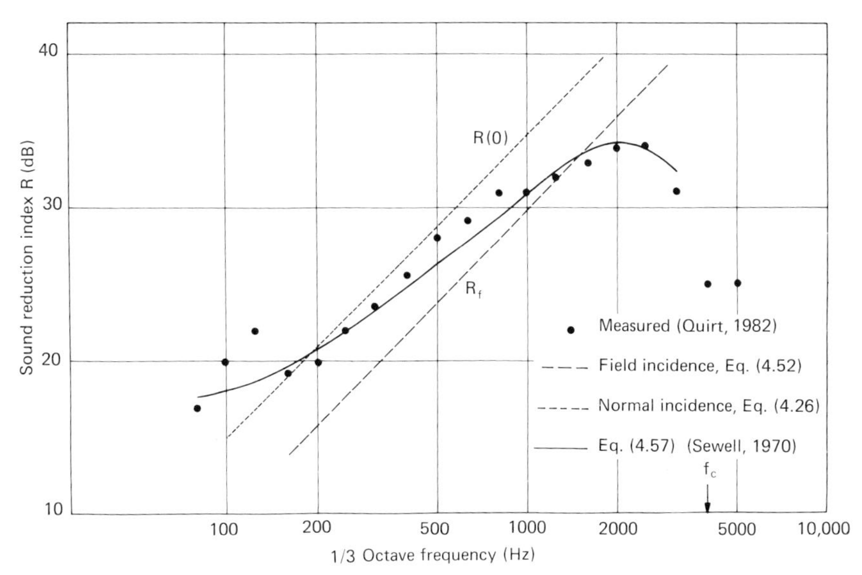

Fig. 77. Sound reduction index of 3-mm-thick glazing.

An expression for diffuse-field sound reduction index at frequencies above the critical frequency was derived by Cremer (1942):

$$R_d = R(0) + 10 \log_{10}(f/f_c - 1) + 10 \log_{10} \eta - 2 \quad \text{dB}. \tag{4.53}$$

The dominant influence of coincidence transmission is seen in the presence of the loss factor term. The general form of the theoretical diffuse incidence sound reduction index curve for infinite partitions is shown in Fig. 76. Deviations from this curve are observed in experimental results obtained on bounded panels. An example of a measured sound reduction index curve that exhibits a distinct coincidence dip is presented in Fig. 77; the dip is deep because the damping of glazing has generally fairly small values unless special edge treatment is applied.

4.4 Transmission of Diffuse Sound through a Bounded Partition in a Baffle

The two main factors that can cause the diffuse-field transmission performance of a real, bounded panel in a rigid baffle to differ significantly from the theoretical performance of an unbounded partition are (1) the existence of standing-wave modes and associated resonance frequencies, and (2) diffraction by the aperture in the baffle that contains the panel. An additional influence of room boundaries on transmission through partitions that separate rooms is discussed in Section 4.5.

As we have seen in Chapter 2, the radiation efficiency of modes of bounded plates vibrating at frequencies below the critical frequency is very much influenced by the presence of the boundaries, but is generally less than unity: it will be shown in Chapter 5 that the response of a mode to acoustic excitation is proportional to its radiation efficiency. We have also seen that the radiation efficiency of an infinite partition, excited by plane waves obliquely incident at angle ϕ, is $\sigma = \sec \phi$, which generally exceeds unity, and that the response is mass controlled below the critical frequency. Therefore, in comparing the transmission coefficients at subcritical frequencies of bounded and unbounded partitions of the same thickness and material, it would seem that the relatively low values of the radiation ratios of the bounded-plate modes would be offset by the enhanced response produced by modal resonance, which is absent in the infinite partition. However, it transpires from analyses too involved to be presented here that resonant "amplification" is generally insufficient to make up for the low values of modal radiation efficiencies associated with modal *resonance* frequencies, except in very lightly damped structures.

Application of the multi-mode, diffuse-field response equation presented in Chapter 5 [Eq. (5.37)] shows that the ratio of the mass-controlled, field incidence, infinite-partition transmission coefficient τ_∞, to the average bounded-plate, resonant-mode transmission coefficient τ_r, for homogeneous plates of equal densities and thicknesses is approximately

$$\tau_\infty/\tau_r \simeq (hc_l'\omega/\sqrt{3}c^2)(\eta_{tot}/\sigma^2), \tag{4.54}$$

in which h is the plate thickness, σ the modal average radiation efficiency in a frequency band centred on ω, and η_{tot} the sum of the plate loss factors due to frictional mechanisms and to sound radiation. Note the sensitivity to the magnitude of the radiation efficiency; this arises because both response and radiation are proportional to σ. Well below the critical frequency, an approximate expression for the modal average radiation efficiency of a simply supported rectangular plate is

$$\sigma \simeq (P\lambda_c/\pi^2 A)(f/f_c)^{1/2}, \tag{4.55}$$

where P is the panel perimeter length, A the panel area, and λ_c the wavelength at f_c, which is by definition equal in the plate and the fluid. Substitution of this expression into Eq. (4.54) yields

$$\tau_\infty/\tau_r \simeq 200(A/P\lambda_c)^2\eta_{tot}, \tag{4.56}$$

which is frequency independent and typically much greater than unity, except in the case of very lightly damped, lightweight, stiff panels of small surface area. At frequencies approaching f_c, σ increases very rapidly toward unity and τ_r approaches τ_∞.

It is an experimentally observed fact that the subcritical transmission coefficients of many simple homogeneous partitions, as measured when they are inserted between reverberation rooms, approximate reasonably closely to that given by the infinite-partition field incidence formula, Eq. (4.51). It is clear, therefore, that a model based upon transmission by the mechanism of excitation and radiation of modes *at resonance* is not adequate. Further, convincing evidence for the dominance of a non-resonant transmission mechanism is provided by the observation that the subcritical sound reduction indices of many partitions is not significantly altered by an increase in their total damping. One may infer from this that a form of response and radiation that is not sensitive to the action of damping mechanisms is responsible for the major part of the sound transmission process in such cases.

There are two forms of qualitative explanation of the origin of this behaviour. As the basis of the first we observe that the response of a uniform bounded plate to an applied force field may be considered to be the sum of the response to that field of an unbounded plate having the same dynamic properties, plus the response of the unbounded plate to the action of boundary force fields that are conceived as forcing it to satisfy the actual boundary conditions of the bounded plate: the result of the action of these forces is seen in boundary wave reflection. A similar concept was introduced in Chapter 1 in connection with the impedance analysis of a bounded beam, in which the incidence of outgoing bending waves upon the beam boundaries causes reflected waves to be generated. Such a decomposition of response is not straightforward in the case of fluid-loaded plates because regions of the imaginary unbounded plate outside the boundaries can communicate acoustically with regions within the boundary, a process that is not physically valid. However, the influence of fluid loading on the vibration of practical structures transmitting airborne sound is generally negligible.

Hence we may visualise the response of a bounded plate as comprising two components: (1) the infinite-plate response component, which is "forced" to travel at the trace wave speed $c/\sin\phi$ of the incident wave; (2) the waves caused by the incidence of this forced wave on the actual boundaries: the latter waves, which are *free* bending waves travelling at their natural, or free, wave speeds, are multiply reflected by the various boundaries, and those components having frequencies equal to the natural frequencies of the bounded plate modes interfere constructively to create resonant motion in these modes. We may, at least qualitatively, consider the transmission processes associated with free- and forced-wave components to coexist independently, one controlled by damping and one not. As already shown by Eq. (4.56), the forced-wave process (corresponding to τ_∞) tends to transmit more energy than the free-wave process (corresponding to τ_r), in agreement with experimental results.

In Eq. (4.56), the panel dimensions factor $(A/P\lambda_c)$ appears because it controls subcritical modal radiation efficiencies. There is an additional geometric effect associated with the ratio of panel dimensions to the wavelength of forced waves. Equation (4.56) is based upon the field incidence transmission coefficient; the associated transmission analysis assumed that an infinitely extended forced bending wave exists, having a radiation efficiency equal to sec ϕ. In fact, forced-wave motion actually exists physically only within the boundaries of the plate, and therefore the number of forced wavelengths between boundaries is limited. This "windowing" effect, already met in connection with bounded panel radiation, spreads the wavenumber spectrum of forced vibration around the line value $k \sin \phi$ of the equivalent infinitely extended wave train. Some of these spectral components are subsonic and do not radiate. Those forced-wave components produced by near grazing incidence ($\phi \to \pi/2$), which dominate forced transmission, leak most strongly into the subsonic wavenumber range ($k_x > k$). The effect is akin to that which allows the subsonic free-wave components to radiate at all, as already seen in Chapter 2; however, it acts in a reverse manner. We see, therefore, that the effect of reducing the size of a panel is to enhance free-wave, resonant transmission, and to reduce forced-wave transmission. Since the latter normally dominates in large panels, the low-frequency sound reduction index of small panels is likely to be greater and more affected by damping than that of larger panels of the same material.

An alternative explanation of the dominance at subcritical frequency of non-resonant transmission mechanisms is based upon the fact that the radiation efficiency of a given mode, which has a subcritical natural frequency, increases with the frequency of modal vibration. It transpires from detailed analysis that the greatest contribution of a mode to sound energy transmission occurs at frequencies far above its resonance frequency, in which case its response, and the resulting transmission, is mass controlled.

The effect of panel size on forced-wave transmission is quantified in an analysis by Sewell (1970). His expression for the transmission coefficient of a panel of area A, at a frequency corresponding to acoustic wavenumber k, may be expressed in terms of a sound reduction index for subcritical frequencies ($\omega < \omega_c$) as

$$R_{nr} \simeq R(0) - 10 \log_{10}[\ln(kA^{1/2})] + 20 \log_{10}[1 - (\omega/\omega_c)^2] \quad \text{dB}, \qquad (4.57)$$

where the subscript nr indicates non-resonant. Sewell's original formula contains a panel shape factor and another area factor, neither of which is usually significant. This formula has been compared by Quirt (1982) with the field incidence formula and found to be superior at low frequencies (Fig. 77).

4.5 Transmission of Sound through a Partition between Two Rooms

In Section 4.4 it was assumed that the panel was located in an aperture in an infinite rigid-plane baffle and that the incident sound field was diffuse. It is of considerable practical interest to know how the sound reduction index formulas obtained using this model compare with those derived on the basis of a model of a rectangular partition that forms the complete dividing wall between rectangular rooms.

A comprehensive analysis of this problem has been presented by Josse and Lamure (1964). The essential difference between their model and the diffuse-field model is that the sound field is represented in terms of a series of room acoustic modes. The response of the panel and the transmission of energy into the receiving room depend upon two coupling factors: the first is a function of the closeness of the natural frequencies of the acoustic modes of the rooms and of the panel vibration modes; the second depends upon the degree of spatial matching between the acoustic mode pressure distributions over the panel and the distributions of panel mode displacements. The analysis is not based upon a fully coupled system formulation, since the exciting pressure field on the surface of the panel is assumed to be that which would exist in the absence of panel motion; this assumption is not likely to lead to significant errors in most practical problems of airborne sound transmission in buildings. (Why do you think this is true?)

The panel modes' normal velocities are assumed to have the non-dimensional form

$$\phi_{\mathrm{N}}(x, y) = \sin(m\pi x/a) \sin(n\pi y/b), \tag{4.58}$$

where m and n are integers and a and b are the panel side lengths. The acoustic-mode pressure is assumed to take the non-dimensional form

$$\psi_{\mathrm{N1}}(x_1, y_1, z_1) = \cos(m_1\pi x/a) \cos(n_1\pi y/b) \cos(p_1\pi z/c). \tag{4.59}$$

The response of a panel mode of natural frequency ω_{N} to excitation by an acoustic mode driven at frequency ω by the source is proportional to a factor β, which derives from an integral over the panel surface of the product of ϕ_N and ψ_{N1}:

$$\beta = \frac{mn}{[(m^2 - m_1^2)(n^2 - n_1^2)]} \times \frac{j\omega}{[j\eta\omega_N^2 + (\omega_N^2 - \omega^2)]}, \tag{4.60}$$

where η is the mechanical loss factor of the panel mode. We see that the response is very sensitive to the degree of spatial matching ($m \sim m_1, n \sim n_1$) and frequency matching ($\omega \sim \omega_N$): in fact, β remains finite with $m = m_1$ and $n = n_1$, the original integral taking a special form.

Three classes of modal matching can be identified:

(1) Geometric matching: $m \simeq m_1$, $n \simeq n_1$, $\omega \not\simeq \omega_N$, no resonance.

(2) Semi-coincidence: $m \simeq m_1$, $n \not\simeq n_1$, $\omega \simeq \omega_N$, resonance,
$m \not\simeq m_1$, $n \simeq n_1$, $\omega \simeq \omega_N$, resonance.

(3) Coincidence: $m \simeq m_1$, $n \simeq n_1$, $\omega \simeq \omega_N$, resonance.

It is shown that the last condition cannot exist below the critical frequency ω_c of the panel. An essentially statistical analysis of contributions of the various classes of coupling to panel response yields the following results for the response in frequency bands centred on ω.

(1) Geometric matching; subcritical frequencies, $\omega < \omega_c$:

$$\langle \overline{v^2} \rangle = \frac{2\langle \overline{p_1^2} \rangle}{m^2\omega^2}, \tag{4.61}$$

where $\langle \overline{v^2} \rangle$ is the space-averaged mean-square vibration velocity of the panel and $\langle \overline{p_1^2} \rangle$ the space-averaged mean-square pressure in the source room. The response is seen to be equal to that of a limp panel of mass m to a surface pressure of mean-square value $2\langle \overline{p_1^2} \rangle$, which is the average value on the surface of a large reverberant room containing a broadband sound field of space-average mean-square value $\langle \overline{p_1^2} \rangle$.

(2) Semi-coincidence, subcritical frequencies, $\omega < \omega_c$:

$$\langle \overline{v^2} \rangle = \frac{2\langle \overline{p_1^2} \rangle}{m^2\omega^2} \left[\frac{2c}{\eta} \left(\frac{1}{a} + \frac{1}{b} \right) \left(\frac{1 + \omega/\omega_c}{(\omega\omega_c)^{1/2}} \right) \right]. \tag{4.62}$$

The factor in square brackets represents the influence of panel resonances on response, as the presence of the loss factor suggests.

(3) Coincidence; supercritical frequencies, $\omega > \omega_c$:

$$\langle \overline{v^2} \rangle = \frac{2\langle \overline{p_1^2} \rangle \pi\omega/\omega_c}{m^2\omega^2 4\eta(1 - \omega_c/\omega)^{1/2}}. \tag{4.63}$$

The contribution of semi-coincident modes of class (2) at super-critical frequencies is generally negligible compared with that of the coincident modes. The influence of damping is again clearly evident.

(4) Response in frequency band $\Delta\omega$ centred on ω_c, $\Delta\omega > c/b, c/a$:

$$\langle\overline{v^2}\rangle = \frac{2\langle\overline{p_1^2}\rangle\pi}{m^2\omega^2 4\eta}\left(\frac{\omega_c}{\Delta\omega}\right)^{1/2}. \tag{4.64}$$

Note that the response depends upon the bandwidth $\Delta\omega$ because it peaks sharply in the vicinity of the critical frequency.

The corresponding sound reduction indices are as follows:

(1) Subcritical frequencies; $\omega < \omega_c$:

$$R = R(0) - 10\log_{10}\left\{\left[\frac{3}{2} + \ln\left(\frac{2\omega}{\Delta\omega}\right)\right] + \frac{16c^2}{\eta\omega_c}\frac{1}{(\omega\omega_c)^{1/2}}\left(\frac{a^2+b^2}{a^2b^2}\right)\left[1 + \frac{2\omega}{\omega_c} + 3\left(\frac{\omega}{\omega_c}\right)^2\right]\right\} \quad \text{dB}. \tag{4.65}$$

The term in the braces which contains η represents the resonant contribution to transmission; it is seen to increase in significance as the partition size decreases, a characteristic observed in the case of a baffled panel. A valuable feature of Eq. (4.65) is that it offers a means of estimating the minimum value of η that will significantly affect sound transmission.

Evaluation of Eq. (4.65) for typical practical cases shows that resonant transmission is relatively more important (3–6 dB) in this case of a division between two rooms than in the case of a panel in a baffle.

The non-resonant component of sound reduction index is

$$R = \begin{cases} R(0) - 5.6 \quad \text{dB} & (1/3 \text{ octave band}), \\ R(0) - 4.0 \quad \text{dB} & (1/1 \text{ octave band}). \end{cases}$$

The expressions are close to the empirical field incidence formula.

(2) Supercritical frequencies, $\omega > \omega_c$:

$$R = R(0) + 10\log_{10}(2\eta/\pi)(\omega/\omega_c - 1) \quad \text{dB}. \tag{4.66}$$

This formula agrees with Cremer's [Eq. (4.53)] and indicates a 9 dB per octave increase with frequency.

(3) Frequency band centred on ω_c, $\Delta\omega/\omega_c > (k_c a)^{-1}, (k_c b)^{-1}$:

$$R = R(0) + 10\log_{10}(2\eta/\pi)(\Delta\omega/\omega_c) \quad \text{dB}. \tag{4.67}$$

Experimental verification of Eqs (4.65)–(4.67) is provided by the results shown in Fig. 78 of measurement on a 5-cm-thick brick wall plastered on both sides to a thickness of 1.5 cm.

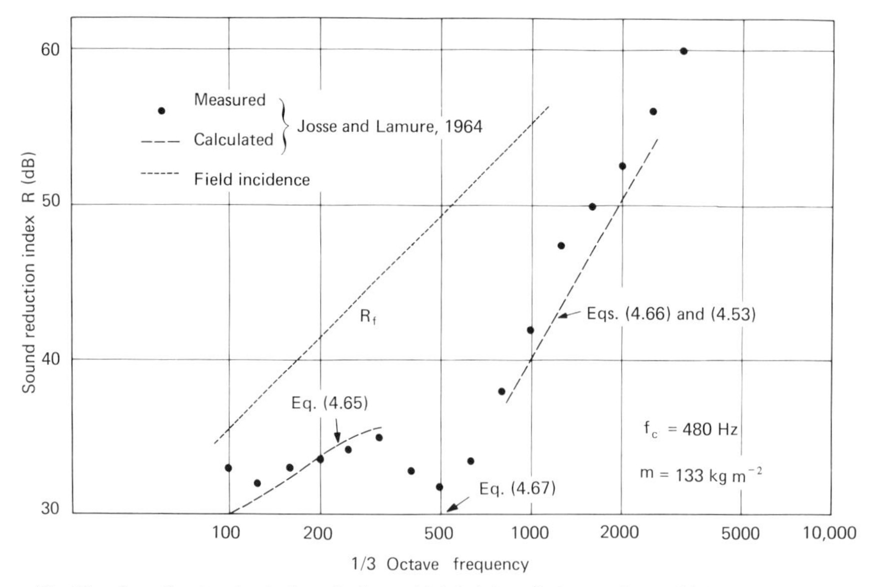

Fig. 78. Sound reduction index of a 5-cm-thick brick wall plastered to a thickness of 1.5 cm.

4.6 Double-Leaf Partitions

Theoretical and experimental analyses of sound transmission through single-leaf partitions show that the sound reduction index at a given frequency generally increases by 5–6 dB per doubling of mass, provided that no significant flanking or coincidence-controlled transmission occurs. In practice, it is often not only necessary for structures to have low weight, but also to provide high transmission loss: examples include the walls of aircraft fuselages, partition walls in tall buildings, and movable walls between television studios. This requirement can clearly not be met by single-leaf partitions.

The most common solution to this problem is to employ constructions comprising two separate leaves separated by an air space or cavity. It would be most convenient if the sound reduction index of the combination were to equal the sum of those of the two leaves when used as single-leaf partitions. Unfortunately the air in the cavity dynamically couples the two leaves, with the result that the sound reduction index of the combination may fall below this ideal value, sometimes by a large amount. In the following sections, various idealised models are theoretically analysed in order to illustrate the general sound transmission characteristics of double-leaf partitions and the dependence of these characteristics on the physical parameters of the systems.

It is clear from even a superficial review of the available literature that theoretical analysis of the sound transmission behaviour of double-leaf partitions is far less well developed than that of single-leaf partitions, and that consequently greater reliance must be placed upon empirical information. The reason is not hard to find; the complexity of construction and the correspondingly larger number of parameters, some of which are difficult to evaluate, militate against the refinement of theoretical treatments. In particular, it is difficult to include the effects of mechanical connections between leaves, and of non-uniformly distributed mechanical damping mechanisms, in mathematical models. The following analyses are offered, therefore, more as vehicles for the discussion of the general physical mechanisms involved, than as means of accurate quantitative assessments of the performance of practical structures.

4.7 Transmission of Normally Incident Plane Waves through an Unbounded Double-Leaf Partition

The idealised model is shown in Fig. 79. Uniform, non-flexible partitions of mass per unit area m_1 and m_2, separated by a distance d, are mounted upon viscously damped, elastic suspensions, having stiffness and damping

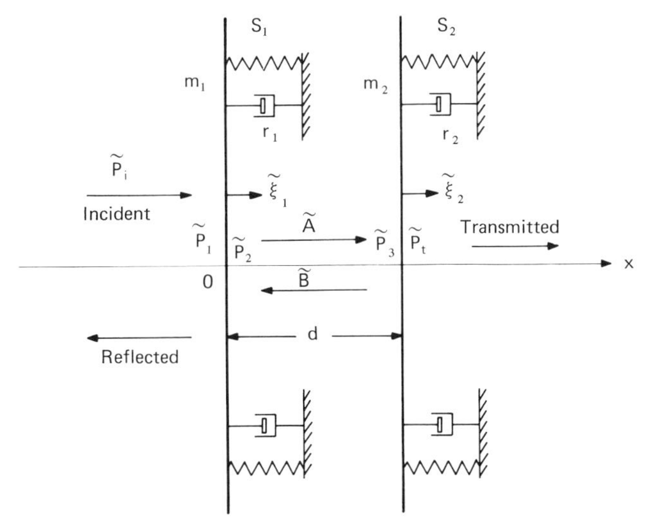

Fig. 79. Idealised model of normal incidence sound transmission through a double-leaf partition.

coefficients per unit area of s_1, s_2 and r_1, r_2, respectively. It is assumed initially that the fluid in the cavity behaves isentropically, without energy dissipation, and that the pressure–density relationship is adiabatic, as in the free air outside. A plane wave of frequency ω is incident normally upon leaf 1.

The cavity wave coefficients $\tilde{A}$ and $\tilde{B}$ are related to the leaf displacements $\tilde{\xi}_1$ and $\tilde{\xi}_2$ and cavity pressures $\tilde{p}_2$ and $\tilde{p}_3$ as follows:

$$\tilde{p}_2 = \tilde{A} + \tilde{B}, \tag{4.68}$$

$$\tilde{p}_3 = \tilde{A}\exp(-jkd) + \tilde{B}\exp(jkd), \tag{4.69}$$

$$j\omega\tilde{\xi}_1 = (\tilde{A} - \tilde{B})/\rho_0 c, \tag{4.70}$$

$$j\omega\tilde{\xi}_2 = [\tilde{A}\exp(-jkd) - \tilde{B}\exp(jkd)]/\rho_0 c. \tag{4.71}$$

The equations of motion of the leaves are

$$j\omega\tilde{\xi}_1\tilde{z}_1 = \tilde{p}_1 - \tilde{p}_2, \tag{4.72}$$

$$j\omega\tilde{\xi}_2\tilde{z}_2 = \tilde{p}_3 - \tilde{p}_t, \tag{4.73}$$

in which

$$\tilde{z}_1 = j\omega m_1 + r_1 - js_1/\omega = m_1(j\omega + \eta_1\omega_1) - js_1/\omega, \tag{4.74}$$

$$\tilde{z}_2 = j\omega m_2 + r_2 - js_2/\omega = m_2(j\omega + \eta_2\omega_2) - js_2/\omega, \tag{4.75}$$

where η_1 and η_2 are the respective mechanical loss factors, and ω_1 and ω_2 the *in vacuo* natural frequencies of the two leaves. The pressure $\tilde{p}_1$ is related to the pressure $\tilde{p}_i$ of the incident wave by

$$\tilde{p}_1 = 2\tilde{p}_i - j\omega\rho_0 c\tilde{\xi}_1, \tag{4.76}$$

and the transmitted wave has pressure $\tilde{p}_t$ given by

$$\tilde{p}_t = j\omega\rho_0 c\tilde{\xi}_2. \tag{4.77}$$

Let us assume first that the cavity width is very small compared with an acoustic wavelength, in which case $kd \ll 1$. Equations (4.68) and (4.69) indicate that, in this case, $\tilde{p}_3 \simeq \tilde{p}_2 = \tilde{p}_c$; in other words, we may assume that the cavity pressure is uniform. Equations (4.70) and (4.71) may be combined to give

$$j\omega\rho_0 c\tilde{\xi}_2 = j\omega\tilde{\xi} - jkd(\tilde{A} + \tilde{B})$$

or

$$(\rho_0 c^2/d)(\tilde{\xi}_1 - \tilde{\xi}_2) = \tilde{p}_c, \tag{4.78}$$

which indicates that the air acts as a spring of stiffness $s' = \rho_0 c^2/d$. Equations (4.72)–(4.78) may be combined to yield the leaf displacement ratio

$$\tilde{\xi}_1/\tilde{\xi}_2 = [j\omega(\tilde{z}_2 + \rho_0 c) + \rho_0 c^2/d]/(\rho_0 c^2/d), \tag{4.79}$$

and the pressure amplitude transmission coefficient

$$\frac{\tilde{p}_t}{\tilde{p}_i} = -\frac{2j(\rho_0 c)^2/kd}{[\tilde{z}_2 + \rho_0 c - j\rho_0 c/kd][\tilde{z}_1 + \rho_0 c - j\rho_0 c/kd] + (\rho_0 c/kd)^2}. \quad (4.80)$$

Comparison of the terms in square brackets in the denominator of Eq. (4.80) with Eqs. (4.74) and (4.75) shows that the impedance of each leaf is combined with an acoustic radiation (damping) term $\rho_0 c$ and an acoustic stiffness term $\rho_0 c/kd$. Now the mechanical stiffness s_1 of leaf 1 may be equated to $\omega_1^2 m_1$, where ω_1 is the *in vacuo*, undamped resonance frequency of leaf 1 on its mounting. Hence the ratio of mechanical to acoustic stiffness is

$$\delta_1 = \frac{s_1/\omega_1}{\rho_0 c/k_1 d} = \frac{\omega_1 m_1 k_1 d}{\rho_0 c}. \quad (4.81)$$

The same form of relationship can be written for leaf 2. If the model is considered to represent an approximation to normal incidence sound transmission through a bounded panel, ω_1 and ω_2 can be taken as the fundamental, *in vacuo*, natural frequencies of each panel. The products $\omega_1 m_1$ and $\omega_2 m_2$ are proportional to the square of the ratios of the panel thicknesses to the typical panel dimension, and it turns out that for many lightweight double-leaf partitions of practical dimensions δ_1 is less than unity, so that the acoustic stiffness predominates.

If acoustic damping, mechanical damping, and stiffness are neglected, the maximum transmission coefficient $\tau = |\tilde{p}_t/\tilde{p}_i|^2$ occurs at a frequency such that

$$(-\omega^2 m_1 + \rho_0 c^2/d)(-\omega^2 m_2 + \rho_0 c^2/d) = (\rho_0 c^2/d)^2.$$

The solution is

$$\omega_0 = \left[\left(\frac{\rho_0 c^2}{d}\right)\left(\frac{m_1 + m_2}{m_1 m_2}\right)\right]^{1/2}. \quad (4.82)$$

This is termed the *mass–air–mass resonance frequency*, which is seen to decrease with increase of the leaf separation d. This frequency is a minimum when $m_1 = m_2$: we shall symbolise it by ω_{0m}.

At low frequencies, such that $kd \ll 1$, the transmission behaviour may be classified as follows:

(1) Frequencies below the mass–air–mass resonance frequency, $\omega < \omega_0$: In this case, $\omega^2 m_2 m_1 < (m_1 + m_2)(\rho_0 c^2/d)$, the damping terms have negligible influence, and Eq. (4.80) becomes

$$\tilde{p}_t/\tilde{p}_i \simeq -2j\rho_0 c/\omega(m_1 + m_2). \quad (4.83)$$

Hence

$$\tau \simeq (2\rho_0 c/\omega m_t)^2,$$

and

$$R = R(0, m_t) \quad \text{dB}, \tag{4.84}$$

where $m_t = m_1 + m_2$. Comparison with Eq. (4.20) shows that the partition behaves like a single-leaf partition having a mass equal to the sum of the masses of the two leaves: damping has negligible effect.

(2) Frequencies close to the mass–air–mass resonance frequency, $\omega \simeq \omega_0$: In this case, the pressure transmission coefficient is

$$\tilde{p}_t/\tilde{p}_i \simeq -2\rho_0 c/(\eta_1\omega_1 m_2 + \eta_2\omega_2 m_1 + K\rho_0 c), \tag{4.85}$$

where the factor K equals $(m_1/m_2) + (m_2/m_1)$. This result suggests that, if mechanical damping is low, it is preferable to minimise transmission at resonance. This can be done by maximising K by making m_1/m_2 or $m_2/m_1 \gg 1$: in these cases $\omega_0 > \omega_{0m}$. However, later analysis shows that benefit near ω_0 is gained at the expense of performance at higher frequencies. In the special case of leaves of equal mass m, *in vacuo* fundamental natural frequency ω', and loss factor η,

$$\frac{\tilde{p}_t}{\tilde{p}_i} \simeq -\frac{2}{2\eta(m\omega'/\rho_0 c) + 2}. \tag{4.86}$$

If, in addition, the mechanical damping is sufficiently large to make η much greater than $\rho_0 c/\omega' m$, which is generally greater than $\rho_0 c/\omega_0 m$, then

$$\tilde{p}_t/\tilde{p}_i \simeq -2\rho_0 c/2m\omega'\eta. \tag{4.87}$$

The sound reduction index is hence

$$R = R(0, m_t, \omega') + 20 \log_{10} \eta \quad \text{dB}, \tag{4.88}$$

where $R(0, m_t, \omega')$ is based upon the total mass and the *in vacuo* fundamental natural frequency of the leaves: the transmission is damping controlled. If $\eta \ll \rho_0 c/\omega' m$ then τ is close to unity and virtually all the incident sound energy is transmitted. As already stated, the transmission peak caused by resonance is made less severe by using leaves of different weight.

(3) Frequencies above the mass–air–mass resonance frequency, $\omega > \omega_0$: In this case, $\omega^2 m_2 m_1 > (m_1 + m_2)(\rho_0 c^2/d)$ and

$$\frac{\tilde{p}_t}{\tilde{p}_i} \simeq \frac{2j(\rho_0 c)^2/kd}{\omega^2 m_1 m_2}. \tag{4.89}$$

Substitution from Eq. (4.82) for $\rho_0 c^2/d$ yields

$$\frac{\tilde{p}_t}{\tilde{p}_i} \simeq \frac{2j\rho_0 c}{\omega(m_1 + m_2)}\left(\frac{\omega_0}{\omega}\right)^2.$$

Hence

$$R \simeq R(0, m_t) + 40 \log_{10}(\omega/\omega_0) \quad \text{dB}, \tag{4.90a}$$

which may also be expressed as

$$R = R(0, m_1) + R(0, m_2) + 20 \log_{10}(2kd) \quad \text{dB}, \tag{4.90b}$$

where $R(0, m_t)$ is based upon the total mass of the partition. The sound reduction index therefore rises at 18 dB/octave from the value it would have at the resonance frequency if simply controlled by total mass. The great improvement over the performance below the resonance frequency, as indicated by the term $40 \log_{10}(\omega/\omega_0)$, is typical of transmission through inertial layers coupled by a resilient layer: the physical explanation is that leaf 2 acts as a mass driven through a spring by the motion of leaf 1, above the system resonance frequency. The transmission characteristics in the low frequency range, for which $kd \ll 1$, are presented in graphical form in Fig. 80.

The behaviour of the system at higher frequencies, for which it may not be assumed that $kd \ll 1$, may be analysed by solving Eqs. (4.68)–(4.77) for arbitrary kd. The general solution for the ratio of transmitted to incident

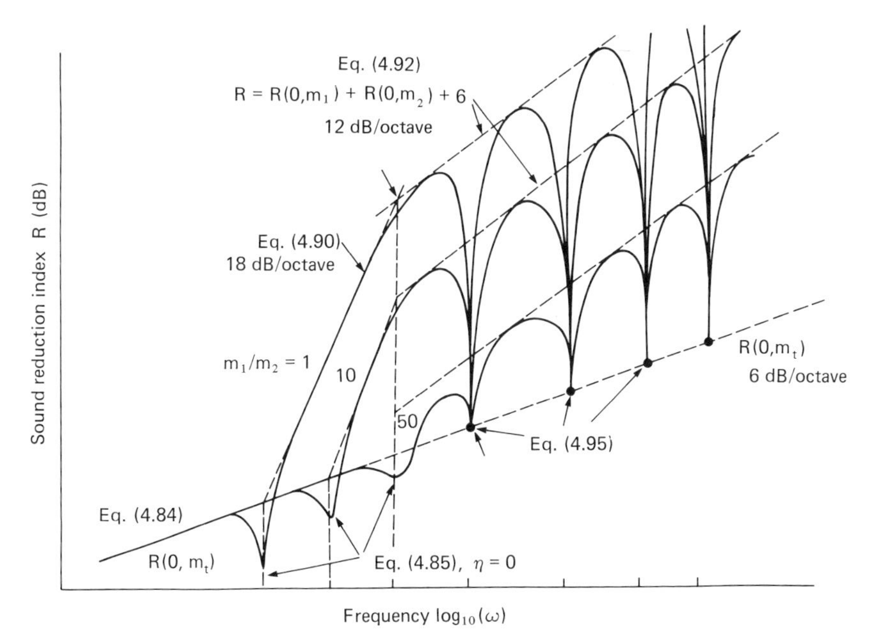

Fig. 80. Illustration of the theoretical effect on normal-incidence sound reduction index of varying leaf mass ratios, while keeping the total mass constant.

pressures is

$$\frac{\tilde{p}_t}{\tilde{p}_i} = -\frac{2j\rho_0^2 c^2 \sin kd}{\tilde{z}'_1 \tilde{z}'_2 \sin^2 kd + \rho_0^2 c^2}, \tag{4.91}$$

where $\tilde{z}' = \tilde{z} + \rho_0 c(1 - j \cot kd)$. Note that Eq. (4.91) reduces to Eq. (4.80) if $kd \ll 1$. The variation of this ratio with frequency is complicated; it varies between minima, which correspond to acoustic anti-resonances of the cavity, when $kd = (2n - 1)\pi/2$, and maxima, at resonances when $kd = n\pi$, n being any non-zero positive integer. At the anti-resonance frequencies the ratio takes the approximate form

$$\frac{\tilde{p}_t}{\tilde{p}_i} = -\frac{2j\rho_0^2 c^2}{\omega^2 m_1 m_2}, \tag{4.92a}$$

which gives

$$R \simeq R(0, m_1) + R(0, m_2) + 6 \quad \text{dB}, \tag{4.92b}$$

which, unlike the case at ω_0, is maximised by making $m_1 = m_2$. This is greater than the sum of the mass-controlled sound reduction indices of the two leaves considered as single partitions. At the resonance frequencies, the solution is indeterminate in the absence of energy losses in the cavity, and the displacements of the two leaves are predicted to be equal. An alternative approach to the solution is to consider the acoustic impedance imposed on the first leaf by the combination of the fluid in the cavity and second leaf. The general expression for the specific acoustic impedance of a column of fluid terminated by a specific acoustic impedance $\tilde{z}_l$ is

$$\tilde{z}_0 = \rho_0 c \frac{\tilde{z}_l + j\rho_0 c \tan kl}{\rho_0 c + j\tilde{z}_l \tan kl}. \tag{4.93}$$

(Prove.) At resonance, $\tan kl = 0$ and

$$\tilde{z}_0 = \tilde{z}_l. \tag{4.94}$$

Hence the loading on leaf 1 is the same as if leaf 2 were directly attached to it. The corresponding sound reduction index is

$$R = R(0, m_t) \quad \text{dB}, \tag{4.95}$$

which is that given by a single leaf of mass $m_t = m_1 + m_2$. The general variation of the sound reduction index with frequency is shown in Fig. 80. The asymptotic frequency average value of R is approximately equal to the sum of $R(0, m_1)$ and $R(0, m_2)$, which increases at 12 dB/octave. If this line is extrapolated to low frequencies it will intersect the line given by Eq. (4.90), at a frequency given by $kd = 1/2$ which corresponds approximately to one-sixth the lowest cavity acoustic resonance frequency.

4.8 The Effect of Cavity Absorption

In practice, sound absorbing materials are placed in the cavity of double-leaf constructions in order to minimise resonance effects. Because absorption will cause the sound wave amplitude to decay with distance, the acoustic wavenumber in the cavity must be assumed to be complex. Let the propagation constant γ be given by

$$\gamma = \alpha + j\beta, \tag{4.96}$$

where β is the phase constant, and α the attenuation constant. The expression for the pressure field in the cavity becomes

$$\tilde{p}(x) = \tilde{A} \exp(-\gamma x) + \tilde{B} \exp(\gamma x). \tag{4.97}$$

The particle velocity is given by

$$\begin{aligned} \tilde{u}(x) &= (j/\omega\rho_0)\,\partial\tilde{p}(x)/\partial x \\ &= (j\gamma/\omega\rho_0)[-\tilde{A} \exp(-\gamma x) + \tilde{B} \exp(\gamma x)]. \end{aligned} \tag{4.98}$$

The equations of motion and continuity of particle velocity are the same as Eqs. (4.68)–(4.77) with jkd replaced by γd. The general solution for ratio of transmitted to incident sound pressure is

$$\frac{\tilde{p}_t}{\tilde{p}_i} = -\frac{4j\omega^3\rho_0^2 c\gamma/\sinh \gamma d}{\gamma^2\tilde{\alpha}_1\tilde{\alpha}_2 - \omega^2\rho_0\gamma(\tilde{\alpha}_1 + \tilde{\alpha}_2)\coth \gamma d + \omega_0^4\rho_0^2}, \tag{4.99}$$

where $\tilde{\alpha}_1 = j\omega(\tilde{z}_1 + \rho_0 c)$ and $\tilde{\alpha}_2 = j\omega(\tilde{z}_2 + \rho_0 c)$. This equation reduces to Eq. (4.91) when $\alpha = 0$ and $\beta = k$.

The mass–air–mass resonance frequency may be significantly changed by the presence of a porous absorbent material in the cavity because the bulk modulus, together with effective density and phase speed can be very different from those in air. The attenuation constant α is largely determined by a property of the absorbent material termed the *flow resistivity* (Bies, 1971). This quantity is defined as the pressure drop per unit thickness per unit velocity which is produced by a steady low-speed flow of fluid through the material: it is assumed that at low frequencies and small amplitudes this resistance applies also to oscillatory flow. The attenuation constant α is very approximately related to the flow resistivity r by $\alpha \simeq r/2\rho_0 c$.

On the basis of the assumptions that $\alpha d < 1$, which is generally true for typical values of flow resistance at frequencies close to the mass–air–mass resonance frequency, and also that $\beta d \ll 1$ (since β is of the order of k), Eq. (4.99) shows that a low-frequency transmission peak, corresponding to damped mass–air–mass resonance, will occur when $\beta^2 = (\omega_0/c)^2 + \alpha^2$. The corresponding solution for the pressure coefficient is

$$\frac{\tilde{p}_t}{\tilde{p}_i} \simeq -\frac{2\rho_0 c}{\omega(m_1+m_2)(2\alpha\beta c^2/\omega_0^2)+(\eta_1\omega_1 m_2+\eta_2\omega_2 m_1)+K\rho_0 c}, \tag{4.100}$$

or, using the approximate relationship between α and r,

$$\frac{\tilde{p}_t}{\tilde{p}_i} \simeq -\frac{2\rho_0 c}{\omega(m_1+m_2)(rd/\rho_0 c)(\beta/k_0^2 d)+(\eta_1\omega_1 m_2+\eta_2\omega_2 m_1)+K\rho_0 c}, \tag{4.101}$$

in which $k_0 = \omega_0/c$ and $K = (m_1/m_2) + (m_2/m_1)$. If, for the sake of qualitative interpretation of this result, we let $\beta \simeq k_0$ and $\omega \simeq \omega_0$, then

$$\frac{\tilde{p}_t}{\tilde{p}_i} \simeq -\frac{2\rho_0 c}{m_1(r/\rho_0+\eta_2\omega_2)+m_2(r/\rho_0+\eta_1\omega_1)+K\rho_0 c}, \tag{4.102}$$

which is independent of ω_0 and d, as in the case without absorption [Eq. (4.85)]. Comparison of Eqs. (4.102) and (4.85) shows that the influence of the mechanical damping remains the same, and that it is significant compared with the cavity absorption effect only if $\eta_1\omega_1 \geqslant r/\rho_0$ and/or $\eta_2\omega_2 \geqslant r/\rho_0$. In the special case of leaves of equal mass and no mechanical damping, cavity absorption is only significant if $r \geqslant \rho_0^2 c/m$ or $\alpha \geqslant \rho_0/2m$. In this case,

$$R = R(0, m_t, \omega_0) + 20\log_{10}(r/\rho_0\omega_0) \quad \text{dB}, \tag{4.103}$$

where the first term corresponds to that of a single leaf of mass equal to the sum of the masses of the two leaves, as in Eq. (4.84) for frequencies below the mass–air–mass resonance frequency.

In practice, a flow resistivity high enough to make $r/\rho_0\omega_0$ greater than unity may be realised, in which case Eq. (4.103) suggests that the sound reduction index at resonance is greater than that indicated by Eq. (4.84) for the non-resonant case. There are various reasons for this apparent anomaly: first, the presence of absorption to some extent decouples the motions of the leaves even below the resonance frequency, so that off-resonance transmission is also reduced; second, $r/2\rho_0 c$ is usually considerably greater than the effective value of α at the low frequencies typical of mass–air–mass resonance; third, it was assumed that $\alpha d \ll 1$, which approximates to $rd/2\rho_0 c \ll 1$, which restricts the range of flow resistances for which the analysis is valid.

If the attenuation constant is so large that αd approaches 1, then the wave pressure amplitude is reduced by a factor of e, or nearly three, during one journey between leaves. The leaves become effectively decoupled, and the low-frequency mass–air–mass resonance phenomenon cannot occur. In this case, the sound reduction index is substantially larger than the foregoing analyses suggest. It is clearly profitable to install a cavity absorbent with the highest possible flow resistance, provided it does not produce significant mechanical coupling between the leaves, which is difficult to avoid in practice.

At frequencies well above the mass–air–mass resonance frequency there can occur acoustic resonances of the cavity, as previously discussed. The introduction of absorption into the cavity is particularly effective in suppressing these resonances. If $\alpha d < 1$, resonance behaviour is still observable, but the sound reduction index exceeds the minimum value indicated by Eq. (4.95). If $\alpha d > 1$, then acoustic resonance of the cavity is virtually eliminated; the solution of Eq. (4.99) for the maximum pressure transmission ratio is

$$\frac{\tilde{p}_t}{\tilde{p}_i} \simeq -j\left(\frac{2\rho_0 c}{\omega m_1}\right)\left(\frac{2\rho_0 c}{\omega m_2}\right)\left(\frac{k}{\beta}\right)\exp(-\alpha d), \qquad (4.104)$$

which gives

$$R = R(0, m_1) + R(0, m_2) + 8.6\alpha d + 20\log_{10}(\beta/k) \quad \text{dB}, \qquad (4.105)$$

where 8.6α (dB m^{-1}) corresponds to the attenuation of waves travelling through the absorptive material. The physical interpretation of Eq. (4.105) is that the incident acoustic wave is progressively attenuated by passage through the first leaf, the absorbent, and the second leaf, and provided that $\alpha d > 1$, there is effectively no acoustic coupling between the two leaves.

4.9 Transmission of Obliquely Incident Plane Waves through an Unbounded Double-Leaf Partition

So far, only normally incident sound has been considered. When a plane wave is incident at an oblique angle ϕ upon leaf 1 it sets up a bending wave travelling in the plane of the leaf with the trace wavenumber $k_z = k \sin\phi$. Provided that the leaves and cavity are unbounded and uniform, waves travelling with the same wavenumber component parallel to the plane of leaf 1 are set up in the fluid in the cavity, in leaf 2, and in the fluid external to the partition, as shown in Fig. 81. In satisfaction of the acoustic-wave equation, and in the absence of cavity absorption, the wavevector components of the cavity wave in the direction normal to the planes of the leaves have magnitudes given by

$$k_x = k(1 - \sin^2\phi)^{1/2} = k\cos\phi.$$

Hence the pressure wave system in the cavity takes the form

$$\tilde{p}(x, z) = [\tilde{A}\exp(-jk_x x) + \tilde{B}\exp(jk_x x)]\exp(-jk_z z), \qquad (4.106)$$

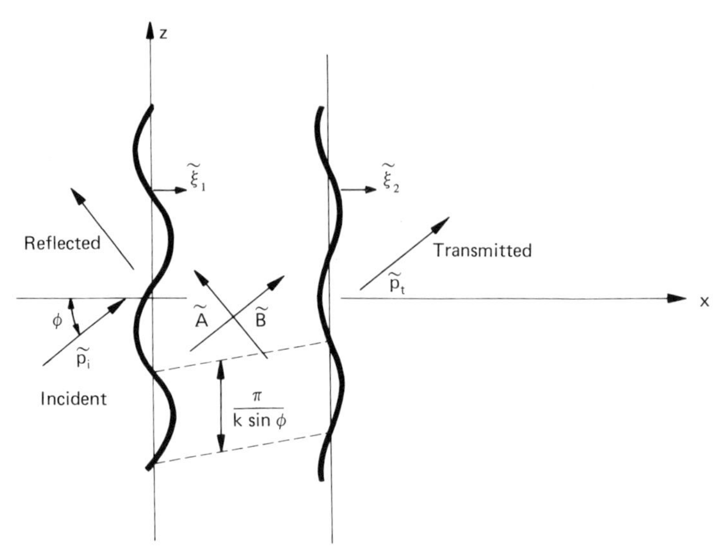

Fig. 81. Transmission of obliquely incident sound through an unbounded double-leaf partition.

and the corresponding particle velocity normal to the planes of the leaves is

$$\tilde{u}_x(x, z) = \frac{\cos \phi}{\rho_0 c} [\tilde{A} \exp(-jk_x x) - \tilde{B} \exp(jk_x x)] \exp(-jk_z z). \quad (4.107)$$

The physical interpretation of Eqs. (4.106) and (4.107) is that the acoustic impedance of the cavity is increased by a factor sec ϕ compared with the normal incidence value. The fluid-loading (radiation) impedance produced by the fluid external to the partition is increased by the same factor to $\rho_0 c \sec \phi$.

The relevant impedances of the leaves are the wave impedances corresponding to a progressive bending wave of wavenumber $k \sin \phi$ as given by Eq. (4.30) with $\phi_1 = \phi$. It is assumed in the following analysis that the fluids inside the cavity and outside the partition are the same, so that no refraction occurs. We have already seen, in the analysis of single-leaf transmission, that at frequencies well below the critical frequency of a leaf the inertial component of wave impedance greatly exceeds the stiffness component, and therefore in this frequency range the latter may be neglected. Equations (4.68)–(4.71) become

$$\tilde{p}_2 = \tilde{A} + \tilde{B}, \quad (4.108)$$

$$\tilde{p}_3 = \tilde{A} \exp(-jkd \cos \phi) + \tilde{B} \exp(jkd \cos \phi), \quad (4.109)$$

$$j\omega\tilde{\xi}_1 = (\tilde{A} - \tilde{B})/(\rho_0 c \sec\phi), \tag{4.110}$$

$$j\omega\tilde{\xi}_2 = [\tilde{A}\exp(-jkd\cos\phi) - \tilde{B}\exp(jkd\cos\phi)]/(\rho_0 c \sec\phi). \tag{4.111}$$

Equations (4.74)–(4.77) become

$$\tilde{z}_1 = j\omega m_1 + r_1, \tag{4.112}$$

$$\tilde{z}_2 = j\omega m_2 + r_2, \tag{4.113}$$

and

$$\tilde{p}_1 = 2\tilde{p}_i - j\omega\rho_0 c \sec\phi\tilde{\xi}_1, \tag{4.114}$$

$$\tilde{p}_t = j\omega\rho_0 c \sec\phi\tilde{\xi}_2. \tag{4.115}$$

The general solution for the ratio of transmitted to incident pressures is

$$\frac{\tilde{p}_t}{\tilde{p}_i} = -\frac{2j\rho_0^2 c^2 \sec^2\phi \sin(kd\cos\phi)}{\tilde{z}_1'\tilde{z}_2' \sin^2(kd\cos\phi) + \rho_0^2 c^2 \sec\phi}, \tag{4.116}$$

where $\tilde{z}' = \tilde{z} + \rho_0 c \sec\phi[1 - j\cot(kd\cos\phi)]$.

When $kd\cos\phi \ll 1$, Eq. (4.116) reduces to

$$\frac{\tilde{p}_t}{\tilde{p}_i} = -\frac{2j\rho_0^2 c^2 \sec^2\phi/(kd\cos\phi)}{[\tilde{z}_1 + \rho_0 c\sec\phi - j\rho_0 c/(kd\cos^2\phi)][\tilde{z}_2 + \rho_0 c\sec\phi - j\rho_0 c/(kd\cos^2\phi)] + \cdots + [\rho_0 c/(kd\cos^2\phi)]^2}. \tag{4.117}$$

Comparison with the equivalent normal-incidence, low-frequency result, Eq. (4.80), shows that the effective stiffness of the cavity has increased by a factor $\sec^2\phi$. Hence the oblique-incidence mass–air–mass resonance frequency is greater than the normal-incidence value by a factor $\sec\phi$.

The low-frequency transmission behaviour may be classified as follows:

(1) Frequencies below the oblique incidence mass–air–mass resonance frequency, $\omega < \omega_0 \sec\phi$: The pressure transmission coefficient is

$$\tilde{p}_t/\tilde{p}_i \simeq -2j\rho_0 c/\omega(m_1 + m_2)\cos\phi, \tag{4.118}$$

which is the same as the oblique-incidence expression for a single leaf of mass m_t. Hence

$$R(\phi) = R(\phi, m_t) \quad \text{dB}. \tag{4.119}$$

(2) Frequencies close to the oblique-incidence mass–air–mass resonance frequency, $\omega \simeq \omega_0 \sec\phi$: In this case, we must take mechanical damping of the leaves into account; the result is

$$\tilde{p}_t/\tilde{p}_i \simeq -2\rho_0 c\sec\phi/(\eta_1\omega_1 m_2 + \eta_2\omega_2 m_1 + K\rho_0 c\sec\phi), \tag{4.120}$$

where $K = (m_1/m_2) + (m_2/m_1)$, as in the normal-incidence case. The influence of mechanical damping, in comparison with that of different leaf weights, is seen to decrease with increasing angle of incidence. It is seen that mass–air–mass resonance can take place at all frequencies above ω_0, the frequency increasing with ϕ.

(3) Frequencies above the oblique-incidence mass–air–mass resonance frequency, $\omega > \omega_0 \sec \phi$: In this case leaf inertia dominates and the result takes the same form as for normal incidence:

$$\frac{\tilde{p}_t}{\tilde{p}_i} \simeq \frac{2j\rho_0 c}{\omega(m_1 + m_2)\cos\phi}\left(\frac{\omega_0 \sec\phi}{\omega}\right)^2 \tag{4.121}$$

and

$$R(\phi) \simeq R(\phi, m_t) + 40\log_{10}[(\omega/\omega_0)\cos\phi] \quad \text{dB} \tag{4.122}$$

or, alternatively,

$$R(\phi) \simeq R(\phi, m_1) + R(\phi, m_2) + 20\log_{10}(2\,kd\cos\phi) \quad \text{dB}. \tag{4.123}$$

The behaviour at higher frequencies, for which it may not be assumed that $kd \ll 1$, may be analysed by solving Eq. (4.116) for arbitrary kd. As in the normal-incidence case, transmission maxima produced by acoustic resonance of the cavity alternate with transmission minima caused by anti-resonance. These frequencies are higher than the corresponding values for normal incidence by the factor $\sec\phi$. At the anti-resonance frequencies given by $kd\cos\phi = (2n-1)\pi/2$, the pressure transmission coefficient minimum is

$$\tilde{p}_t/\tilde{p}_i \simeq 2j\rho_0^2 c^2/\omega^2 m_1 m_2 \cos^2\phi, \tag{4.124}$$

and

$$R(\phi) = R(\phi, m_1) + R(\phi, m_2) + 6 \quad \text{dB}. \tag{4.125}$$

Because the anti-resonance frequencies increase in proportion to $\sec\phi$, the sound reduction index maxima for any particular value of n are actually independent of angle of incidence and are given by Eq. (4.92).

At the resonance frequencies, the panels move as one and the sound reduction index minimum corresponds to Eq. (4.46):

$$R(\phi) = R(\phi, m_t) \quad \text{dB}. \tag{4.126}$$

The value is the same for all angles of incidence because $\omega\cos\phi$ is a constant. This conclusion is extremely important because at every frequency above the lowest cavity resonance frequency $\omega = \pi c/d$, there is an angle of incidence for which resonant transmission occurs: the same is true of mass–air–mass resonance above ω_0. Hence, in a diffuse field, resonance phenomena effectively control the maximum achieved sound reduction index. Therefore, it is vital to suppress these resonances, by inserting absorbent material into the cavity and/or by dividing up the cavity in order to suppress lateral wave motion.

The individual leaves of a double-leaf partition, of course, exhibit coincidence effects in response to the imposed sound fields. Mathematically, the leaf impedance terms in the foregoing equations for obliquely incident sound would be modified to include the bending stiffness and appropriate damping terms. A complete analysis of this problem is extremely complex, because the combined leaf–cavity fluid system is a waveguide that carries waves involving coupled motion of the two media (see Chapter 6). However, it is intuitively obvious that coincidence effects in partitions consisting of nominally identical leaves are likely to be rather more severe in a frequency range around the critical frequency than those in partitions having dissimilar leaves; empirical data shows this to be the case. Where the two leaves are effectively decoupled by cavity absorbent, the decrease in sound reduction index below the mass-controlled value caused by coincidence effects in the two leaves can be approximated by the arithmetic sum of the individual coincidence dips in R of the two leaves when tested in isolation (Sharp, 1978). As with single-leaf coincidence, mechanical damping of the leaves largely controls the severity of coincidence effects.

A generalised form of oblique incidence sound reduction index for a double-leaf partition that does not contain absorbent is presented in Fig. 82. Note that the maxima and minima are independent of ϕ.

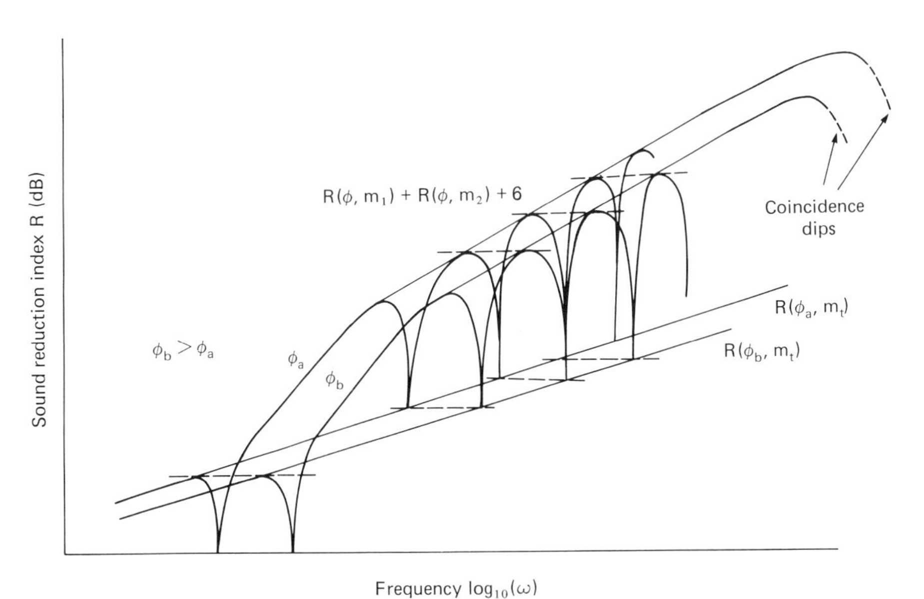

Fig. 82. Oblique incidence sound reduction index curves of a double-leaf partition at two angles of incidence: no cavity absorption.

The general analysis of transmission of obliquely incident plane sound waves through an infinite double-leaf partition containing absorbent material is complicated by the refraction effects caused by the difference of phase velocities of waves in the free air and in the absorbent: it is too involved to be of value in this book. However, if the cavity is filled with absorbent material of substantial flow resistivity, wave motion parallel to the leaves is strongly inhibited and the behaviour approximates to that for normal incidence. The reason is that, whereas the phase change undergone by the acoustic wave during its return journey across the cavity relative to that undergone by the leaf between transmission and return points, is equal to $\beta d \cos \phi$, the attenuation of the pressure amplitude is equal to $\exp(-\alpha d \sec \phi)$, which increases with angle of incidence. It is, in fact, found that the insertion into a cavity of sheets of absorbent material considerably thinner than the cavity width can produce substantial improvements in the performance of a lightweight, double-leaf partition. The mechanism is probably the attenuation of waves in the cavity having relatively large wavevector components parallel to the leaves (caused by highly oblique incidence): however, such minimal treatment does not significantly reduce the adverse influence of the mass–air–mass resonance phenomenon. In the case of double glazing, absorption can only be provided in the reveals at the edges of the cavity, which does little to influence mass–air–mass resonance effects. Figure 83 shows some diffuse

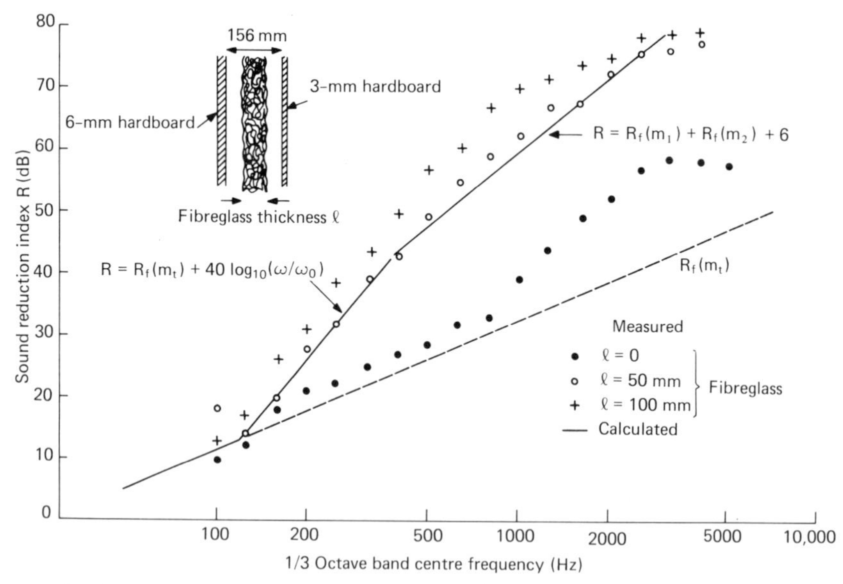

Fig. 83. Performance of a mechanically isolated double-leaf partition (Sharp, 1978).

incidence performance curves of a double-leaf partition in which the leaves are not mechanically coupled. Note the substitution of R_f in Eqs. (4.84), (4.90), and (4.92). A useful evaluation of simple formulas for the performance of double-leaf partitions without mechanical connections is provided by Pagliarini and Pompoli (1983). Quirt (1983) assesses double glazing formulas.

4.10 Mechanical Coupling of Double Partition Leaves

In the model that has been analysed so far, the two leaves are coupled only by the fluid in the cavity. In practice it is necessary to mount the leaves on one or more frames. In the simplest form of construction used in buildings, sheets of board are attached to either side of a common timber frame, which effectively constrains the two leaves to undergo equal transverse displacements at the positions of the frame members or studs, as shown in Fig. 84. There exists no comprehensive theoretical analysis of sound transmission through double partitions, although some useful approximate theory is available (Sharp, 1978).

In order approximately to evaluate the influence of such connections on sound transmission, we may take advantage of some earlier results relating to scattering of waves in plates by discontinuities and to sound radiation generated by line forces. It is assumed that the dynamic behaviour of each stud is independent of any other, which is reasonable on a frequency-average basis, and that the studs simply translate as rigid bodies in their own planes. By these assumptions we restrict our attention to bending waves incident normally on the stud–leaf connection lines, and we neglect the bending stiffness of the studs. These limitations are fairly severe, but they allow some of the more important features of the leaf-coupling process to be understood.

An incident plane wave having a frequency below the critical frequency of the leaves drives a bending wave in leaf 1, which is incident normally upon a stud. The impedance discontinuity scatters the bending wave and

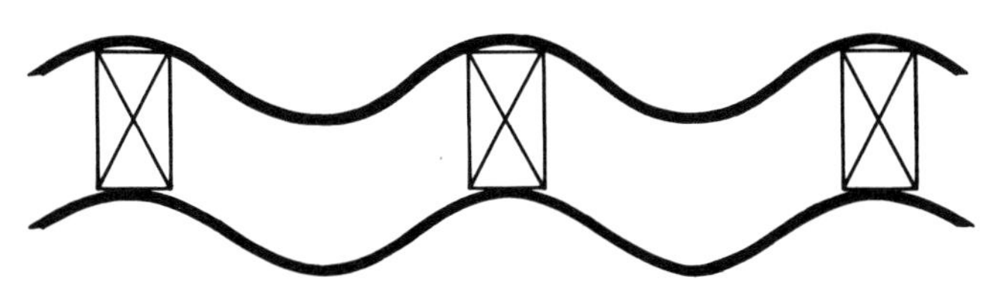

Fig. 84. Timber stud partition.

leaf 2 is driven by the stud. The sound radiated from leaf 2 due to this excitation is considered to be the transmitted field, the acoustic path through the cavity being assumed inactive. The total mechanical impedance of a stud, as "seen" by leaf 1, is equal to the inertial impedance of the stud itself plus the line force impedance of leaf 2. The incident bending-wave velocity amplitude is given approximately by

$$j\omega\tilde{\xi}_1 \simeq 2\tilde{p}_i/j\omega m, \tag{4.127}$$

in which fluid-loading effects have been neglected. The velocity of the stud is determined as in Eq. (2.105) and is equal to

$$j\omega\tilde{\xi}_s = j\omega\tilde{\xi}_1\tilde{Z}_F/(2\tilde{Z}_F + j\omega m_s), \tag{4.128}$$

where $\tilde{Z}_F$ is the line force impedance of each leaf and m_s the mass per unit length of the stud. The effective line force per unit length applied to the second leaf is

$$\tilde{F}_2 = j\omega\tilde{\xi}_s\tilde{Z}_F, \tag{4.129}$$

and the sound power radiated per unit length of the stud is, for frequencies much less than the critical frequency of the leaves, given by Eq. (2.94):

$$\bar{P} = |\tilde{F}_2|^2\rho_0/4m^2\omega. \tag{4.130}$$

If there are n studs per unit width of the partition, the ratio of transmitted to incident sound power per unit area is, for a sound wave incident at angle ϕ,

$$\tau = 2n\bar{P}\rho_0 c/|\tilde{p}_i|^2 \cos\phi. \tag{4.131}$$

The line force impedance of an infinite thin plate is, by analogy with the beam impedance equation (1.88),

$$\tilde{Z}_F = 2D^{1/4}\omega^{1/2}m^{3/4}(1 + j). \tag{4.132}$$

In general the plate impedance will exceed the stud impedance, except at high audio frequencies when a rigid stud model is inappropriate anyway; hence the stud impedance term may be neglected in the denominator of Eq. (4.128). Subsequent substitution of the plate impedance expression (4.132) into Eqs. (4.128), (4.130), and (4.131) yields the following expression for sound power transmission coefficient associated with the presence of the studs:

$$\tau_s = \frac{2n\rho_0^2 c c_l'}{\sqrt{3}\omega^2\rho_s^2 h \cos\phi}, \tag{4.133}$$

in which ρ_s, c_l', and h are the density, longitudinal wave speed, and thickness of the leaf material, respectively.

It is clearly of interest to compare this expression with that for transmission through a double wall by the mechanism of cavity acoustic coupling between the leaves. Except near the mass–air–mass frequency, the minima in the sound reduction coefficient are associated with resonances of the cavity and are given by Eq. (4.126), with $2m = m_t$. Hence the minimum ratio of stud-transmitted power to air-transmitted power, in the frequency range of cavity resonances, is given approximately by

$$\frac{\tau_s}{\tau_a} = \frac{\tau_s}{(\rho_0 c/\omega m \cos\phi)^2} = \frac{2nc_l' h \cos\phi}{\sqrt{3}c},$$

$$= \frac{0.7nc\cos\phi}{f_c}. \qquad (4.134)$$

This ratio is independent of frequency. Substitution of $\phi = 45^\circ$ and values of n and f_c typical of double-leaf constructions used in building yields maximum values of this ratio on the order of unity. The implication is that the sound reduction index is not likely substantially to exceed that given by the mass law for the total mass of the partition. (This is confirmed by the data in Fig. 87.) This is not entirely unexpected since both the acoustic resonances and studs force the two panels to vibrate with similar velocities, although the radiation characteristics are different in the two cases: the stud excitation causes flexural near-field radiation (and also, in practice, radiation by subsequent scattering of the waves excited by one stud by incidence upon other studs), whereas the acoustic coupling causes supersonic travelling-wave radiation. This difference is responsible for the presence of $\cos\phi$ in the transmission ratio; it arises from the travelling-wave radiation ratio $\sec\phi$. The presence of f_c in the denominator indicates that transmission via studs is more important for partitions having low critical frequencies.

Equation (4.134) gives only an approximate estimate of stud transmission because most of the incident acoustic waves will not be incident in a plane normal to a stud axis, and therefore cannot, in principle, cause stud translation. (Why?) In addition, studs possess stiffness and therefore can resonate, or even exhibit coincidence interactions with the leaves. However, the resulting transmission may not differ greatly from that given by Eq. (4.134) because it is based on an assumption of negligible stud impedance and, in addition, acoustically induced plane flexural waves incident upon a yielding stud at angles other than normal will necessarily produce supersonic bending wavenumbers in a stud. Results of a theoretical analysis of normal incidence sound transmission through a stud partition, shown in Fig. 85, indicate that resonances of the inter-stud leaf panels can also reduce the minimum sound reduction index to that given by the mass law (Lin and Garrelick 1977). This

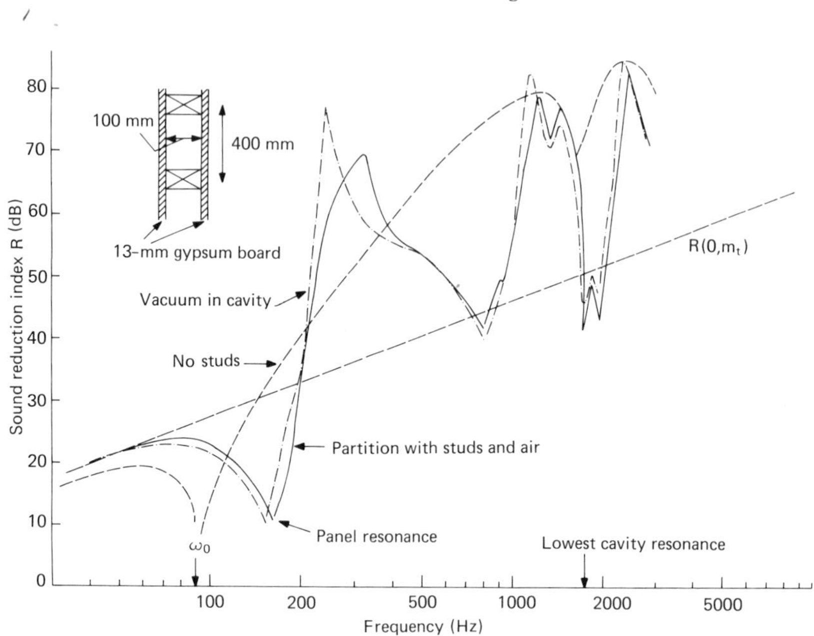

Fig. 85. Theoretical normal-incidence performance of a stud partition (Lin and Garrelick, 1977).

result suggests that damping of the leaves may be important even well below the critical frequency.

Our elementary analysis of transmission due to mechanical line connection between the leaves has yielded a result from which conclusions of practical importance can be drawn. Since the insertion of absorbent into a cavity raises the frequency-average sound reduction index well above the minimum values associated with cavity resonance, the presence of rigid line connections will normally short circuit the absorbent and hence severely degrade the partition performance, except at frequencies near the mass–air–mass resonance frequency. The short-circuiting effect is less serious in the case of leaves with a high critical frequency, as indicated by Eq. (4.134). This is confirmed by the results shown in Fig. 86, in which the improvement in sound reduction index afforded by the introduction of cavity absorbent is shown for two constructions in which the leaf critical frequencies are very different (Ingemansson and Kihlman, 1959).

If a very flexible connection is used between the studs and one of the leaves, the behaviour of the leaf in response to stud movement can be modeled approximately by a base-excited mass–spring system: Elementary theory shows the velocity ratio of base and mass to decrease in proportion

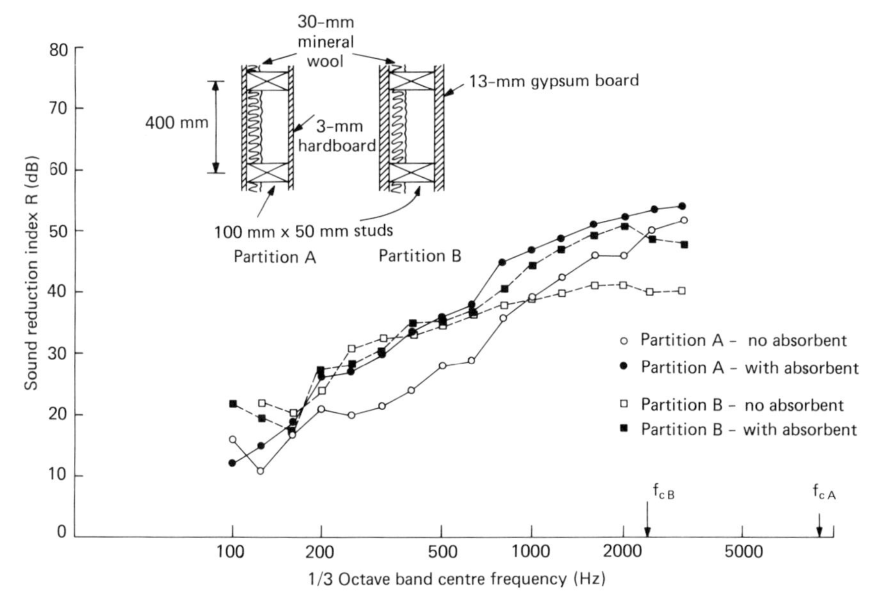

Fig. 86. Improvement made by cavity absorption with leaves of different critical frequency (Ingemansson and Kihlman, 1959).

to the square of frequency above the fundamental stud–connection–leaf resonance frequency. The sound power transmission coefficient is proportional to the square of the velocity ratio and thus to the fourth power of frequency. It is therefore essential that short circuiting of the absorbent cavity by mechanical connection should be reduced by the use of flexible stud connections or by making the studs themselves flexible. For example, it is common in modern stud wall constructions in buildings to use studs formed from 0.5-mm-thick steel sheet, having a cross section so designed as to provide the necessary static stiffness and stability, but to provide a high degree of dynamic flexibility at audio frequencies. Figure 87 shows that the short-circuiting effect is greatly reduced by this form of construction (Northwood, 1970). Useful expressions are given by Gu and Wang (1983).

Improvements to the mid- and high-frequency sound insulation performance of existing building partitions, which is based upon similar principles, can be effected by attaching thin-sheet materials via flexible-point or -line connections (Cremer *et al.*, 1973). As shown by Eq. (4.134) it is important that only very flexible sheets, having low critical frequencies, should be used, in order to minimise near-field radiation; also, the number of connections per unit width should be reduced as far as possible.

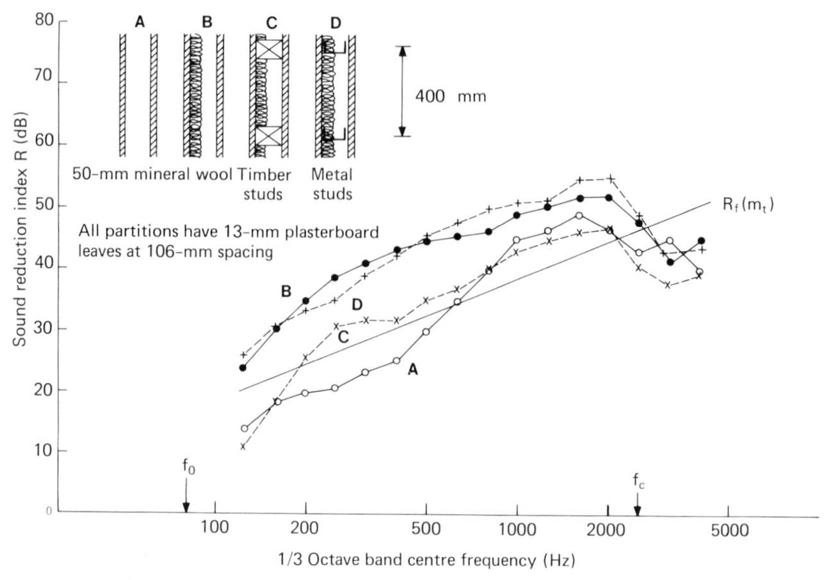

Fig. 87. Influence of stud type on performance (Northwood, 1970).

4.11 Close-Fitting Enclosures

A common method of reducing sound radiation from machinery or industrial plant is partially or fully to cover the radiating surfaces with a sheet of impervious material; such covering is sometimes known as cladding, especially in the case of pipework. The cavity formed between the surface and its enclosure is usually relatively shallow compared with an acoustic wavelength over a substantial fraction of the audio frequency range, and it normally contains sound-absorbent material. Theoretical predictions of the performance of such enclosures have not been conspicuously successful to date, and designers still rely heavily on empirical data. The reasons are threefold: (1) the enclosure and source surfaces are strongly coupled by the intervening fluid, so that the radiation impedance of the source is affected by the dynamic behaviour of the enclosure; (2) the geometries of sources are often such that the cavity wavefields are very complex in form and difficult to model deterministically; and (3) the dimensions of the cavities are not sufficiently large for statistical models of the cavity sound fields to be applied with confidence.

In view of the lack of reliable theoretical treatments of the problem, we shall confine our attention to a simple one-dimensional model that exhibits

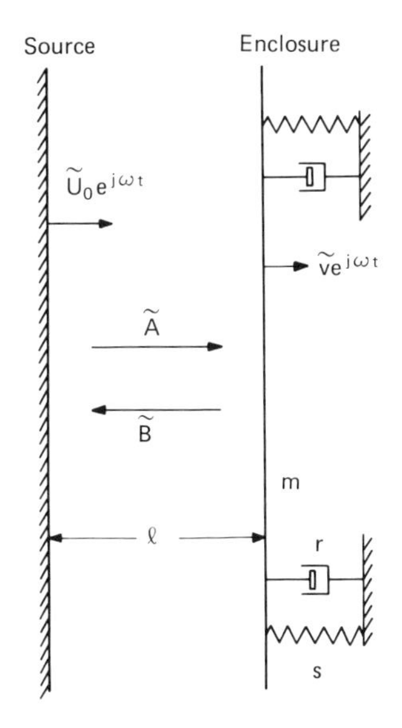

Fig. 88. One-dimensional model of vibrating source and enclosure.

some but not all of the mechanisms that operate in practical cases. The major difference between this model and that of a double partition is that the motion of the primary source surface is assumed to be inexorable, i.e., unaffected by the presence of the enclosure. This assumption is reasonable because the internal impedance of a machinery structure is generally much greater than that of its enclosure.

The model is shown in Fig. 88. The complex amplitude of pressure at the surface of the source is given by

$$\tilde{p}_0 = \tilde{A} + \tilde{B}, \tag{4.135}$$

and the associated particle velocity is given by

$$\rho_0 c \tilde{u}_0 = \tilde{A} - \tilde{B}. \tag{4.136}$$

The pressure in the cavity that drives the panel is

$$\tilde{p}_1 = \tilde{A} \exp(-jkl) + \tilde{B} \exp(jkl), \tag{4.137}$$

and the associated particle velocity, which equals the panel velocity, is given by

$$\rho_0 c \tilde{v} = \tilde{A} \exp(-jkl) - \tilde{B} \exp(jkl). \tag{4.138}$$

Let the specific impedance of the panel that represents the enclosure be represented by

$$\tilde{z}_p = j(\omega m - s/\omega) + r, \tag{4.139}$$

to which must be added in series the radiation impedance, which we assume to be equal to $\rho_0 c$; we denote the total impedance by $\tilde{z}_t$. The equation of motion of the panel is hence

$$\tilde{z}_t \tilde{v} = \tilde{p}_1. \tag{4.140}$$

Substituting for $\tilde{B}$ from Eq. (4.136) and using Eq. (4.137), Eq. (4.140) becomes

$$\tilde{z}_t \tilde{v} = 2\tilde{A} \cos kl - \rho_0 c \tilde{u}_0 \exp(jkl). \tag{4.141}$$

Equations (4.136) and (4.138) allow us to obtain a second equation relating $\tilde{A}$ and $\tilde{v}$:

$$\rho_0 c \tilde{v} = -2j\tilde{A} \sin kl + \rho_0 c \tilde{u}_0 \exp(jkl). \tag{4.142}$$

Hence we can eliminate $\tilde{A}$ in order to relate $\tilde{v}$ and $\tilde{u}_0$:

$$\frac{\tilde{z}_t \tilde{v} + \rho_0 c \tilde{u}_0 \exp(jkl)}{2 \cos kl} = \frac{\rho_0 c \tilde{u}_0 \exp(jkl) - \rho_0 c \tilde{v}}{2j \sin kl}, \tag{4.143}$$

of which the solution is

$$\frac{\tilde{v}}{\tilde{u}_0} = \frac{1}{\cos kl + j(\tilde{z}_t/\rho_0 c) \sin kl}. \tag{4.144}$$

The ratio of sound power radiated by the panel enclosure to that radiated in the absence of the enclosure is

$$\frac{\bar{P}_e}{\bar{P}} = \left|\frac{\tilde{v}}{\tilde{u}_0}\right|^2 = \left\{\left[\cos kl - \frac{\sin kl(m\omega - s/\omega)}{\rho_0 c}\right]^2 + \sin^2 kl\left[1 + \frac{r}{\rho_0 c}\right]^2\right\}^{-1}. \tag{4.145}$$

The *insertion loss* (IL) is actually a logarithmic measure of the difference of sound pressure levels with and without the enclosure. In this one-dimensional case

$$\mathrm{IL} = 10 \log_{10}(\bar{P}/\bar{P}_e) \quad \mathrm{dB}.$$

It may immediately be seen that the insertion loss is zero whenever $\sin kl = 0$, irrespective of the mechanical damping of the enclosure. This occurs at frequencies when the cavity width is equal to an integer number of half-wavelengths; the impedance at the source surface then equals the impedance of the panel plus the radiation impedance, and the panel velocity equals the source surface velocity. This situation is similar to that of the double partition at normal incidence, when the impedances of the two partitions simply add. The difference here is that, according to our assumption, the source surface motion is inexorable, which is equivalent to assuming that the load impedance is very much less than the internal impedance of the source. Hence the panel velocity $\tilde{v}$ equals the source velocity $\tilde{u}_0$, and the presence of the panel has no effect.

The insertion loss also takes minimum values when

$$\tan kl = \frac{\rho_0 c}{m\omega - s/\omega}. \tag{4.146}$$

In practice the lowest frequency at which this occurs is normally such that $kl \ll 1$ and $\tan kl \simeq kl$. Hence

$$\omega_1^2 \simeq \rho_0 c^2/ml + \omega_0^2, \tag{4.147}$$

where $\omega_0^2 = s/m$: the fluid cavity bulk stiffness $\rho_0 c^2/l$ is added to the mechanical stiffness, as in the case of mass–air–mass resonance of the double partition. In this case the insertion loss becomes

$$\mathrm{IL} = 20 \log_{10}(1 + r/\rho_0 c) + 10 \log_{10}(\rho_0 l/m + k_0^2 l^2) \quad \mathrm{dB}, \tag{4.148}$$

where $k_0 = \omega_0/c$, and $k_0 l \ll 1$ because $kl \ll 1$.

It is clearly beneficial to make the *in vacuo* natural frequency of the panel as high as possible because the second term will normally be negative. The specific panel damping factor $r/\rho_0 c$ may be written as $\eta\omega_0 m/\rho_0 c$, and damping is therefore only significant if $\eta \gg \rho_0 c/\omega_0 m$. If the mechanical damping is rather low, the minimum insertion loss is normally negative; more power is radiated from the enclosure at this resonance frequency than from the unenclosed source! How can this be? It has nothing to do with the surface area of the enclosure in comparison with that of the source, because they are equal. The answer is revealed by recalling that the basic expression for acoustic power radiation from a vibrating surface is $\bar{P} = \overline{pv}$, which for a surface vibrating in s.h.m. is $\bar{P} = \frac{1}{2}\mathrm{Re}[\tilde{p}\tilde{v}^*] = \frac{1}{2}|\tilde{v}|^2 \,\mathrm{Re}(\tilde{z}_\mathrm{r})$, where $\tilde{z}_\mathrm{r}$ is the impedance seen by the surface. Since the source vibration is inexorable, the real part of the impedance presented by the fluid plus enclosure must, in this case, exceed that for the unbounded fluid. Reference to Eq. (4.93) indicates that $\mathrm{Re}(\tilde{z}_\mathrm{r})$ for the source is maximised when Eq. (4.146) is satisfied. The resonant behaviour of the enclosure/airspace combination creates high acoustic pressures in the air space. However, not all this power is radiated from the enclosure; some is dissipated by enclosure motion, which is why the enclosure damping is an important factor in controling the minimum insertion loss.

Equation (4.148) clearly indicates that a combination of high stiffness, high damping, and low mass is required for good low-frequency enclosure performance. These requirements are very different from those for good performance of a single-leaf partition at frequencies below the critical frequency, although the maxima in insertion loss correspond to the normal incidence sound reduction index for a panel of twice the mass per unit area.

The mechanical stiffness of the enclosure is only significant, however, if it exceeds the acoustic stiffness of the fluid, i.e., $\omega_0^2 m > \rho_0 c^2/l$. Although increasing the cavity width l will reduce the severity of this insertion loss minimum, it decreases the frequencies of standing-wave resonance in the cavity, and therefore may not always be beneficial, at least in theory.

Higher-frequency minima in insertion loss occur whenever Eq. (4.146) is satisfied, but the values of these minima are greater than that at the lowest resonance frequency, as can be shown by substituting the corresponding values of sin kl in Eq. (4.145). Let tan $kl = \alpha$, then

$$\sin^2 kl = \alpha^2/(1 + \alpha^2).$$

Assuming that these higher resonances occur well above the *in vacuo* natural frequency of the enclosure, then $\alpha \simeq \rho_0 c/\omega m$ and the insertion loss minima are given by

$$\begin{aligned} \mathrm{IL} = {} & 20 \log_{10}(\rho_0 c/\omega m) - 10 \log_{10}[1 + (\rho_0 c/\omega m)^2] \\ & + 20 \log_{10}(1 + r/\rho_0 c) \quad \mathrm{dB}, \end{aligned} \tag{4.149}$$

which can be negative. In fact, the frequencies at which the minima occur are very close to those at which sin kl and therefore IL are zero. The presence of sound-absorbent material in the cavity will improve the insertion loss at these higher resonances but is not likely to be very effective at the lowest resonance frequency given by Eq. (4.147). A theoretical insertion loss curve is shown in Fig. 89.

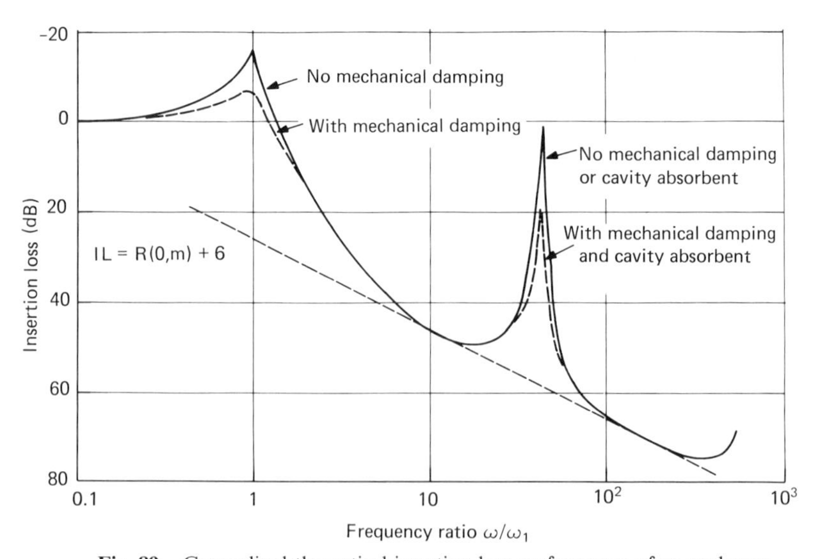

Fig. 89. Generalised theoretical insertion loss performance of an enclosure.

The aforegoing analysis is based upon a very simplistic model of a source and enclosure. A number of attempts have been made to develop more realistic models, particularly with respect to the three-dimensional nature of acoustic fields in real enclosure cavities; unfortunately none have been conspicuously successful (Tweed and Tree, 1978). Low-frequency mass–cavity resonance dips appear in most published results, although the frequencies are often well below 100 Hz and therefore may not be of great consequence. In the case of sources that are small in comparison with the enclosure dimensions, there is evidence of resonance due to the acoustic coupling of opposite walls, which is given by Eq. (4.147) with l as the enclosure half-width dimension. The lower-order cavity standing-wave resonances can be discerned in some results also. In the absence of absorbent material in the cavity the insertion loss of an enclosure above the lowest standing-wave resonance usually rises by about 3 dB per octave, and falls substantially below the sound reduction index of the material when used as a partition. When even a moderate amount of absorbent material (50 mm) is introduced, the insertion loss above the lowest standing-wave resonance frequency generally increases to values close to the field incidence sound reduction index, or even to the normal incidence value in the case of small cavities and pistonlike source motion. In the frequency range encompassing the mass–cavity resonance, and the lowest standing-wave resonance, the one-dimensional theory appears to give fairly accurate predictions of enclosure behaviour.

The insertion loss of an enclosure is severely degraded by non-resilient connection to the vibrating source. Holes in an enclosure are usually necessary and also degrade insertion loss; however, isolated holes only effectively transmit sound of wavelengths smaller than the peripheral length of the hole.

4.12 Transmission of Sound through Stiffened, Composite, and Non-Uniform Panels

For reasons of weight and static stiffness, structural panels are often mounted on frames stiffened by ribs or corrugations or are fabricated as sandwich constructions. The effect of these features on sound transmission can be considerable and is frequently detrimental to the sound reduction index when compared with a uniform, unstiffened panel of the same average mass per unit area.

In considering the transmission of diffuse incidence sound through a bounded panel in Section 4.4, a ratio of the mass-controlled, sub-critical transmission coefficient τ_{∞} to resonant-mode transmission coefficient τ_r was introduced: the ratio τ_r/τ_{∞} is proportional to the square of the radiation efficiency (σ^2) and inversely proportional to the loss factor (η). For most practical, uniform, homogeneous panels below the critical frequency, the loss factor and radiation efficiencies are such as to make the ratio τ_{∞}/τ_r much greater than unity. However, if the radiation efficiency is increased by modification of the plain panel, the ratio will decrease, resonance-controlled transmission may predominate, and the sound reduction index will then fall below the mass-controlled value. A consequence of major practical importance is that the panel damping will, to a greater or lesser extent, control the overall sound reduction index, in contrast to the mass-controlled case.

The effects of support constraints and impedance discontinuities on the radiation efficiency of a panel have been briefly discussed in Chapters 2 and 3. One way of qualitatively understanding the effect of attaching line stiffeners or ribs to a panel is to consider the panel to consist of an assemblage of smaller panels, each bounded by the adjacent ribs. The radiation efficiency of small panels has been shown to exceed that of larger panels of the same material; hence, the radiation efficiency of the assemblage exceeds that of the unstiffened panel. In order for this argument to be fully justified, the panel vibrations on the two sides of a rib should be incoherent (phase unrelated). In practice, vibrational waves are transmitted across such features, and therefore this assumption cannot strictly be correct. An alternative approach, which allows for such transmission, is based upon the derivation of dispersion relationships for bending waves that interact with the ribs. The wave reflections produced by the ribs alter the dispersion relationship in such a way that free waves having wavenumber components of supersonic phase velocity can propagate at frequencies below the uniform-plate critical frequency (Mead, 1975); these components increase the sub-critical radiation efficiency and may cause the panel to be excited in a coincident manner by incident sound waves at frequencies below critical. In practice, the effect on sound transmission is as though the critical frequency had been lowered by one or two octaves, the degree of change being dependent upon the spacing and stiffness (bending and torsional) of the ribs. Figure 90 shows sound reduction index curves of stiffened panels and of their unstiffened counterpart, in which the influence of damping can clearly be seen to be much greater for the former than for the latter (von Venzke *et al.*, 1973).

A common means of increasing the static bending stiffness and buckling resistance of thin panels is to corrugate them. This makes them orthotropic, giving very great differences of bending stiffness in the two orthogonal directions. We note in passing that waves in deeply convoluted plates and shells (e.g., bellows) and in plates with castellated corrugations are not governed

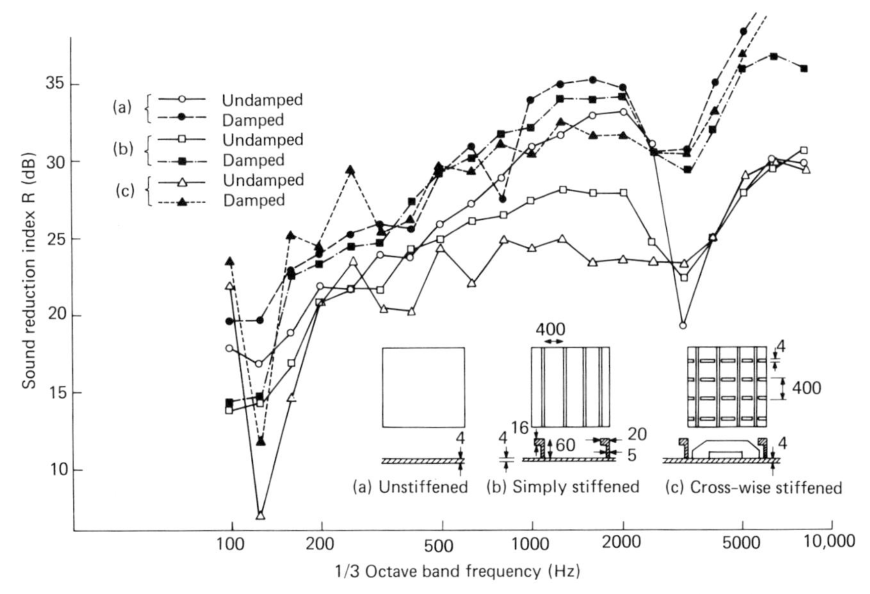

Fig. 90. Effects of stiffening and damping on the sound reduction index of aluminium plates (von Venzke *et al.*, 1973).

by the ordinary plate bending-wave equation. Assuming that an orthotropic bending model is appropriate, it is immediately clear that the critical frequency for bending waves travelling in the direction normal to the corrugations is decreased in comparison to that of a uniform flat plate of equal thickness. It is therefore not surprising that the effect of corrugations is generally to spread the effect of coincidence on sound transmission over a wider frequency range. An analysis of sound transmission through orthotropic plates (Heckl, 1960) yields the following expression for the diffuse-field transmission coefficient:

$$\tau_d \simeq \begin{cases} \dfrac{\rho_0 c}{\pi \omega m} \dfrac{f_{c_1}}{f} \left[\ln\left(\dfrac{4f}{f_{c_1}} \right) \right]^2, & f_{c_1} < f < f_{c_2}, \quad (4.150a) \\[2ex] \dfrac{\pi \rho_0 c}{\omega m} \dfrac{(f_{c_1} f_{c_2})^{1/2}}{f}, & f > f_{c_2}, \quad (4.150b) \end{cases}$$

where f_{c_1} and f_{c_2} are the critical frequencies based upon the maximum and minimum bending stiffness. Figure 91 compares Eq. (4.150a) with some results of sound transmission tests on corrugated panels made by Cederfeldt (1974).

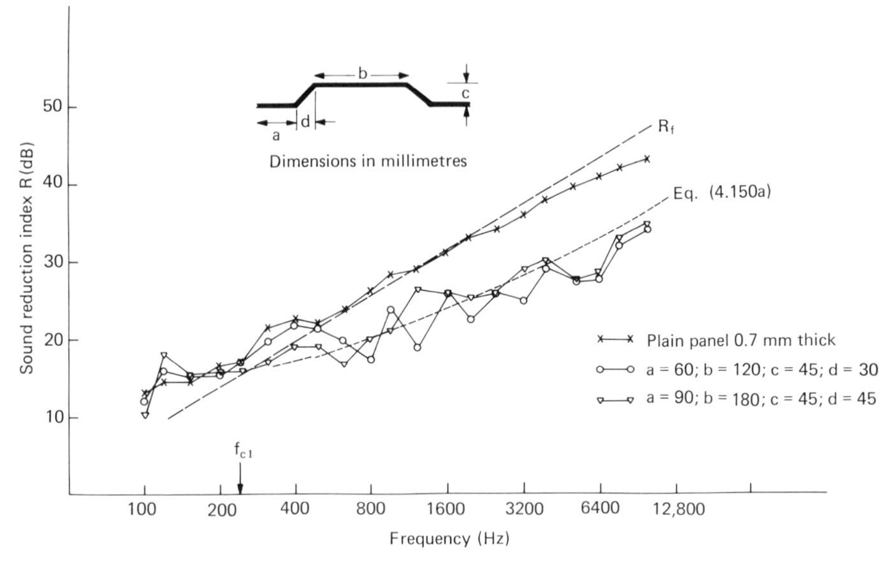

Fig. 91. Sound reduction indices of two corrugated steel panels (Cederfeldt, 1974)

Composite panels frequently take the form of sandwich constructions, especially where a combination of high static strength, stability, and low weight is necessary. Honeycomb panels, which consist of a resin-impregnated paper, plastic, or metal foil core sandwiched between very thin metal or fibre-reinforced plastic face plates, are widely used in aerospace structures. The form of the transverse wave dispersion curve is determined partly by the bending stiffness of the whole construction and partly by the shear stiffness of the core, except at very low frequencies, where the section bending stiffness predominates, and at very high frequencies, where the individual face-plate–bending stiffnesses become significant. A generalised dispersion curve is shown in Fig. 51.

The relationship between the wavenumber of transverse waves in a honeycomb panel and the wavenumbers corresponding to pure bending and to pure shear is (Kurtze and Watters, 1959)

$$1+\left(\frac{k_\mathrm{s}}{k_\mathrm{b}}\right)^2\left(\frac{k}{k_\mathrm{b}}\right)^2-\left(\frac{k}{k_\mathrm{b}}\right)^4-\left(\frac{k_\mathrm{b}}{k_{\mathrm{b_f}}}\right)^4\left(\frac{k_\mathrm{s}}{k_\mathrm{b}}\right)^2\left(\frac{k}{k_\mathrm{b}}\right)^6=0, \qquad (4.151)$$

where k_s is the shear wavenumber in the absence of transverse bending forces, k_b the overall cross-section bending wavenumber in the absence of shear distortion, and $k_{\mathrm{b_f}}$ the wavenumber for faceplate bending alone. With core depth d, shear modulus G, total mass per unit area m, face plate thickness

h ($\ll d$), and elastic modulus E, these wavenumbers are given by

$$k_s^2 = m\omega^2/Gd \qquad \text{(non-dispersive)}, \tag{4.152}$$

$$k_b^4 = m\omega^2/D_1 \qquad \text{(dispersive)}, \tag{4.153}$$

$$k_{b_f}^4 = m\omega^2/2D_2 \qquad \text{(dispersive)}, \tag{4.154}$$

where

$$D_1 = Ed^2h/2(1 - \nu^2) \qquad \text{and} \qquad D_2 = Eh^3/12(1 - \nu^2).$$

Because $h \ll d$, $D_1 \gg D_2$, and $k_b^4 \ll k_{b_f}^4$ for most practical constructions, the last term on the left-hand side of Eq. (4.151) can be neglected, since k_s/k_b and k/k_b are generally of the order of unity in the frequency range of interest. The frequency at which k_s equals k_b, which is shown in Fig. 51, is then given by

$$\omega_{bs}^2 = G^2d^2/mD_1, \tag{4.155}$$

at which frequency

$$k \simeq 1.25k_b = 1.25k_s. \tag{4.156}$$

Below this frequency $k_b > k_s$, and the solution of Eq. (4.151) tends to $k = k_b$, so that shear distortion has little influence; at higher frequencies $k_b < k_s$, and the solution tends to $k = k_s$, so that shear distortion dominates.

The free wavenumber in the shear-controlled region is approximately $k_s = \omega(m/Gd)^{1/2}$. Hence if the phase speed $(Gd/m)^{1/2}$ is less than the speed of sound c, coincidence should not occur until the very high frequencies at which face plate bending becomes significant and the wave becomes dispersive. Hence a honeycomb panel having this property might be expected to exhibit mass-controlled transmission behaviour over the whole of the frequency range of practical interest; in addition, it would be expected to form a good low-frequency enclosure structure because of its combination of low mass and high transverse stiffness at low frequencies.

In practice, a number of other factors combine to affect adversely the sound reduction index of many honeycomb panel constructions. If the shear wave phase speed is not very much lower than the speed of sound, panel boundary effects can significantly raise the radiation efficiency, and resonance-controlled transmission can become important, especially because the inherent mechanical damping of this type of structure is not normally very high. On the other hand, if the shear stiffness is too low, the static stiffness and stability characteristics may be unsatisfactory. Studies of the transmission characteristics of the types of honeycomb structure used in aircraft and helicopters have shown that the construction must be carefully optimised to obtain acceptable performance. Figure 92 shows some typical sound transmission curves for this type of honeycomb panel.

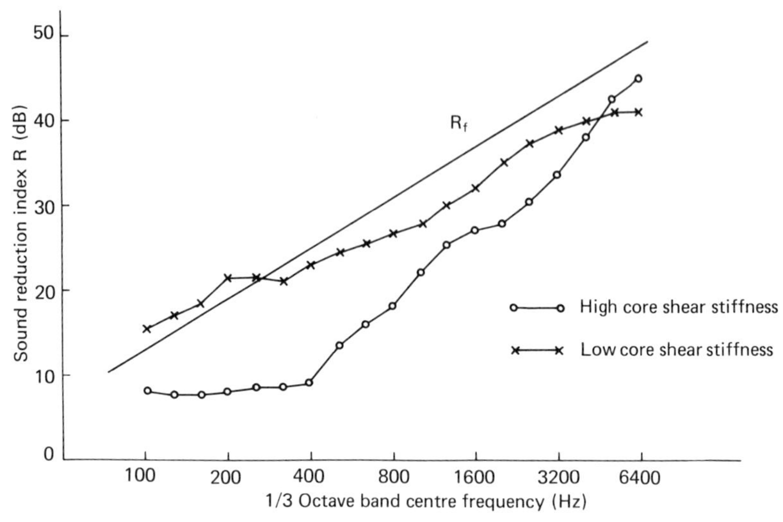

Fig. 92. Measured sound reduction index of honeycomb sandwich panels.

Honeycomb panels usually contain cores that are very stiff in compression normal to the plane of the panel. Sandwich panels containing cores of compressible materials, such as plastic foams, exhibit a transverse, dilatational resonance between the masses of the face plates and the compressional stiffness of the core. This phenomenon is analogous to the mass–air–mass resonance phenomenon of double-leaf partitions, but tends to occur at relatively higher frequencies in practice. The effect is to produce a sound reduction index curve below the resonance frequency that rises very little with frequency and lies well below the mass law line. A typical performance curve, is shown in Fig. 93. Performance data for a variety of sandwich panels containing rigid plastic foam or paper honeycomb cores are presented by Jones (1981). The adverse effects of resonance do not seem to be greatly mitigated by the use of a highly damped core, and such constructions are generally to be avoided. A form of sandwich construction that is widely used for partitions in ships takes the form of steel face plates, a few millimetres thick, separated by a core of "cross-cut" mineral wool having the fibres oriented normally to the plane of the plate. The compressional stiffness is sufficiently high to avoid adverse dilatational effects in the frequency range of interest, but the shear stiffness is low and the damping is moderately high. There is, of course, an upper limit to the core thickness, for which the dilatational effects are not of concern. Einarsson and Söderquist (1982) describe a means of combining a honeycomb sandwich with a plain sheet to provide good insulation performance with minimum thickness.

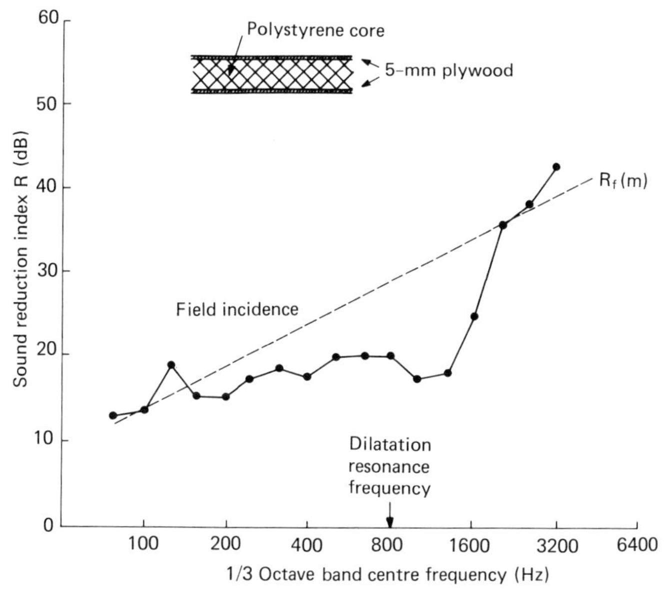

Fig. 93. The effect of dilatation resonance on sound reduction index.

4.13 Transmission of Sound through Thin-Walled, Circular Cylindrical Shells

The mechanisms of sound transmission through thin-walled, circular cylindrical shells are of practical interest in relation to two rather different classes of structure. One class contains large-diameter, very thin walled shells, characteristic of aerospace vehicles and large ducts carrying low-pressure gas flows; the other includes smaller diameter, thicker-walled structures, commonly described as pipes, which are widely used in industrial installations and in fluid transportation networks. In general, it is of concern to keep sound energy from entering the first class of structures and from escaping from those in the second class.

The two features that cause the sound transmission characteristics of thin-walled, circular cylindrical shells to differ from those of flat partitions and panels are the curvature of the walls and the constraint imposed by the cylinder walls on the sound field in the contained fluid. Whereas, in a uniform flat plate, small-amplitude flexural waves and in-plane longitudinal and shear waves are all uncoupled and can propagate independently, the curvature of the walls of a cylinder couples the radial, axial, and tangential motions so that the three equations of motion in these directions contain contributions from all three displacements and from their derivatives (Leissa, 1973). The solutions of these equations represent three classes of propagating wave, which are characterised by the ratios of the displacements u, v, and w in the axial, tangential, and radial directions, respectively. The "flexural" class is characterised by the fact that u/w and v/w are generally much less than unity. This class is of most interest in audio-frequency acoustic-coupling problems because an inviscid fluid can only exchange energy with a shell via radial motion, and also the wave impedances of the flexural class are generally relatively well matched to those of sound waves in the audio-frequency range.

As with the process of sound transmission through flat partitions, the sound transmission characteristics of cylinders are strongly dependent upon the relative phase speeds of waves in the solid and fluid media. When waves propagate within a medium confined by parallel boundaries, the constraints imposed by these boundaries upon the forms of wave motion possible at any frequency lead to so-called waveguide behaviour, of which a commonplace example is observable while sitting in the bath and disturbing the water surface: Multiple reflections at the boundaries transform waves spreading out cylindrically from the source into guided waves propagating in the direction of the waveguide axis; these guided waves involve characteristic (modal) distributions of the wave variable across the waveguide, which can be seen as interference patterns. When considering transmission through flat plates, we did not explicitly consider modal waveguide behaviour in the plates, although it is inherent to modal behaviour of rectangular plates, the boundaries of which form two orthogonal waveguides. Had we chosen to consider a waveguide model formed by a flat strip of uniform width l, located between boundaries that provide simple supports (Fig. 94), Eq. (1.41) could be used, together with the boundary conditions, to yield the dispersion relationship for flexural modes

$$k_{zp}^2 = k_b^2 - k_x^2 = k_b^2 - (p\pi/l)^2 = \omega(m/D)^{1/2} - (p\pi/l)^2, \tag{4.157}$$

where k_{zp} is the wavenumber corresponding to the propagation of the characteristic waveguide modes in which the wave variables take the form of a standing pattern $\sin(p\pi x/l)$ across the width of the strip. [Derive Eq. (4.157).] This relationship is represented qualitatively in Fig. 94. Equation (4.157) shows that real (propagating) solutions exist for each value of p only at frequencies greater than that for which $k_{zp} = 0$. The frequencies at which $k_{zp} = 0$ are the resonance frequencies of a simply supported beam of length l made from the plate material, and they are known as the cutoff frequencies of the waveguide modes of order p. Below its cutoff frequency, a mode cannot effectively propagate wave energy and its amplitude decays exponentially away from the point of excitation.

A cylindrical shell clearly represents a waveguide for waves in both the solid and fluid media. In fact, only coupled waves can propagate in a fluid-

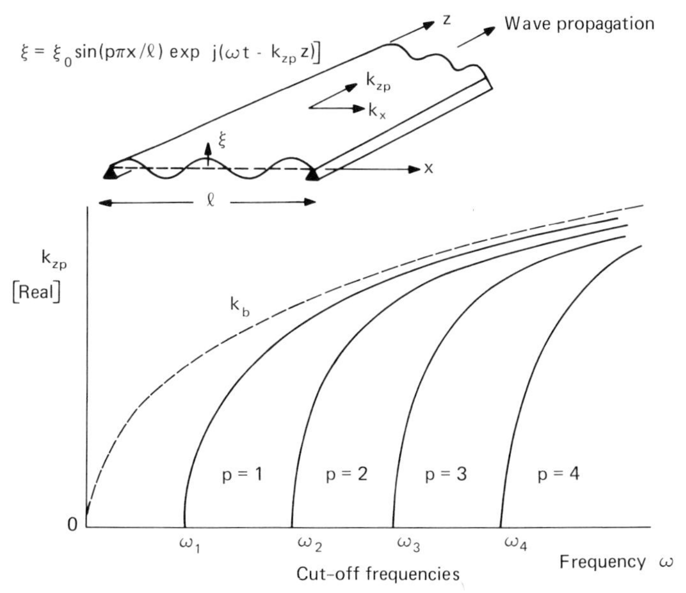

Fig. 94. Waveguide behaviour of a flat-plate strip (simply supported).

filled cylinder, but in cases of gas-filled metal cylinders these waves in general correspond closely to the uncoupled *in vacuo* shell waves and to the rigid-wall duct acoustic waves (Fuller and Fahy, 1981). Because of wall curvature and the requirement that all the kinematic and dynamic variables be continuous around the circumference, radial (and the other two) displacements of the wall produce tangential and axial strains of the median surface of the wall, together with associated stresses (Leissa, 1973). These "membrane stresses" generate radial force components that depend upon the local curvature, in a manner similar to the generation of transverse forces by curvature of a tensioned membrane. [There is, however, an essential difference between these two cases: in a cylinder the tangential membrane stress is, in part, proportional to transverse (radial) displacement, and the associated curvature is determined mainly by the shell geometry; in the case of a tensioned membrane, the stress is produced by externally applied forces and is independent of (small) transverse displacements, and the curvature is proportional to local transverse displacement.] Radial displacements of the shell wall also give rise to bending stresses, in exactly the same manner as in flat plates; the resulting radial forces are independent of shell radius, but strongly dependent upon wall thickness through the second moment of area of the wall cross section. Hence the total radial force is produced by a combination of membrane and bending forces.

4.14 Flexural Wave Propagation in a Circular Cylindrical Shell

Flexural-type waves propagating in a uniform cylindrical shell may be characterised by axial and circumferential wavenumbers k_z and k_s, as shown in Fig. 95. The wavefronts of free propagating waves of wavenumber k_{cs} form a spiral pattern like a barber's pole, the angle between the wavevector and a generator of the cylinder being given by $\theta = \tan^{-1}(k_s/k_z)$. As with the flat strip, characteristic waveguide modes may be formed by the superposition of helically propagating waves having the same axial wavenumber components k_z and oppositely directed circumferential wavenumber components $\pm k_s$. The closure of the shell in the circumferential coordinate direction requires that the wave variables be continuous around the circumference and that the characteristic circumferential patterns take the form ${}^{\sin}_{\cos}(k_s s)$, where $s = a\phi$ and $k_s = n/a$ $(n = 0, 1, 2, \ldots, \infty)$, so that an integer number of complete circumferential wavelengths, $\lambda_s = 2\pi/k_s = 2\pi a/n$, fit around the circumference: this is unlike the flat strip, for which an integer number of *half*-wavelengths is necessary. The wavenumber relationship corresponding to Eq. (4.157) is

$$k_z^2 = k_{cs}^2 - k_s^2 = k_{cs}^2 - (n/a)^2, \tag{4.158}$$

where k_{cs} is the wavenumber of the two propagating helical wave components.

The two most important cylindrical-shell parameters are the non-dimensional frequency $\Omega = \omega a/c_l' = \omega/\omega_r$ and the non-dimensional thickness parameter $\beta = h/\sqrt{12}a$, where h is the wall thickness, a the mean radius of the shell, c_l' the longitudinal wavespeed in a plate of the shell material, and $\omega_r = c_l'/a$ (rad s^{-1}) the ring frequency. The physical significance of these

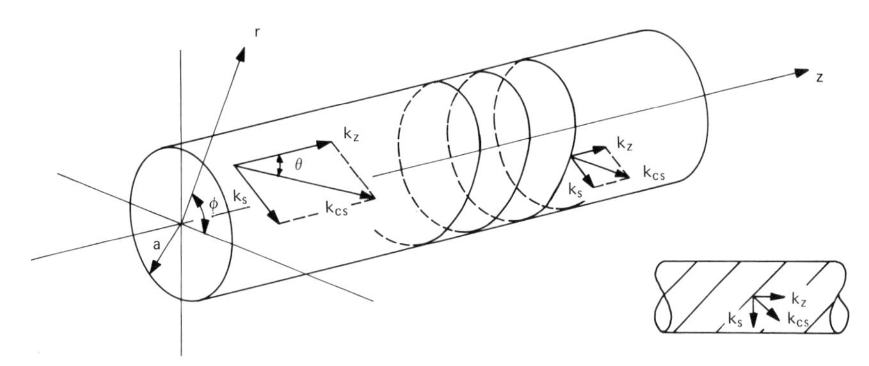

Fig. 95. Cylindrical-shell coordinates and wavenumbers.

parameters is as follows: The ring frequency ω_r is the lowest frequency at which an $n = 0$ axisymmetric-mode resonance can occur. The longitudinal wavelength in the shell wall equals the shell circumference when $\Omega = 1$, and an $n = 0$ breathing, or hoop, resonance occurs at this frequency: note that ω_r is dependent upon cylinder radius, but not on wall thickness. Later we shall see that the ring frequency separates frequency regions in which wall curvature effects are more, and less, important. Large-diameter shells, such as aircraft fuselages, have ring frequencies of the order of 300 Hz, whereas the ring frequencies of industrial pipes tend to be in the range 1–5 kHz.

The free flexural wavenumber in a flat plate of thickness equal to that of the shell wall is

$$k_b = \omega^{1/2}(m/D)^{1/4} = \omega^{1/2}(12)^{1/4}/h^{1/2}c_l'^{1/2}. \tag{4.159}$$

This wavenumber can be expressed in a non-dimensional form involving the cylindrical-shell parameters Ω and β and the radius a as

$$k_b a = \Omega^{1/2}\beta^{-1/2}. \tag{4.160}$$

This non-dimensional wavenumber forms a useful reference value in the subsequent analysis.

A form of bending wave that dominates the low-frequency vibration behaviour of slender cylinders, e.g., pipelines, is the beam-bending wave. There is no cross-sectional distortion and the circumferential variation of radial displacement corresponds to the order $n = 1$. The beam-bending wavenumber is that derived for bars in Section 1.6:

$$k_{bb} = \omega^{1/2}\,(m'/EI)^{1/4}, \tag{4.161}$$

in which $m' = 2\pi ah\rho$ and $I = \pi a^3 h$. The non-dimensional form equivalent to that in Eq. (4.160) is

$$k_{bb}a = (2)^{1/4}\Omega^{1/2} \simeq 1.2\Omega^{1/2}, \tag{4.162}$$

which does not contain β and is therefore independent of wall thickness, unlike $k_b a$.

There are many thin-shell equations of varying degree of complexity, the differences arising largely from differences in the assumed strain-displacement relationships (Leissa, 1973). For the purposes of introducing the reader to those features of cylindrical shell behaviour that cause the sound transmission characteristics to differ markedly from those of a flat plate, frequency and dispersion relationships derived from relatively simple forms of shell equations will be used.

One approximate form of thin-shell equation (Heckl, 1962b) yields the following flexural-wave dispersion relationship between the non-dimensional axial wavenumber $k_z a$ and the non-dimensional frequency Ω, for given non-dimensional circumferential wavenumber $k_s a = n$:

$$\Omega^2 = (1 - \nu^2)\{(k_z a)^2/[(k_z a)^2 + n^2]\}^2 + \beta^2\{[(k_z a)^2 + n^2]^2 + [n^2(4 - \nu) - 2 - \nu]/2(1 - \nu)\}, \tag{4.163}$$

in which ν is Poisson's ratio. This expression is accurate for thin shells ($\beta^2 \ll 10^{-1}$) and for values of $n \geqslant 2$. The first term on the right-hand side of the equation is associated with membrane strain energy and the second, which contains β^2, is associated with strain energy of wall flexure. The cross-sectional resonance frequencies, or cutoff frequencies of an infinitely long cylinder, are given by Eq. (4.163) with $k_z = 0$, which corresponds to an infinite axial wavelength. It is most important to observe that these frequencies are determined purely by strain energy of wall flexure; they correspond to Rayleigh's inextensional mode frequencies, which were derived by assuming that the median surface of the shell wall did not strain. Evaluation of the resulting equation for Ω^2,

$$\Omega_n^2 = \beta^2 n^4\left[1 - \frac{1}{2}\left(\frac{1}{1 - \nu}\right)\left(\frac{4 - \nu}{n^2} - \frac{2 + \nu}{n^4}\right)\right], \tag{4.164}$$

yields the approximate values for the cutoff frequencies of the lower-order modes which are presented in Table V.

It is seen from Table V that, except for the lowest-order modes, the cutoff frequencies are reasonably well approximated by thc formula

$$\Omega_n \simeq \beta n^2. \tag{4.165}$$

Equation (4.165) indicates that the number of cutoff or cross-sectional resonance frequencies below the ring frequency is given approximately by

$$n_r \simeq \beta^{-1/2}. \tag{4.166}$$

TABLE V

Cutoff Frequencies of Flexural Modes of Thin-Walled Circular Cylindrical Shells

Circumferential mode order n	$\Omega_n/\beta n^2$	Ω_n/β
2	0.67	2.68
3	0.85	7.65
4	0.91	14.56
5	0.95	23.75
6	0.96	34.56
7	0.97	47.10

The dispersion relationship, Eq. (4.163), may be expressed in graphical form in three ways, depending upon which of the non-dimensional variables $k_z a$, $k_s a = n$, or Ω forms the curve parameter. As already stated, the relationship between the internal acoustic and shell wavenumbers is of crucial importance in determining the coupling of the media; this is true also for the interaction with external fluid, and therefore a two-dimensional wavenumber diagram with Ω as the parameter is found to be useful. Experimental evidence suggests that the level of excitation of long pipes by sound in a contained fluid is largely determined by an axial wavenumber coincidence phenomenon, involving internal fluid and shell waveguide modes. In this case, curves of axial wavenumber versus Ω, for given n, are revealing: both types of curve are presented below.

In order to illustrate the form of the shell wavenumber diagram, Eq. (4.163) is simplified by the omission of the less-important flexure term; this omission will not significantly distort the qualitative picture of the mechanisms of sound transmission discussed below. In Fig. 96, the loci of the non-dimensional $k_z \sim k_s$ relationship are presented for constant values of Ω: the radius vector represents the shell wavenumber k_{cs}, the angle θ indicating the direction of component wave propagation relative to a generator. The particular non-dimensional form of wavenumber is chosen so that the curves are universal, as the simplified form of Eq. (4.163) shows:

$$\frac{(k_z a\beta^{1/2})^4}{[(k_z a\beta^{1/2})^2 + (n\beta^{1/2})^2]^2} + [(k_z a\beta^{1/2})^2 + (n\beta^{1/2})^2]^2 \simeq \Omega^2. \qquad (4.167)$$

Also shown in Fig. 96 are the constant-frequency (Ω) loci for a flat plate of the same thickness and material as the cylinder walls. The broken line corresponds to equality of the first (membrane) term and second (bending) term in Eq. (4.167). It may therefore be considered to enclose the region in which membrane effects are predominant and in which the loci for $\Omega < 1$ correspond approximately to the straight-line forms appropriate to a membrane wall cylinder of vanishing wall thickness ($\beta \to 0$) in which bending effects are negligible: consider $k_z a \sim k_s a$ for fixed Ω in Eq. (4.163), when $\beta = 0$. One striking effect of membrane stresses is to exclude the possibility of purely axial propagation below the ring frequency.

Although Fig. 96 is universally applicable, it must be realised that the number of shell waveguide modes that can propagate in the frequency range below the ring frequency is dependent upon the shell thickness parameter, through Eq. (4.165), and that there is really not a continuum of circumferential wavenumbers, because of the requirement for continuity of the distribution of wave variables around the circumference. Hence only the intersection points between vertical lines drawn at $n\beta^{1/2}$ and the loci are physically significant. The number of shell modes that are substantially affected by membrane

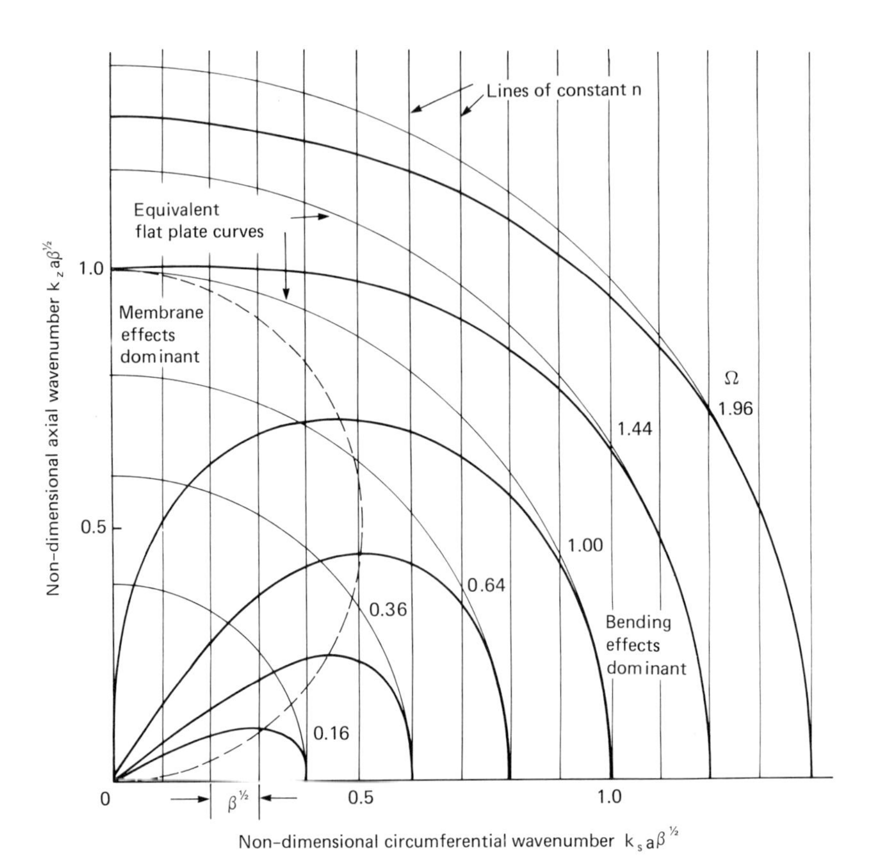

Fig. 96. Universal constant-frequency loci for flexural waves in thin-walled cylindrical shells ($n > 1$).

effects decreases with increase in the wall thickness, as would be expected. From Fig. 96 we may take the formula $n_m \beta^{1/2} = 0.5$ as an indication of this number: alternatively, we can assume that any mode with a cutoff frequency below $\Omega = 0.25$ is subject to membrane effects; for example, if $h/a = 0.003$, $n_m = 17$, and if $h/a = 0.05$, $n_m = 4$.

An alternative form of Fig. 96 is shown in Fig. 97, in which the axial wavenumbers corresponding to particular values of n are plotted against Ω: the curves correspond to cross plots along the vertical lines, corresponding to fixed values of n in Fig. 96. The effects of membrane stresses in increasing the curves correspond to cross plots along the vertical lines corresponding with the equivalent flat-strip (pure-bending) curves plotted on the same figure.

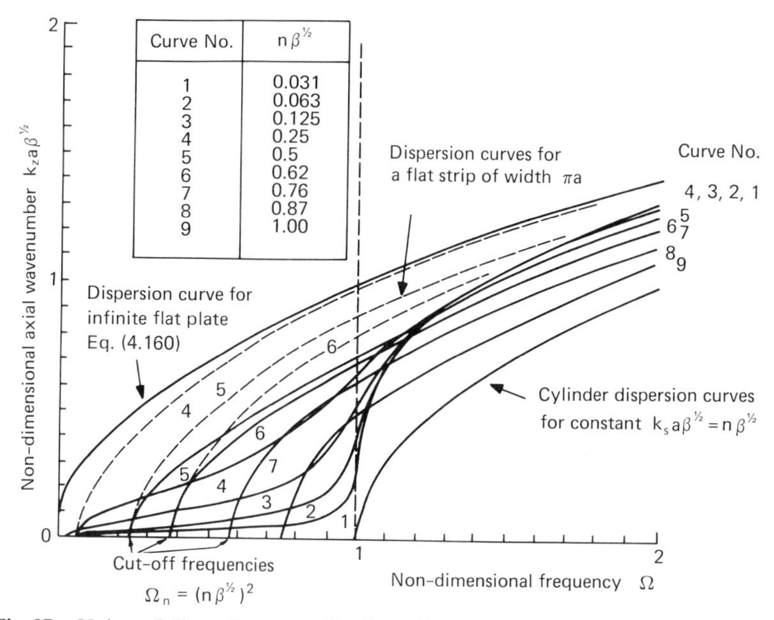

Fig. 97. Universal dispersion curves for flexural waves in thin-walled cylindrical shells.

4.15 Coupling between Shell Modes and Acoustic Duct Modes

The acoustic coupling between the fluid contained in a cylinder and the shell is very much dependent upon the relative axial phase speeds (or axial wavenumbers) of the waveguide modes in the two media. Acoustic duct theory (Morse, 1948) shows that the modes proper to a rigid-walled cylindrical waveguide take the form

$$p_{mn}(r, \phi, z) = \tilde{P}_{mn} {}^{\cos}_{\sin}(n\phi) J_n(k_r r) \exp(-jk_z z), \tag{4.168}$$

where J_n is a Bessel function and the radial wavenumber k_r is determined by the zero normal-particle wall boundary condition as solutions of the equation $[J_n'(k_r r)]_{r=a} = 0$. The characteristic solutions of the equation are multi-valued for given n and therefore they are superscripted as k_r^{np}, in which n indicates the number of diametral pressure nodes and p the number of concentric circular pressure nodes (Fig. 98).

The radial and axial wavenumbers satisfy the acoustic wave equation

$$k_z^2 + (k_r^{np})^2 = k^2. \tag{4.169}$$

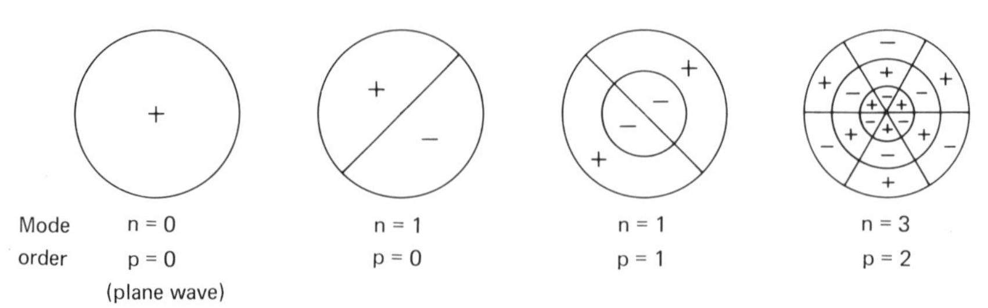

Fig. 98. Cross-sectional distribution of pressure phase and nodal surfaces of acoustic modes of a circular cylindrical waveguide.

The cutoff frequencies below which a particular mode cannot propagate freely and carry energy in an infinitely long duct are given by

$$k_z^2 = k^2 - (k_r^{np})^2 = 0, \tag{4.170a}$$

or

$$ka = k_r^{np}a. \tag{4.170b}$$

Values of $k_r^{np}a$ are given in Table VI. In terms of the ring frequency of the shell, Eq. (4.170b) may be expressed as

$$\Omega_{np} = (k_r^{np}a)(c_i/c_l'), \tag{4.171}$$

where c_i is the speed of sound in the contained fluid. For instance the lowest cutoff frequency is given by

$$\Omega_{10} = 1.84c_i/c_l', \tag{4.172}$$

TABLE VI

Cutoff Frequencies of Acoustic Modes of Hard-Wall Ducts of Circular Cross Section[a]

	p				
n	0	1	2	3	4
0	0	3.83	7.02	10.17	13.32
1	1.84	5.33	8.53	11.71	14.86
2	3.05	6.71	9.97	13.17	16.35
3	4.20	8.02	11.35	14.59	17.79
4	5.32	9.28	12.68	15.96	19.20
5	6.92	10.63	13.99	17.31	20.58
6	7.50	11.73	15.27	18.64	21.93
7	8.58	12.93	16.53	19.94	23.27
8	9.65	14.12	17.77	21.23	24.59

[a] Values of $k_r^{np}a$ are tabulated: see Eq. (4.170b).

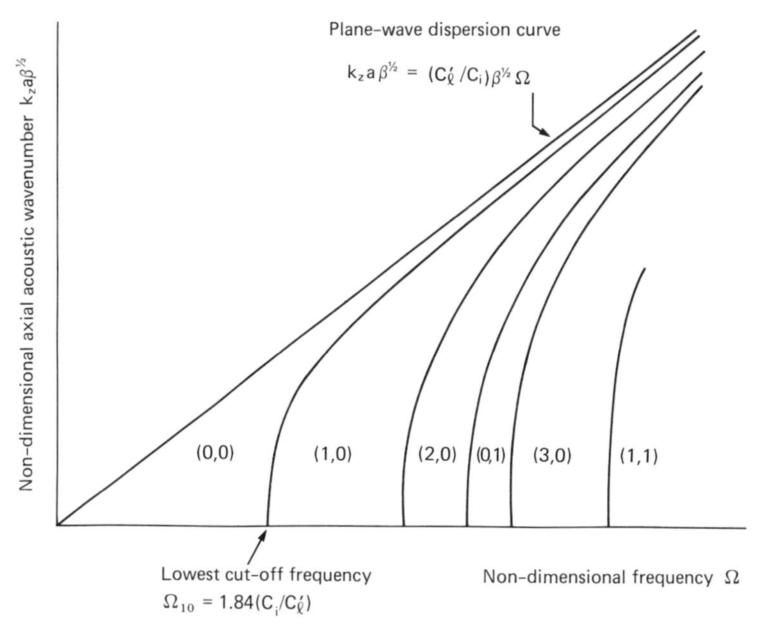

Fig. 99. Dispersion curves for circular cylindrical acoustic waveguide.

which for steel or aluminium alloy, and air at room temperature becomes

$$\Omega_{10} = 0.125. \tag{4.173}$$

The dispersion relationship [Eq. (4.169)] may be represented graphically in the same form as Fig. 97 for the shell modes (Fig. 99): the appropriate values of β and c_i/c_l' must be employed, because unfortunately there is no universal form of combined structural and acoustic wavenumber diagram.

Strictly speaking the acoustic duct modes and the shell wall modes do not exist independently in a fluid-filled duct (see Chapter 6) but in cases of coupling between metal ducts and low-pressure gases the coupled modes resemble closely their uncoupled components and for the purposes of approximate analysis, and qualitative understanding of the coupling characteristics, they may be assumed to retain their uncoupled characteristics. Hence the dispersion diagrams for the shell waves and the fluid waves may be superposed as shown in Fig. 100: only waves of equal circumferential order n may couple. Figure 100 shows that equality of fluid and shell axial wavenumbers can occur; this is a *coincidence* condition.

In most practical cases, coincidence between the lower-order shell modes and the acoustic modes of low radial order p occurs at frequencies close to the acoustic mode cutoff frequencies because membrane effects keep the

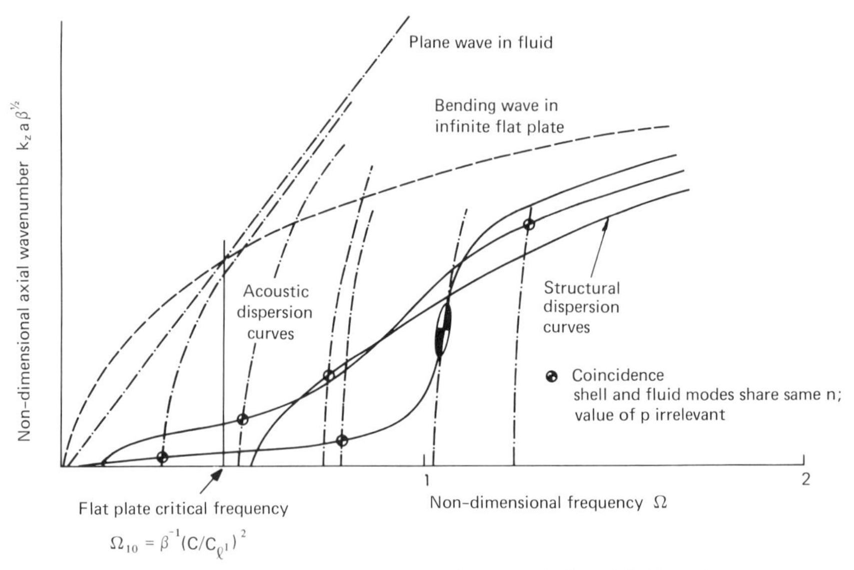

Fig. 100. Illustration of coincidence between shell and fluid modes.

slope of the structural wavenumber curve low, whereas the acoustic wavenumber curves rise rapidly with Ω. The lowest possible coincidence frequency, corresponding to coincidence between the beam-bending mode and the (1, 0) acoustic mode, corresponds closely to the cutoff frequency of this mode, which is given by Eq. (4.172).

Between this frequency and the ring frequency, a number of further such coincidences can occur; any one shell mode can be coincident with all the acoustic modes of equal circumferential order (n) and increasing radial order (p). However, the accuracy of equating coincidence frequency to the corresponding acoustic mode cutoff frequency decreases as membrane effects become less dominant, i.e., as n increases and/or as the ring frequency is approached. Figure 97 indicates that the low $n\beta^{1/2}$ structural wavenumber curves rise rapidly in the vicinity of $\Omega = 1$, eventually to become pure bending wavenumber curves. Acoustic and structural wavenumber curves tend to run nearly parallel, and multiple coincidence between two curves is possible. The high density of coincidences appears to be responsible for the reduction in transmission loss normally observed with thin-walled cylinders around the ring frequency (Fig. 101). In thicker-walled shells, the wavenumber rise described above is not so steep, and a ring frequency dip is not so apparent.

In order to distinguish between shell waves of supersonic and subsonic wavenumber k_{cs} at any frequency, for the purpose of considering sound radiation into the *external* fluid, acoustic wavenumber loci corresponding

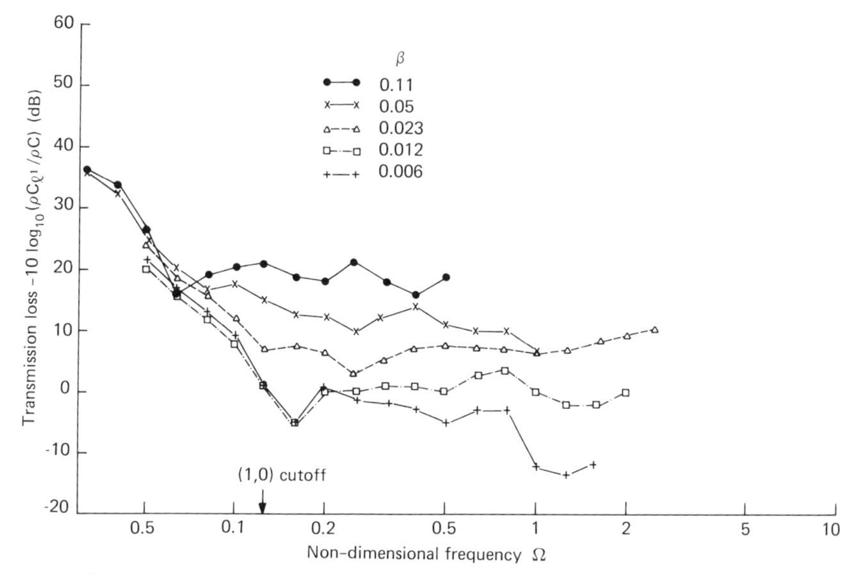

Fig. 101. Normalised transmission loss of pipes (Holmer and Heymann, 1980).

to any value of Ω may be superimposed upon the structural wavenumber diagram (Fig. 96) in the form where Ω is the parameter, as shown in Fig. 102. Because the external fluid is considered to extend to infinity, it is assumed that any direction of acoustic propagation is possible: note that the speed of sound c_0 in the external fluid is used.

It is clear from Fig. 102 that the effect of membrane stress, in turning the constant-frequency loci toward the origin, is to increase the phase velocity of waves having wavenumber combinations falling into the membrane-dominated region, so that supersonic shell modes, having $k_{cs} < k$, can propagate at frequencies below the equivalent flat-plate critical frequency ω_c. By contrast, below ω_c all *flat-plate* wavevectors describe a circular locus of radius larger than that of the circular acoustic-wavevector locus, because $k_b > k$. At and above the ring frequency the membrane effects become increasingly weak, because circumferential membrane resonance occurs at $\Omega \simeq 1$, and the shell wave loci suddenly "unwind" in the region $n\beta^{1/2} \leqslant 0.5$, and tend toward platelike loci as frequency increases further.

If the ring frequency is well below the critical frequency, the unwinding of the loci at $\Omega \simeq 1$ pulls them completely outside the circular acoustic-wavevector locus, and all the shell waves become subsonic: only when the critical frequency is reached do all the shell waves become supersonic, as in flat plates. On the other hand, if the ring frequency is well above the critical frequency, all shell waves will already be supersonic before unwinding occurs

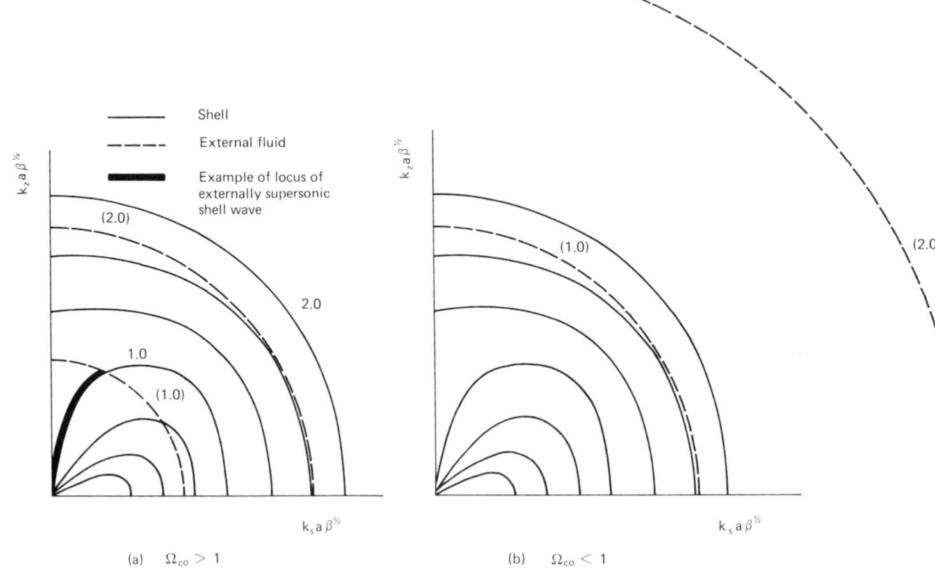

Fig. 102. Shell and external fluid wavevector loci for $\Omega_{co} > 1$ and $\Omega_{co} < 1$.

and will remain so as frequency increases. The non-dimensional form of the critical frequency is

$$\omega_{co}/\omega_r = \Omega_{co} = \beta^{-1}(c_0/c_l')^2, \tag{4.174}$$

and hence the two regimes are divided as follows:

$$\Omega_{co} \begin{cases} < 1, & \beta > (c_0/c_l')^2, \quad (4.175a) \\ > 1, & \beta < (c_0/c_l')^2. \quad (4.175b) \end{cases}$$

For air at 20°C, $(c_0/c_l')^2 \simeq 4.6 \times 10^{-3}$: values of β typical for aircraft fuselages in air satisfy condition (4.175b) whereas industrial pipes tend to satisfy condition (4.175a).

4.16 Transmission Characteristics

Theoretical and experimental analyses of sound transmission through partitions show that in the frequency ranges where coincidence can occur it controls the magnitude of the sound transmission coefficient. We have seen that, in practice, coincidence is generally possible between cylindrical-shell modes and the acoustic modes of a contained fluid below the shell ring frequency, although the incidence of coincidence decreases with increase of the speed of sound in the fluid [Eq. (4.171)]. Above the ring frequency this form of "internal" coincidence may or may not occur, depending upon the ratio of critical frequency (based upon the contained fluid sound speed) to ring frequency, given by

$$\omega_{c_i}/\omega_r = \Omega_{c_i} = \beta^{-1}(c_i/c_l')^2. \tag{4.176}$$

If Ω_{c_i} is considerably greater than unity, internal coincidence is not possible between the ring frequency and the critical frequency ω_{c_i}, in the range given by

$$1 < \Omega < \beta^{-1}(c_i/c_l')^2. \tag{4.177}$$

In the case of an aircraft fuselage made of aluminium alloy, with a diameter of 5 m and a wall thickness of 3 mm, the ring frequency is 320 Hz and the internal critical frequency is approximately 4 kHz. However, an industrial steel pipe of 500-mm diameter and 13-mm wall thickness, which contains air at 20°C, has a ring frequency of 3.2 kHz and an internal critical frequency of 1 kHz.

The transmission loss is also dependent upon the external radiation efficiency of the shell waves. If the fluids inside and outside the cylinder have the same speed of sound, shell waves excited by internal coincidence necessarily have radiation efficiencies close to unity, and their very high axial phase velocities (low axial wavenumbers) produce radiated fields having predominantly radial wavevectors corresponding to almost two-dimensional radiation. If the speed of sound in the contained fluid exceeds that in the external fluid, as is very often the case, this characteristic is even more pronounced. Hence, by the principle of reciprocity, explained in Chapter 5, coincidence-controlled transmission into a cylindrical shell below the ring frequency is likely to occur only when waves are incident at almost 90° to the cylinder axis. As the ring frequency is approached, the range of angles for which coincidence transmission is possible widens as the shell axial wavenumbers increase.

The predominant form of transmission above the ring frequency depends upon both the internal and external non-dimensional critical frequencies Ω_{c_i} and Ω_{co}, which in turn depend upon β, c_i/c_l', and c_0/c_l'. If both Ω_{c_i} and Ω_{co} are less than unity, supercritical flat-plate transmission theory should apply, although it should be remembered that the result depends upon the assumed angular distribution of incident acoustic intensity. It is likely that the energy of acoustic fields in a contained fluid that is excited by sources internal to the cylinder is concentrated in the many modes that are excited well above their cutoff frequencies, and therefore have axial wavenumber close to the acoustic wavenumber based upon c_i. Hence flat-plate transmission theory for near-grazing waves is likely to be more relevant than diffuse-field theory, and the transmitted intensity will then be concentrated in the axial direction: stiffness control will predominate.

In measurements of the sound transmission characteristics of circular cylindrical-shell structures, a transmission coefficient equal to the ratio of power radiated per unit surface area of shell to power passing axially through unit area of cross section has been used (Holmer and Heymann, 1980). The associated transmission loss index has been found to be characterised by a "floor" between the lowest internal coincidence frequency $\Omega_{10} = 1.84 c_i / c_l'$ and the ring frequency $\Omega = 1$, in which coincidence minima can be discerned (Fig. 101). These minima have been examined in detail by Bull and Norton (1982), who also discuss the effect of variation of β and of mean internal flow on coincidence. With cylinders of very small β, an absolute minimum in the index in the region of the ring frequency is observed, which is not so apparent in cylinders of higher β, as expected from our earlier discussion. Above the ring frequency the transmission loss index is observed to rise fairly rapidly, as flat-plate theory would suggest, although there is as yet insufficient quantitative evidence to establish the frequency dependence.

Below the lowest coincidence frequency, experimental data show a rapidly increasing transmission loss index as the frequency falls. The physical mechanisms responsible for this behaviour have not yet been firmly established. If an infinitely long, uniform axisymmetric model is assumed, transmission by stiffness-controlled breathing mode response to plane acoustic waves in the contained fluid produces a transmission index far higher than that measured, unless elaborate precautions are taken to prevent the excitation of resonant modes of higher circumferential order. The latter may be generated by direct mechanical excitation, by evanescent modes in the fluid in the vicinity of the source, or by non-axisymmetric features such as bends or circumferential variations of wall thickness (Rennison, 1977). The inference is that in practical systems beam bending and higher modes are largely responsible for the observed "transmission." Beam-bending waves in an infinitely long pipe can radiate acoustic power above their critical frequency, which is given by Eq. (4.162) as

$$\omega a / c_0 = k_{bb} a = 1.2 \Omega^{1/2}$$

or

$$\Omega = 1.44 (c_0 / c_l')^2, \tag{4.178}$$

which is independent of pipe diameter.

As an example, this frequency for a 300-mm-diameter steel pipe in air at 20° is 17 Hz. Well above this frequency the axial phase velocity is so high compared with the speed of sound in the fluid that the pipe may be assumed to translate uniformly. The corresponding radiation efficiency is given by Eq. (2.110) for $k_0 a \ll 1$ as

$$\sigma = \tfrac{1}{2}\pi(k_0 a)^3. \tag{4.179}$$

Measurements of radiation efficiency on industrial pipes below Ω_{10} suggest that this is a more reasonable assumption than that based upon uniform cylinder pulsation, which hoop-stiffness–controlled response would produce. However, the laboratory studies discussed in Chapter 2 indicate that higher circumferential mode resonances can reduce the radiation efficiency below that given by Eq. (4.179). Quantitative theoretical estimates of the transmission loss index below Ω_{10} are therefore unavailable at the time of writing.

Quantitative theoretical estimates of the transmission index in the floor region above Ω_{10} are also of uncertain accuracy, and heavy reliance must be placed on empirical data for the moment. However, the aforegoing qualitative discussion of the coincidence mechanism suggests that the floor level is likely to fall in proportion to the logarithm of density (in frequency) of the possible incidence of coincidence and rise in proportion to the logarithm of the square of the mass per unit area of the cylinder wall (6 dB per doubling of mass). Any effects of mean flow are most likely to influence the transmission process by varying the coincidence density below the ring frequency and by affecting the flat-plate transmission process above the ring frequency. It might be expected that structural damping would significantly affect any coincidence-controlled transmission mechanism, as it does in flat plates. The very limited experimental evidence available indicates some effect of damping (Holmer and Heymann, 1980; Rennison, 1977); however, theoretical studies suggest that damping may not control near-coincidence transmission in practical systems (Bull and Norton, 1982).

In this discussion of transmission through cylindrical shells the role of resonances of modes of finite-length pipes has not been mentioned. It is not clear at present whether they need to be explicitly considered. Flat-plate transmission theory at supercritical frequencies, in which coincidence effects play a dominant part, seems to work quite well, although plate resonances are not explicitly considered. The modal density of a simply supported shell of length l can be obtained from Fig. 97, as indicated in Chapter 1, by drawing horizontal lines at values of k_z given by $m\pi/l$. Intersections of lines of integer values of m and n indicate modal natural frequencies. At frequencies below the ring frequency, membrane effects on low-n modes are seen to reduce the modal density to values below that of a flat plate of area equal to the shell area $2\pi al$. In the vicinity of the ring frequency, the relaxation of membrane effects produces a peak in the modal density. At higher frequencies the modal density asymptotes to the flat-plate value. Further research is necessary to elucidate the influence of modal density on transmission.

Problems

4.1 A piston of 100-mm diameter and 0.02-kg mass is elastically mounted in a rigid co-planar baffle so that it has an *in vacuo* natural frequency of 50 Hz. By considering the normal incidence of plane waves upon the piston, together with the scattered and transmitted fields (fields radiated due to piston vibration), derive an equation of motion for the piston. Hence evaluate the sound power transmission coefficient in air as a function of frequency between 10 and 500 Hz. Compare the sound reduction index of the piston with the mass law and explain the physical reasons for the differences. Hint: Use small-*ka* approximations for fluid loading and radiated field. How will τ change with angle of incidence of the excitation field?

4.2 Determine the magnitude of the loss factor necessary to reduce the resonant transmission coefficient of the system described in 4.1 by a factor of ten.

4.3 The sound of an over-flying aircraft is assumed to fall as plane waves on the surface of a deep lake. Determine the sound-pressure level, relative to the incident-field sound-pressure level, produced at a water depth of 500 mm by the 1-kHz component of the incident sound, when the aircraft is at an angle of elevation of 72°. Is it appropriate to evaluate an intensity transmission coefficient τ in this case?

4.4 A plane sound wave in air is obliquely incident upon a large horizontal metal sheet. The temperature of the sheet is raised by some means to higher than ambient temperature. Qualitatively investigate the influence of increase of temperature on the sound transmission coefficient (a) at sub-critical frequencies and (b) at super-critical frequencies.

4.5 A simply supported, baffled, rectangular plate has an elastic modulus of 10^9 Nm^{-2}, a density of 700 kg m^{-3}, and a thickness of 4 mm. The dimensions of the plate are 1 × 1.2 m. Determine the value of the plate loss factor that makes the resonance-controlled transmission coefficient for 500-Hz octave band diffuse incidence sound in air approximately equal to the mass-controlled transmission coefficent. Equation (4.56) applies. Also try Eq. (4.65).

4.6 Evaluate δ_1 in Eq. (4.81) for a double-glazed window construction consisting of two panes of 6-mm-thick glass mounted in a 1500 × 700 mm frame at a separation distance of 19 mm. Also evaluate the mass–air–mass resonance frequency of this construction.

4.7 Using the model of a double-leaf partition illustrated in Fig. 79, derive an expression for the ratio of leaf displacements $\tilde{\xi}_1/\tilde{\xi}_2$ at the mass–air–mass resonance frequency in terms of the leaf mass ratio m_1/m_2.

4.8 By incorporating a resistance term in the one-dimensional fluid momentum equation and deriving a modified wave equation, demonstrate that the attenuation constant α is approximately related to the flow resistivity r by $\alpha \simeq r/2\rho_0 c$.

4.9 Compare the transverse force line impedances of the leaves and studs of typical lightweight double-leaf building partitions. You are expected to seek the necessary information concerning materials and dimensions from appropriate sources.

4.10 By reference to the simple enclosure model illustrated in Fig. 88 and the associated analysis, find an expression for the ratio of power radiated by the enclosed source to power radiated by the enclosure, in terms only of the specific damping ratio of the enclosure. Hint: Estimate the power dissipated by the enclosure damping. Show that, for practical values of enclosure damping, the maximum value of high-frequency IL is equal to the normal incidence sound reduction index for a panel of twice the mass per unit area of the enclosure.

4.11 Evaluate β for circular cylindrical shells typical of industrial pipes, ventilation ducts, and aircraft fuselages. Determine the ratio of diameter to wall thickness of a steel cylinder that makes the ring frequency equal to the critical frequency in air at 20°C. Repeat the calculation for 100°C.

5 Acoustically Induced Vibration of Structures

5.1 Practical Aspects of Acoustically Induced Vibration

Any solid structure exposed to a sound field in a contiguous fluid medium will respond to some degree to the fluctuating pressures acting at the interface between the two media. It is common experience that heavy road traffic can produce visually observable low-frequency displacements in window panes of houses adjacent to a highway; concert goers are particularly thrilled when they can actually feel, through their seats, the response of the auditorium structure to a fortissimo passage; viewers at air displays frequently experience strong chest cavity vibration when a low-flying pilot turns on the afterburners; and heavy rock devotees love the body vibrations induced by the very high powered loudspeaker systems used by performing groups.

The process of transmission of airborne sound energy through partitions involves vibrational response to sound pressures acting on one side, together with the consequent radiation of sound from the other side. This process, which has been described and analysed in Chapter 4, generally involves extremely small transverse displacement amplitudes of the partitions, which therefore do not constitute a threat to the integrity of the structures. There are, however, cases of considerable practical significance in which acoustically induced vibration levels are of such a magnitude as to require the engineer to assess the likely effects on the structural and operational integrity of the system concerned. Acoustically induced fatigue failure first came to prominence in the 1950s owing to the high levels of jet noise to which aircraft structures were exposed; in the 1960s aerospace engineers became concerned with the high levels of vibration induced in spacecraft structures and on-board equipment packages due to rocket noise at launch; and acoustically induced fatigue failure of nuclear reactor and industrial plant components came to light at about the same time.

Other examples of acoustically induced vibration of practical significance include the degradation of receiving sonar array performance by the response of the surrounding hull structure to the incoming sound field; low-frequency, airborne sound-induced vibration of complete building structures by sources such as large reciprocating compressors; vibration of the tank tops (engine support structures) of ships by airborne noise radiated by the underside of the engines; and the process of absorption of sound by structures in buildings, such as lightweight ceilings, glazing, and panelling. The mechanisms of structural vibration caused by explosive blasts and by sonic booms are qualitatively similar, but these airborne disturbances differ from those considered in this book because they are of finite amplitude and are therefore non-linear in nature.

It will be shown in this chapter that the characteristics of response of a structure to incident sound are closely linked to its sound radiation characteristics. This feature is the result of an extremely important fundamental property of sound sources and sound fields, which is expressed in the *principle of reciprocity*, together with its various ramifications. This principle will be discussed only briefly, and in a special case, in this chapter. Similarly, only brief mention will be made here of the general equation that relates sound fields in fluid volumes, and the conditions on bounding surfaces of arbitrary geometry, which is known as the Kirchhoff–Helmholtz integral equation (see Section 3.2). In many cases, a simpler form of this equation, applicable to plane bounding surfaces and known as the Rayleigh integral, Eq. (3.1), is sufficiently accurate for practical purposes: further treatment of the general integral equation is contained in Chapters 6 and 7.

5.2 Decomposition of a Sound Field

Before entering upon any mathematical analysis of the response of structures to sound, it is helpful to discuss certain fundamental principles of the interaction process in qualitative terms. Consider a rigid body of arbitrary shape upon which a sound field is incident. At points on the surface of the body the sound field particle velocity normal to the surface must be zero. This condition may be satisfied by imagining the surface of the body to undergo a form of motion such that its normal surface velocity is equal and opposite to that which would exist in the sound field in that normal direction, but in the absence of the body. The unobstructed incident sound field and that "radiated" by the body in its imagined motion are then superimposed: the latter field component is correctly termed the "scattered" field. This form of thinking is helpful because we already have some physical feel for the nature of sound fields radiated by vibrating bodies.

In the simplest case, imagine a plane sound wave normally incident upon an infinitely extended rigid plane surface. We know that the pressure at the surface of a uniformly vibrating infinite-plane surface, on the side to which positive velocity is directed, is $p(t) = \rho_0 c v(t)$. Hence the total pressure on the rigid surface is

$$p_{\mathrm{bl}}(t) = p_{\mathrm{i}}(t) + \rho_0 c[p_{\mathrm{i}}(t)/\rho_0 c] = 2p_{\mathrm{i}}(t), \tag{5.1}$$

in which p_{bl} symbolises the total surface pressure in the obstructed, or blocked, sound field, and p_{i} the pressure at the position of the surface in the unobstructed sound field. Now imagine that a simple harmonic plane sound wave is incident at an angle ϕ to the rigid surface. In the absence of the surface, the incident-wave particle velocity component normal to position of the absent surface takes the form

$$u_{\mathrm{n}}(x, t) = (\tilde{p}_{\mathrm{i}}/\rho_0 c) \cos \phi \exp[j(\omega t - kx \sin \phi)], \tag{5.2}$$

in which x is the coordinate in the plane of the surface. We now imagine the surface to be present and vibrating with normal surface velocity $(-u_{\mathrm{n}})$. Analysis in Chapters 2 and 3 of the radiation from a vibrating-plane surface shows that the wave impedance presented by a fluid to such a travelling wave is [Eqs. (2.35) and (3.38)]

$$\tilde{z}_{\mathrm{wf}} = \rho_0 c/[1 - (k \sin \phi/k)^2]^{1/2} = \rho_0 c \sec \phi. \tag{5.3}$$

Hence the pressure on the rigid surface takes the form

$$p_{\mathrm{bl}}(x, t) = 2\tilde{p}_{\mathrm{i}} \exp[j(\omega t - k \sin \phi x)] = 2p_{\mathrm{i}}(x, t). \tag{5.4}$$

We see therefore that "pressure doubling" occurs for all angles of incidence, except for 90° when the incident wave is not obstructed.

Imagine now that the surface is not rigid but has uniform elastic and inertial properties. In this special case, the spatial distribution of response of the surface must necessarily match that of the incident pressure field, i.e., it has a surface wavenumber $k_x = k \sin \phi$. The dynamic properties of the surface may be expressed in terms of a structural wave impedance $\tilde{z}_{\mathrm{ws}}(k_x, \omega)$. The surface pressure cannot now be equal to $2p_i(x, t)$, because the surface moves under the action of surface pressure and therefore produces an additional scattered field, which we call the radiated field. Now, the blocked pressure field corresponds to zero motion of the surface, and hence it is valid to superpose the blocked field and the radiated field i.e., that produced by the absence of motion plus that produced by motion. As far as the structure is concerned, it moves in response to the total surface pressure that arises from the sum of these two field components. Hence we may write a surface response equation as

$$\tilde{v}(k_x, \omega)\tilde{z}_{ws}(k_x, \omega) = -\tilde{p}_{bl}(k_x, \omega) - \tilde{p}_{rad}(k_x, \omega), \tag{5.5}$$

in which positive v is directed out from the structure into the fluid, and the fluid is assumed to exist only on one side of the structure.

As in the case of our imagined surface motion, $\tilde{p}_{rad}(k_x, \omega)$ may be expressed in terms of the surface velocity and the fluid wave impedance:

$$\tilde{v}(k_x, \omega)\tilde{z}_{ws}(k_x, \omega) = -\tilde{p}_{bl}(k_x, \omega) - \tilde{v}(k_x, \omega)\tilde{z}_{wf}(k_x, \omega). \tag{5.6}$$

The structural and fluid wave impedances are seen to add to form a fluid-loaded structure wave impedance, introduced in Chapter 3 [Eq. (3.43)]. The response equation may be rewritten as that of the fluid-loaded structure to the blocked-pressure field

$$\tilde{v}(k_x, \omega)[\tilde{z}_{ws}(k_x, \omega) + \tilde{z}_{wf}(k_x, \omega)] = -\tilde{p}_{bl}(k_x, \omega). \tag{5.7}$$

Although this equation has been developed here specifically in relation to an infinite plane surface of uniform properties exposed to an incident plane wave, in which case both the scattered and radiated fields have a unique wavenumber component k_x, the principles involved apply to any flexible surface of any geometric form. However, it is often difficult to evaluate the appropriate forms of $\tilde{z}_{ws}$ and $\tilde{p}_{bl}$ in cases where the structure is of irregular geometry and non-uniform dynamic properties. In such cases the scattered and vibrational fields do not possess a unique wavenumber even if the blocked field does, and total response must be evaluated by including the contributions of all possible values of k_x, by inverse Fourier transformation of Eq. (5.7). An application of such an analysis is presented in Section 5.3.

One valuable conclusion that can be drawn from our consideration of the components of the total presssure field acting on the surface of an elastic structure is that, if we can reasonably accurately assess the influence of a contiguous fluid on the natural frequencies and mode shapes of a structure, then we may apply the estimated blocked-pressure field to the fluid-loaded structure. The influence of fluid loading on structural vibration characteristics has been discussed in Chapter 3. In many practical cases the fluid-loaded structural mode shapes are altered little from their *in vacuo* forms, even in the presence of a dense fluid such as water, and reasonable estimates can be made of the reduction of natural frequencies by inertial fluid loading.

It should not be assumed that the blocked pressure on the surface of a structure is always close to twice the incident pressure in the absence of the body. This is particularly not the case for bodies that have one or two typical dimensions small compared to an acoustic wavelength, as consideration of the equivalent radiated fields readily shows. For example, consider a long rigid circular cylinder of radius a exposed to plane sound waves incident at 90° to the cylinder axis. If it is assumed that the acoustic wavelength greatly exceeds the cylinder circumference ($ka \ll 1$), then the scattered-field component can be equated approximately to that radiated by the cylinder in undergoing uniform translational vibration with a velocity amplitude $\tilde{u}$ equal and opposite to that in the incident wave. (Why is the small-ka condition necessary for this assumption?) The "scattered" surface pressure is given, correct to first order in ka, by Eq. (3.30) as

$$(\tilde{p}_s)_{r=a} = j\rho_0 cka\tilde{u} \cos\theta, \tag{5.8}$$

where θ is the azimuthal coordinate of the cylinder measured relative to the plane of motion. Now $\tilde{u} = -\tilde{p}_i/\rho_0 c$. Therefore, the ratio of amplitudes of scattered surface pressure to incident pressure is given approximately by

$$|\tilde{p}_s|_{r=a}/|\tilde{p}_i| \simeq ka \cos\theta \ll 1. \tag{5.9}$$

The pressure in the incident wave at the location of the cylinder surface, but in the absence of the cylinder, is given approximately by

$$(\tilde{p}_i)_{r=a} \simeq \tilde{p}_i(1 + jka \cos\theta). \tag{5.10}$$

Thus, although the scattered pressure is small in comparison with the incident pressure, the total force per unit length due to the blocked pressure is actually double that which is produced by the integration of the unobstructed incident field component around the circumference of the cylinder. (Check this conclusion yourself.)

5.3 Response of a Baffled Plate to Plane Waves

The following analysis of the response of a finite plate in a rigid baffle to incident plane waves serves to illustrate the phenomenon of scattering of sound by reflection from a surface of non-uniform specific acoustic impedance, as well as the modal-analysis approach to the problem of estimating the response of flexible structures to sound. It is not always realised that scattering of incident waves caused by spatial variations of surface impedance is at least equal in practical importance to that due to irregularity of surface geometry, particularly when the surface can respond in strongly resonant modes. A practical example of the exploitation of this phenomenon is in the installation of low-frequency resonant panel (membrane) absorbers in recording and broadcasting studios; not only do these panels absorb low-frequency energy and thereby control undesirable room acoustic resonances, but they also scatter incident sound energy into many directions and improve the diffusion of the sound fields in a way that cannot be achieved by shaping the room surfaces because of their small dimensions in comparison with the wavelength of the sound to be controlled.

The following analysis is of a two-dimensional problem because the complexity of the expression is somewhat less than that for a three-dimensional system; however, the principles are the same. Consider an infinitely long, thin, uniform elastic plate of width a set between two semi-infinite rigid-plane baffles; the whole system lies in one common plane (Fig. 103). The edges of the plate are assumed to be simply supported in order to simplify the modal expressions; this is not, however, a necessary condition. A simple harmonic plane wave, of frequency ω and wavenumber k, is assumed to be incident on the plane at an angle ϕ to the normal.

As we have seen, the blocked pressure is equal to twice the incident pressure at the surface of the plane. The transverse normal velocity v of the plate is expressed in terms of a summation over its *in vacuo* normal modes: the primary object of the analysis is to find the velocity amplitudes of these modes.

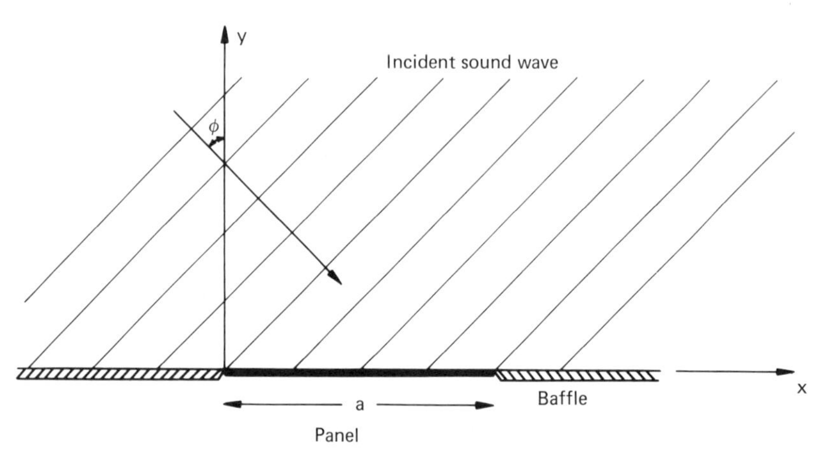

Fig. 103 Panel excited by an incident plane wave.

Let

$$v(x,t) = \sum_p \tilde{v}_p \sin(p\pi x/a) \exp(j\omega t). \tag{5.11}$$

In order to derive an expression for the field radiated by the plate we may express the transverse velocity of the whole plane surface in terms of an infinite set of simple harmonic travelling waves of wavenumber k_x by the Fourier integral transform pair, as in Chapters 2 and 3:

$$\tilde{V}(k_x) = \int_0^a \sum_p \tilde{v}_p \sin(p\pi x/a) \exp(-jk_x x)\, dx, \tag{5.12a}$$

$$v(x) = \frac{1}{2\pi} \int_{-\infty}^{\infty} \tilde{V}(k_x) \exp(jk_x x)\, dk_x. \tag{5.12b}$$

Now the radiated field can be obtained by using the relationship between pressure and surface velocity in the field radiated by a single travelling-wave component [Eqs. (2.34) and (2.47)]:

$$\tilde{P}(k_x, y) = \tilde{V}(k_x)(\omega\rho_0/k_y) \exp(-jk_y y), \tag{5.13}$$

in which $k_y = \pm(k^2 - k_x^2)^{1/2}$, the choice of sign being decided upon physical grounds, as explained in Section 2.5. The radiated-field component expressed by Eq. (5.13) corresponds physically to a plane wave radiated at angle $\phi(k_x) = \cos^{-1}(k_y/k)$, for surface wavenumbers satisfying the condition $k_x < k$, and to a near field, which decays exponentially with distance from the plane, for $k_x > k$. The reason that k_x takes an infinite range of values is that the plate modes exist over only a finite region of the plane, the solution to Eq. (5.12a) for $\tilde{V}_p(k_x)$ corresponding to vibration in a single mode p, being given by Eq. (2.45) as

$$\tilde{V}_p(k_x) = \tilde{v}_p \frac{(p\pi/a)[(-1)^p \exp(-jk_x a) - 1]}{k_x^2 - (p\pi/a)^2}. \tag{5.14}$$

Figure 36 illustrates the dependence of $|\tilde{V}_p(k_x)|^2$ on $k_x/(p\pi/a)$. Equations (5.13) and (5.14) effectively express the scattering behaviour of one vibration mode of the plate; a wave of single $k_x = k \sin\phi$ is incident, and plane waves having k_x within the range $-k < k_x < k$ are radiated at all angles, in the direction opposite to the x-wise component of the incident wavevector as well as in the same direction. It should not, however, be concluded that sound energy is necessarily radiated symmetrically about the normal, because the incident field will force the relative phases of the modal responses to take a certain form that will influence the directional characteristics of the field radiated by all modes when vibrating simultaneously.

The plate responds to the total acoustic-pressure field on its surface, which is the sum of the blocked pressure (twice the incident pressure) and the radiated field. From Eqs. (5.12a), (5.13), and (5.14) the wavenumber spectrum of the radiated field at the surface is given by

$$\tilde{P}(k_x)_{y=0} = \sum_p \tilde{V}_p(k_x)(\omega\rho_0/k_y), \tag{5.15a}$$

and the surface pressure distribution due to modal motion is given by the inverse transform [Eq. (3.53)]

$$\tilde{p}(x, 0) = \frac{1}{2\pi}\int_{-\infty}^{\infty} \sum_p \tilde{V}_p(k_x)(\omega\rho_0/k_y)\exp(jk_x x)\,dk_x. \tag{5.15b}$$

The total surface pressure is equal to the blocked pressure plus the radiated pressure given by Eq. (5.15b). Hence the generalised force per unit length on mode m is given by

$$\tilde{F}_m = \int_0^a 2\tilde{p}_i \sin(m\pi x/a)\exp(-jk\sin\phi\, x)\,dx + \frac{1}{2\pi}\int_0^a\left[\int_{-\infty}^{\infty}\sum_p \tilde{V}_p(k_x)(\omega\rho_0/k_y)\exp(jk_x x)\,dk_x\right]\sin(m\pi x/a)\,dx. \tag{5.16}$$

The result of the first integral, which gives the blocked modal force, is similar in form to Eq. (5.14) with p replaced by m and k_x replaced by $k \sin \phi$. However, unlike k_x, which varies between $-\infty$ and $+\infty$, $k \sin \phi$ is limited to the range between $-k$ and $+k$, the limits corresponding to grazing incidence and zero corresponding to normal incidence. Hence the blocked force can only achieve the peak value at $k \sin \phi = \pm p\pi/a$ if $|k| \geq p\pi/a$, which for resonant vibration

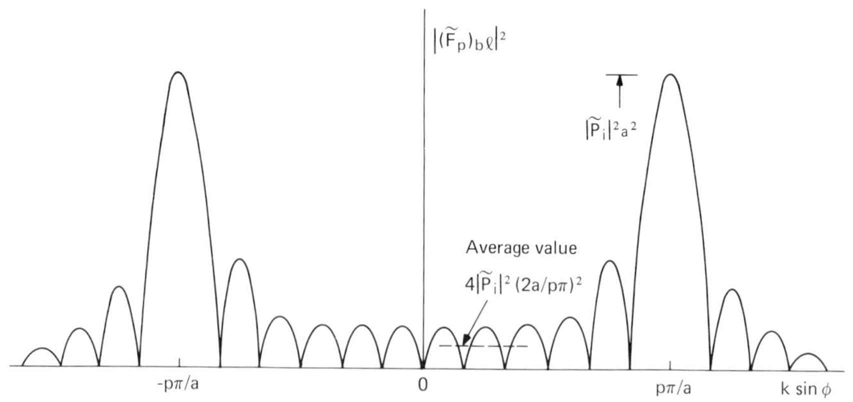

Fig. 104 Wavenumber modulus spectrum of blocked modal force produced by an incident plane sound wave.

means that the modal natural frequency exceeds the critical frequency (Fig. 104). Hence the blocked modal force can peak for lower-order modes at frequencies above their natural frequencies; it requires only a sufficiently high excitation frequency to satisfy the above condition.

This condition is reminiscent of that encountered in Section 2.5 in connection with radiation from plate modes: only the components of $\tilde{V}_p(k_x)$ with $-k < k_x < k$ can radiate sound energy to the far field, and only those same components can be directly excited by incident plane waves. Resonant vibration of all modes having natural frequencies above the critical frequency is most strongly excited by waves incident at angles corresponding to the angles of maximum energy radiation by those modes at resonance. Resonant vibration of all other lower-order modes is excited most strongly by waves close to grazing incidence, with the exception of the fundamental mode, for which the blocked force peaks at $k \sin\phi = 0$.

The panel-mode shape $\sin(m\pi x/a)$ in Eq. (5.16) may be expressed by the Fourier integral transform pair

$$\sin m\pi x/a = \frac{1}{2\pi}\int_{-\infty}^{\infty} \tilde{A}_m(k'_x)\exp(jk'_x x)\,dk'_x,$$

where

$$\tilde{A}_m(k'_x) = \int_0^a \sin(m\pi x/a)\exp(-jk'_x x)\,dx.$$

Thus Eq. (5.16) may be written as

$$\begin{aligned}\tilde{F}_m &= 2\tilde{p}_{\mathrm{i}}\tilde{A}_m(k\sin\phi)\\ &\quad + \frac{1}{4\pi^2}\int_{-\infty}^{\infty}\left\{\left[\int_{-\infty}^{\infty}\sum_p \tilde{V}_p(k_x)(\omega\rho_0/k_y)\exp(jk_x x)\,dk_x\right]\right.\\ &\quad \times\left.\left[\int_{-\infty}^{\infty}\tilde{A}_m(k'_x)\exp(jk'_x x)\,dk'_x\right]\right\}dx, \qquad (5.17)\end{aligned}$$

in which the range of integration over x has been extended from 0 to a to $-\infty$ to ∞ because the form of $\tilde{A}_m(k'_x)$ ensures that the modal displacement does not extend outside the panel. Now, since the modal displacement function is real, $\tilde{A}(-k'_x) = \tilde{A}^*(k'_x)$, and use of the identity

$$\int_{-\infty}^{\infty}\exp[j(k_x - k'_x)x]\,dx = 2\pi\delta(k_x - k'_x)$$

allows Eq. (5.17) to be reduced to

$$\tilde{F}_m = 2\tilde{p}_{\mathrm{i}}\tilde{A}_m(k\sin\phi) + \frac{1}{2\pi}\int_{-\infty}^{\infty}\sum_p \tilde{V}_p\tilde{A}_p(k_x)\tilde{A}_m^*(k_x)(\omega\rho_0/k_y)\,dk_x, \qquad (5.18)$$

in which

$$\tilde{A}_n(\alpha) = \frac{(n\pi/a)[(-1)^n \exp(-j\alpha a) - 1]}{\alpha^2 - (n\pi/a)^2}.$$

Evaluation of the integrals in Eq. (5.18) is left as an exercise for advanced students.

This result indicates qualitatively that each mode m is subject to pressures generated by vibration in all the other modes p through the summation term, which arises because every mode makes a contribution to each k_x component of $\tilde{P}(k_x)_{y=0}$. Physically this represents cross-modal coupling between modes through the scattered-pressure field. The cross-coupling term will normally represent a mixture of reactive and resistive loading through the different forms of k_y in the ranges $-k \le k_x \le k$, $k_x < -k$, and $k_x > k$ [Eqs. (2.35) and (2.36)]. The total response of the structure is formally given by

$$v(x, t) = \exp(j\omega t) \sum_m \tilde{F}_m \sin(m\pi x/a)/\tilde{Z}_m,$$

where $\tilde{Z}_m$ is the *in vacuo* modal impedance.

The general solution for the multi-mode response and scattering by a plate is clearly rather complicated to evaluate, even in this relatively simple case. However, if the modal density of the structure is rather low, it may be reasonable to neglect inter-modal coupling and remove the summation sign from Eq. (5.18), letting $m = p$. The second integral in Eq. (5.18) then represents the combination of modal radiation damping and reactive fluid loading appropriate to each mode at its resonance frequency. Fortunately, reactive fluid loading generally alters mode shapes little from their *in vacuo* forms, even if substantially lowering modal natural frequencies. An example of a general scattering analysis for a thin rod is presented by Liamshev (1958).

The degree of excitation of any mode by an incident plane wave, and the consequent scattering of the wave energy into angles different from the specular direction (rigid-plane surface reflection), is clearly strongly related to the radiation characteristics of the mode. The actual magnitude of resonant modal vibrations is controlled not only by the blocked-pressure force, but by the modal mass and total damping, the latter including radiation damping. The strength of resonant modal scattering increases as the total damping decreases.

The two-dimensional analysis can, of course, be generalised to any arrangement of structure and fluid. However, the blocked-pressure distribution and modal-radiation characteristics may be difficult to evaluate in cases of complex geometry and it may be necessary to employ numerical analyses for the purpose.

5.4 Applications of the Principle of Reciprocity

Our preceding discussions of structural response to incident sound, and its consequent scattering by radiation, have hinted at the existence of a fundamental relationship between the acoustically induced response and radiation characteristics of structures. In the following sections we investigate the origin and nature of the relationships that do in fact exist.

As a preliminary to dealing with flexible bodies immersed in fluids it is helpful to consider further the generation of blocked-acoustic-pressure distributions on rigid bodies by incident sound. The general principle of point-to-point acoustic reciprocity is presented in detail in "Theory of Sound" by Lord Rayleigh (1896). In essence the principle states that the acoustic pressure produced at one point A in a fluid by a simple harmonic point source of volume velocity at another point B in the fluid is the same as the pressure produced at B by the same source located by point A, irrespective of the presence of arbitrary (dynamically linear) boundaries to the fluid. We now imagine one of the points to be adjacent to the rigid surface of a body in the fluid. A point volume velocity source located immediately adjacent to the body surface will produce an acoustic pressure at an observation point in the fluid. On interchanging the source and observation points the same (blocked) pressure will be produced on the body surface.

The elementary source of volume velocity at the body surface can equally well be created by the vibration of a very small element of the body surface. Now we know that the total acoustic pressure produced in a fluid by the vibration of the surface of a body can, by the principle of superposition, be evaluated by integrating over the surface of the body the product of the strength of elemental sources on the surface times an appropriate transfer function between the source points and the observation point. The transfer function will depend upon the shape of the (motionless) body, as well as the source–observation point distance, because the waves emitted by any one elemental surface source will be scattered by the presence of the rest of the (motionless) body (Cremer, 1981). This fact is expressed mathematically by the surface pressure term in the Kirchhoff-Helmholtz integral equation, which is discussed more fully later. Unless the surface of the body takes the form of an infinitely extended plane surface, the transfer function does not take the simple form $C \exp(-jkR)/R$, which is appropriate to a point source in a fluid free of boundaries (free field), and also to the special planar source case to which the Rayleigh equation applies [see Eq. (2.4)]. However, the principle of reciprocity can be invoked to provide a means of bypassing the problem of evaluating the transfer function directly (Smith, 1962).

5.5 Modal Reciprocity: Radiation and Response

Imagine that we are in a position to make a reasonable assumption about the form of motion generated in a body by acoustic excitation, as in the case of a flexibly mounted but otherwise rigid body, or in cases of vibration dominated by response in an individual mode. Let points on the body surface be defined by the position vector $\mathbf{r}_s$ and points in the fluid be defined by $\mathbf{r}$. As we have already seen, the response of a fluid-loaded mode is obtained by evaluating the modal-generalised force caused by the blocked-pressure field on the surface of the structure. This blocked force is defined as

$$\tilde{F}^m_{\text{bl}} = \int_S \psi_m(\mathbf{r}_s)\tilde{p}_{\text{bl}}(\mathbf{r}_s)\, dS, \tag{5.19}$$

where $\psi_m(\mathbf{r}_s)$ describes the assumed mode shape. A point volume velocity source $\tilde{Q}(\mathbf{r}_0)\exp(j\omega t)$ in the fluid volume at $\mathbf{r}_0$ is assumed to generate a simple harmonic excitation field. A reciprocal relationship may be written between the blocked pressure produced on the rigid surface at $\mathbf{r}_s$ by source Q, and the pressure $\tilde{p}(\mathbf{r}_0)$ produced at $\mathbf{r}_0$ by an elemental surface source of volume velocity $\tilde{v}_m\psi_m(\mathbf{r}_s)\,\delta S$, where v_m is the modal coordinate velocity amplitude:

$$\tilde{p}(\mathbf{r}_0)/\tilde{v}_m\psi_m(\mathbf{r}_s)\,\delta S = \tilde{p}_{\text{bl}}(\mathbf{r}_s)/\tilde{Q}(\mathbf{r}_0). \tag{5.20}$$

This relationship is true irrespective of the shape of the body. Hence, from Eq. (5.19) the elemental contribution to $\tilde{F}_{\text{bl}}$ may be written as

$$\delta\tilde{F}^m_{\text{bl}} = \tilde{Q}(\mathbf{r}_0)\tilde{p}(\mathbf{r}_0)/\tilde{v}_m. \tag{5.21}$$

By superposition, the total modal blocked force may be written as

$$\tilde{F}^m_{\text{bl}} = (\tilde{p}(\mathbf{r}_0)/\tilde{v}_m)\tilde{Q}(\mathbf{r}_0), \tag{5.22}$$

where $\tilde{p}(\mathbf{r}_0)$ is the total pressure produced at $\mathbf{r}_0$ by modal vibration of the whole surface with modal velocity amplitude $\tilde{v}_m$.

Although this equation is valid for any position of point source relative to that of the structure, and for any condition of fluid boundaries, it is convenient initially to assume that apart from the presence of the structure, the fluid is unbounded and that the source is in the far acoustic field of the vibrating body. The reason for making these assumptions is that pressure can then be simply related to sound intensity, and also a reference incident pressure can be unambiguously defined.

Modal vibration can be attributed with a radiation resistance thus:

$$\bar{P}^m_{\text{rad}} = \frac{1}{2}|\tilde{v}_m|^2 R^m_{\text{rad}} = \frac{1}{2}\rho_0 c\sigma_m|\tilde{v}_m|^2 \int_S \psi_m(\mathbf{r}_s)^2\, dS, \tag{5.23}$$

where σ_m is the modal radiation efficiency. In the far radiation field, the radial sound intensity is given by

$$I = \tfrac{1}{2}|\tilde{p}|^2/\rho_0 c. \tag{5.24}$$

A source directivity factor $D(\theta, \phi)$ can be defined as the ratio of sound intensity at radius r in the direction (θ, ϕ) to the sound intensity I_0 produced at the same radius by a uniformly directional source of the same power, where $4\pi r^2 I_0 = \bar{P}_{\text{rad}}$. Hence

$$D(\theta, \phi) = \frac{I(\theta, \phi)}{I_0} = \frac{2\pi r^2|\tilde{p}(\theta, \phi)|^2}{\rho_0 c \bar{P}_{\text{rad}}} = \frac{4\pi r^2|\tilde{p}(\theta, \phi)|^2}{\rho_0 c R^m_{\text{rad}}|\tilde{v}_m|^2}. \tag{5.25}$$

Thus if position $\mathbf{r}_0$ has coordinates (r, θ, ϕ), Eq. (5.25) relates the modulus $|\tilde{p}(\mathbf{r}_0)/\tilde{v}_m|$ of the complex ratio appearing in Eq. (5.22) to the radiation directivity factor of the modal source.

A fluid-loaded modal impedance $\tilde{Z}_m$ may be defined by

$$\tilde{Z}_m = \tilde{F}^m_{\text{bl}}/\tilde{v}_m. \tag{5.26}$$

When the fluid-loaded mode is excited at resonance, the imaginary component of $\tilde{Z}_m$ disappears and only the resistive component $\tilde{R}_m$ remains to control the response amplitude. Hence

$$\tilde{v}_m = \tilde{F}^m_{\text{bl}}/R_m = \tilde{F}^m_{\text{bl}}/(R^m_{\text{int}} + R^m_{\text{rad}}), \tag{5.27}$$

in which the total modal resistance has been separated into that associated with energy dissipation within the structure (or radiation of mechanical energy to contiguous systems) and that associated with acoustic energy radiation. Normally only the mean-square modal velocity, and the associated modal energy, is of interest, the phase of the response being of less concern. Hence Eqs. (5.22), (5.24), and (5.25) are combined to give

$$|\tilde{v}_m|^2 = \frac{\rho_0 c D(\theta, \phi)|\tilde{Q}(\mathbf{r}_0)|^2}{4\pi r_0^2} \frac{R^m_{\text{rad}}}{(R^m_{\text{int}} + R^m_{\text{rad}})^2}. \tag{5.28}$$

It is convenient to replace $\tilde{Q}$ by an expression involving the pressure in the incident field at the location of the structure, but in its absence; this is the pressure $\tilde{p}_0$ produced by $\tilde{Q}$ at distance r. Equation (2.2) gives

$$|\tilde{p}_0|^2 = \omega^2 \rho_0^2 |\tilde{Q}|^2/16\pi^2 r^2. \tag{5.29}$$

Substitution into Eq. (5.28) yields

$$\frac{|\tilde{v}_m|^2}{|\tilde{p}_0|^2} = \frac{4\pi D(\theta, \phi)}{\rho_0 c k^2} \frac{R^m_{\text{rad}}}{(R^m_{\text{int}} + R^m_{\text{rad}})^2}. \tag{5.30}$$

The significance of this result is that it shows that the resonant response of a structure in a single, well-defined mode to excitation by the field of a point source, located in the direction (θ, ϕ), can be estimated from a knowledge of the modal radiation resistance and its directional radiation characteristics; the direct-scattering problem has been avoided by use of the reciprocity principle.

Spectral analysis may be used to show that if the source is not of a single frequency, but has a broad random character, the equivalent form of response equation is

$$\frac{\overline{v_m^2}}{G_{p_0}(\omega)} = \frac{2\pi^2 c D(\theta, \phi)}{\rho_0 \omega_m^2 M_m} \frac{R_{\text{rad}}^m}{R_{\text{int}}^m + R_{\text{rad}}^m}, \tag{5.31}$$

where M_m is the modal generalised mass, $G_{p_0}(\omega)$ the single-sided mean-square spectral density of the incident pressure, which is assumed to be effectively uniform over the bandwidth of the mode, and ω_m the natural frequency of the fluid-loaded mode. Note that $M_m \overline{v_m^2}$ is the time-average total energy of vibration of the mode.

In practice it is often likely that one wishes to estimate modal response to a multi-directional sound field because high sound levels that can produce stresses capable of threatening the integrity of a structure are more easily generated within enclosed spaces, because of reverberant build-up. Provided that equivalent point source locations can be identified, the reciprocity Eq. (5.22) applies equally well to this more complex case. However, it is not then possible to use Eq. (5.24), nor is the incident-sound pressure associated with each source readily identified. Nevertheless, it is possible to derive an approximate response formula for the case of excitation by a diffuse field, which is an idealisation in which plane waves are incident upon a structure from all directions with equal probability and random phase.

An incident diffuse field may be modeled as that produced by a uniform distribution over a spherical surface of large radius of point sources of uniform strength and random phase. If the mean-square source strength per unit area is $\bar{q}^2$, an element of the source surface of area δS produces an elemental modal response $\delta|\tilde{v}_m|^2$ given by Eq. (5.28), with $|\tilde{Q}|^2 = 2\bar{q}^2\,\delta S$. Because all the elemental sources are uncorrelated, the total mean-square pressure produced at the location of the absent structure by the whole source surface is given by Eq. (5.29) with $|\tilde{Q}|^2 = 4\pi r^2 \bar{q}^2$. Integration of Eq. (5.28) over the source surface to obtain the total modal response gives

$$\overline{v_m^2} = \frac{\rho_0 c}{4\pi r^2} \frac{R_{\text{rad}}^m}{(R_{\text{int}}^m + R_{\text{rad}}^m)^2} \overline{q^2} \int_{4\pi} D(\theta, \phi) r^2 \, d\theta \, d\phi, \tag{5.32}$$

where dS has been replaced by $r^2 \, d\theta \, d\phi$. By definition of the directivity factor

$$\int_{4\pi} D(\theta, \phi) \, d\theta \, d\phi = 4\pi, \tag{5.33}$$

and since

$$\overline{p_0^2} = \omega^2 \rho_0^2 \overline{q^2}/4\pi, \tag{5.34}$$

the normalised modal response is

$$\frac{\overline{v_m^2}}{\overline{p_0^2}} = \frac{4\pi}{\rho_0 c k^2} \frac{R_{\text{rad}}^m}{(R_{\text{int}}^m + R_{\text{rad}}^m)^2}. \tag{5.35}$$

The response to a broad-band random diffuse field is, by analogy with Eq. (5.31),

$$\frac{\overline{v_m^2}}{G_{p_0}(\omega)} = \frac{2\pi^2 c}{\rho_0 \omega_m^2 M_m} \frac{R_{\text{rad}}^m}{R_{\text{int}}^m + R_{\text{rad}}^m}, \tag{5.36}$$

where again $M_m \overline{v_m^2}$ is the total energy of modal vibration and $G_{p_0}(\omega)$ is assumed to be uniform.

If a number of modes of a structure have their natural frequencies within the bandwidth of a random diffuse field, the total energy of vibration is exactly equal to the sum of the modal energies, provided that the modes form an orthogonal set. In this case the total time-average energy of structural vibration is related to the mean-square pressure $\overline{p_0^2}$ in a bandwidth centered on ω_c by

$$\frac{\bar{E}_s}{\overline{p_0^2}} = \frac{2\pi^2 c n_s(\omega)}{\rho_0 \omega_c^2} \left\langle \frac{R_{\text{rad}}^m}{R_{\text{int}}^m + R_{\text{rad}}^m} \right\rangle_m, \tag{5.37}$$

where $n_s(\omega)$ is the average modal density of the structure, which is the average number of natural frequencies per rad s^{-1}, and the brackets symbolise an average over the modes resonant in the bandwidth. Modal-average radiation efficiency, which is discussed in Chapter 2, is related to modal-average radiation resistance, through Eq. (5.23) by

$$\langle \sigma \rangle_m = \frac{\langle R_{\text{rad}}^m / S_m \rangle_m}{\rho_0 c}, \tag{5.38}$$

where $S_m = \int_S \psi_m^2 \, dS$.

It has been shown by alternative analysis techniques that Eqs. (5.36) and (5.37) apply to any case of diffuse field excitation, irrespective of the actual forms of the acoustic sources producing the field. In enclosures having dimensions much greater than an acoustic wavelength the acoustic modal density is generally so high that even a rather small excitation bandwidth $\Delta\omega$ is sufficient to produce a close approximation to a diffuse field. The acoustic-field energy density is $\overline{p_0^2}/\rho_0 c^2$ and the modal density $n_a(\omega)$ is $V\omega^2/2\pi^2 c^3$, where V is the enclosure volume. Hence the average energy per mode is $2\pi^2 c\overline{p_0^2}/\rho_0\omega^2\,\Delta\omega$. Equation (5.37) may therefore be written

$$\frac{\bar{E}_s}{\bar{E}_a} = \frac{n_s(\omega)}{n_a(\omega)} \left\langle \frac{1}{1 + R_{int}^m/R_{rad}^m} \right\rangle_m . \tag{5.39}$$

The physical interpretation of this equation is that the average ratio of structural energy per mode to acoustic energy per mode is determined by the ratio of modal radiation and internal resistances. If $R_{rad}^m/R_{int}^m \ll 1$ the energy ratio is determined by R_{rad}^m/R_{int}^m, but if $R_{rad}^m/R_{int}^m \gg 1$ the energy ratio is nearly independent of R_{rad}^m and R_{int}^m. This condition represents an upper limit on structural response to excitation by a broad-band diffuse sound field in an enclosure; it is known as equipartition of modal energy in the terminology of statistical energy analysis. A similar conclusion regarding dependence on the resistance ratios may be drawn from broad-band response Eqs. (5.31) and (5.36).

The influence of the resistance ratio on pure tone at resonance is rather different as indicated by Eqs. (5.32) and (5.35). The maximum response is produced by a resistance ratio of unity. We shall discover later how use can be made of this result in designing low-frequency resonant acoustic absorbers.

5.6 Radiation Due to Point Forces and Response to Point Forces

There exists a fundamental relationship between the radiation field of a plate or shell structure subjected to point force excitation, and its response to the acoustic field produced by a point source in a contiguous fluid (Liamshev, 1960). It relates the acoustic pressure at an observation point A in a contiguous fluid produced by the vibration of any (linear) plate or shell structure subject to the action of a single harmonic point mechanical force at point B, and the vibration velocity produced in the structure at point B, in the absence of the mechanical force, by a point source located in the fluid at point A. The direction of the velocity is defined to be the same as that of the force (Fig. 105):

$$\tilde{p}(\mathbf{r}_0)/\tilde{F}(\mathbf{r}_s) = -\tilde{v}(\mathbf{r}_s)/\tilde{Q}(\mathbf{r}_0), \tag{5.40}$$

where $\mathbf{r}_0$ and $\mathbf{r}_s$ are the position vectors of points A and B, respectively.

The following analysis illustrates a particular example of the general relationship. In Section 3.2, the response of a small, fluid-loaded, baffled piston to a mechanically applied, simple harmonic force was examined. Equations

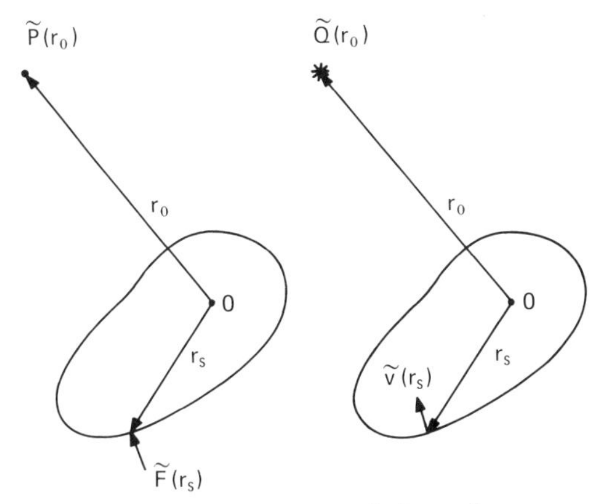

Fig. 105 Reciprocal cases of radiation and response.

(3.7) to (3.9) give

$$\frac{\tilde{v}}{\tilde{F}} = \frac{1}{\tilde{Z}_{\text{tot}}} = \{j[\omega M - S/\omega + (8/3)\rho_0 a^3\omega] + [B + \tfrac{1}{2}\rho_0 c\pi a^2/(ka)^2]\}^{-1}. \qquad (5.41)$$

in which $(M\omega - S/\omega)$ is the reactive mechanical impedance, to which is added the mechanical equivalent of the reactive fluid-loading impedance, and B is the mechanical damping coefficient, to which is added the equivalent acoustic radiation resistance: equivalence is necessary because of the differing definitions of mechanical and acoustic impedance. The effect of fluid loading on the natural and resonance frequencies of the system is clearly shown by Eq. (5.41). We know that the response of an elastic system to acoustic excitation may be evaluated by decomposing the total acoustic field on the surface of the system into its blocked and radiated components, and that this leads to the concept of the fluid-loaded system responding to the blocked-pressure component alone. This approach is most useful when the dominant spatial form of system response can be assumed a priori: the piston is an example of such a system. The mechanical force $\tilde{F}$ may be replaced by the blocked-pressure force due to any form of incident field.

We are currently most interested in the particular type of incident field to which the force–source relationship of Eq. (5.40) applies, namely, the spherical field produced by a point source. The blocked pressure on an infinite baffle may be evaluated most simply by imagining an identical image source to exist on the other side of the baffle in the mirror-image position (Fig. 106). The particle velocity normal to the plane of the (now removed) baffle is zero, and the blocked-pressure amplitude is twice that at the plane

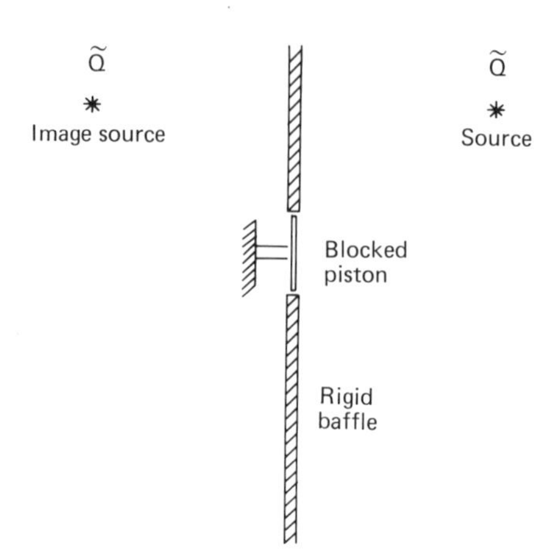

Fig. 106 Source imaged in rigid-plane surface.

due to the real source alone. It is given by

$$\tilde{p}_{\mathrm{bl}}(\mathbf{r}_s) = [j\omega\rho_0\tilde{Q}(\mathbf{r}_0)/2\pi r]\exp(-jkr), \tag{5.42}$$

where $r = |\mathbf{r}_0 - \mathbf{r}_s|$. Since the piston is small ($ka \ll 1$) it may be assumed that the blocked pressure is uniform. With the centre of the piston at the origin of the coordinate system,

$$\tilde{F}_{\mathrm{bl}} = -[j\omega\rho_0\pi a^2\tilde{Q}/2\pi r]\exp(-jkr). \tag{5.43}$$

The piston response to acoustic excitation is

$$\tilde{v}/\tilde{Q} = -(j\omega\rho_0 a^2/2r\tilde{Z}_{\mathrm{tot}})\exp(-jkr). \tag{5.44}$$

When the piston is vibrated by the mechanical force $\tilde{F}$, the radiated field is given by that of a point source in a baffle, because it has been assumed that $ka \ll 1$. Hence, from Eq. (2.3),

$$\tilde{p}/\pi a^2\tilde{v} = (j\omega\rho_0/2\pi r)\exp(-jkr). \tag{5.45}$$

However, $\tilde{v} = \tilde{F}/\tilde{Z}_{\mathrm{tot}}$ and therefore

$$\tilde{p}/\tilde{F} = (j\omega\rho_0 a^2/2r\tilde{Z}_{\mathrm{tot}})\exp(-jkr) = -\tilde{v}Q, \tag{5.46}$$

which confirms Eq. (5.40) in the elementary case.

The general relationship expressed by Eq. (5.40), which holds for any configuration of structure and fluid, subject to the condition of linearity of dynamic behaviour, has considerable practical importance. It is often necessary to investigate the dependence on mechanical forces applied to a structure of the sound field radiated by that structure, but it is frequently very difficult and costly, or even impossible, to replace the forces operating in practice by measurable, controlled forces. However, if the reciprocal experiment is performed, in which acoustic sources, placed in the fluid at positions of interest, are used to excite vibrations in the structural system, the resulting velocities at the points of force operation can be measured relatively easily by small transducers.

A very good example of the application of this technique to the problem of noise radiated by ships, due to machinery operation, is given by Ten Wolde *et al.* (1975). Another field of application is in the diagnosis of the sources of noise in car interiors; an omni-directional acoustic source is placed at the position of the passengers' heads and accelerations are measured on the engine mountings, suspension points, etc. Non-linearity can unfortunately be a problem in this application. A further interesting potential use of the relationship is in determining the optimum positions for the attachment of vibrating machinery to associated structures so as to minimise sound radiation. This is done by finding positions of minimum vibration velocity under acoustic excitation. Of course, reciprocal relationships of the forms discussed above are not limited to simple harmonic time dependence, since they hold for any Fourier components of complex time histories, provided the systems are linear. Recently the reciprocal technique has been used to infer the relationship between the bowing force on a violin string and the radiated sound field.

The literature of sound radiation from plates and shells contains many expressions for the far fields generated by point-excited structures (Junger and Feit, 1972). The responses of the structures to acoustic excitation by point sources in the far field may be evaluated from the same expressions: response to plane-wave excitation may also be evaluated using Liamshev's extension of the point-to-point relationship. Important examples of these relationships are the expression for the response of a uniform plate to point source and to plane-wave excitation; the fluid is considered to exist on one side of the plate only.

In the former case

$$\frac{\tilde{v}}{\tilde{Q}} = \frac{jk \exp(-jkr)}{2\pi r} \frac{\cos\theta}{1 - jkh(\rho_s/\rho_0)(\cos\theta)[1 - (\omega/\omega_c)^3 \sin^4\theta]}, \tag{5.47}$$

where θ is the angle between the normal to the plate and the line joining the point of observation of $\tilde{v}$ to the source point, $h\rho_s$ the plate mass per unit area, and ω_c the critical frequency of the plate. At low frequencies, $\omega/\omega_c \ll 1$ and Eq. (5.47) reduces to

$$\frac{\tilde{v}}{\tilde{Q}} = \frac{jk \exp(-jkr)}{2\pi r} \frac{\cos\theta}{1 - jkh(\rho_s/\rho_0)\cos\theta}. \tag{5.48}$$

Liamshev's plane-wave response theory gives, for $\omega/\omega_c \ll 1$,

$$\frac{\tilde{v}}{\tilde{p}_i} = -\frac{(2/\rho_0 c)\cos\phi}{1 - jkh(\rho_s/\rho_0)\cos\phi}, \tag{5.49}$$

where $\tilde{p}_i$ is the incident plane-wave amplitude. In the limit of very heavy fluid loading ($kh\rho_s/\rho_0 \ll 1$) or high angle of incidence, the response tends to the equivalent free surface particle velocity $-2(\tilde{p}_i/\rho_0 c)\cos\phi$, indicating that the plate impedance is insignificant compared with the fluid impedance.

5.7 Applications of Response Theory

The primary applications of the aforecited theories of acoustically induced response are to problems of flanking-sound transmission in buildings, in which acoustically induced vibration is transmitted along the building structure, thereby bypassing partition structures such as walls and floors, and also to estimates of vibration of aircraft, spacecraft, industrial pipework, and nuclear reactor structures by the intense sound fields to which they are exposed in operation.

One application that has not received much attention is that of the absorption of sound in auditoria by non–load-bearing structures exposed to the sound field; these include lightweight ceilings, gypsum board soffits, stage structures, reflectors, wall paneling, ductwork, and even seats. These are all known to act as low-frequency absorbers, controlling low-frequency reverberation time, but little quantitative information has been published to provide a guide for acoustic designers. In auditoria, sound fields can be considered to be diffuse, and most sounds of interest are transient and have energy distributed over rather broad frequency bands. However, the decay rates of sound fields in auditoria tend to be far less than those of such structures, and therefore the steady-state broad-band, multi-mode response Eq. (5.37) may be used as a reasonable approximation to the short-time average behaviour.

The rate at which energy is dissipated and/or transmitted to connected structures is

$$\overline{P_d} = \sum_m \eta_{\text{int}}^m \omega_m \bar{E}_m, \tag{5.50}$$

in which $\eta_{\text{int}}^m = R_{\text{int}}^m/\omega_m M_m$. The modal energies in response to broad-band, diffuse-field acoustic excitation are given by Eq. (5.36). It is generally the case for all but very lightweight sheets excited at frequencies below critical in air that $R^m_{\text{rad}} \ll R^m_{\text{int}}$; using the approximation, together with Eq. (5.50) gives

$$\overline{P_{\text{d}}} = [2\pi^2 c G_{p_0}(\omega)/\rho_0] \sum_m \eta^m_{\text{rad}}/\omega_m. \tag{5.51}$$

This equation is interesting because it shows the dissipated energy to be dependent only on the radiation loss factors and not on the internal loss factors or damping of the structure. The modal-radiation loss factor is related to the modal-radiation efficiency by

$$\eta^m_{\text{rad}} = \rho_0 c A \sigma_m/\omega_m M. \tag{5.52}$$

Note the appearance of the total structural mass M as opposed to the modal mass. Thus

$$\overline{P_{\text{d}}} = [2\pi^2 c^2 A G_{p_0}(\omega)/M] \sum_m \sigma_m/\omega_m^2. \tag{5.53}$$

In a frequency band of relatively small percentage bandwidth and centre frequency ω_0, this equation may be further simplified to

$$\overline{P_{\text{d}}}/\overline{p_0^2} = [2\pi^2 c^2 A n_{\text{s}}(\omega)/M\omega_0^2]\langle\sigma_m\rangle_m, \tag{5.54}$$

where $\langle\sigma_m\rangle_m$ is the modal-average radiation efficiency appropriate to the band.

In architectural acoustics the measure of the capacity of a surface to dissipate sound energy is the absorption coefficient α, which is defined as the ratio of absorbed to incident intensity. Clearly, its value for any particular surface depends upon the spatial form of the incident field, as well as on frequency and the mechanical and geometric properties of the surface. It is conventional to evaluate the diffuse incidence, or statistical, absorption coefficient, since in practice the actual spatial characteristics of fields are not known. In a diffuse field the intensity incident upon a plane from one side is

$$I = \overline{p_0^2}/4\rho_0 c. \tag{5.55}$$

Therefore,

$$\alpha = \frac{\overline{P_{\text{d}}}/A}{\overline{p_0^2}/4\rho_0 c} = \frac{8\pi^2 \rho_0 c^3 n_{\text{s}}(\omega)}{M\omega_0^2}\langle\sigma_m\rangle_m. \tag{5.56}$$

Since the radiation efficiency of panels at frequencies well below the critical frequency tends to vary approximately as $\omega^{1/2}$ and the average modal density of flat panels is independent of frequency, it is clear from Eq. (5.56) that panel vibration constitutes primarily a low-frequency absorption mechanism. An approximate expression for the low-frequency radiation efficiency of a baffled rectangular panel is

$$\langle\sigma_m\rangle_m = B(1.8hc_l')^{3/2}\omega_0^{1/2}/c^2\pi^2 A(2\pi)^{1/2}, \tag{5.57}$$

where B, A, and h are, respectively, the perimeter, area, and thickness of the panel, and c'_l is the speed of longitudinal waves in the plate. The modal density is

$$n_s(\omega) = \sqrt{3}A/2\pi hc'_l, \tag{5.58}$$

and hence the absorption coefficient becomes

$$\alpha(\omega_0) = 3.3\rho_0 Bc_l'^{1/2}c/A\rho_s h^{1/2}\omega_0^{3/2}. \tag{5.59}$$

This result indicates that one larger panel will absorb less than a number of smaller panels making up the same total area, and that the absorption coefficient decreases by a factor of approximately 3 per octave: absorption by a panel of a given material increases as the thickness is decreased. The value of α for a 2 m × 2 m × 13 mm gypsum board panel at 100 Hz is approximately 0.9. These conclusions are consistent with observations on buildings. The influence of stiffening elements on the absorption coefficient of a given area of panel is clearly to increase the absorption coefficient, because they increase the radiation efficiency.

Equation (5.59) relates to broad-band absorption by multi-modal vibration. Another form of panel absorber commonly used in buildings, particularly in recording studios, is the so-called membrane absorber. This consists of a panel of relatively small area, typically about 1 m square, which is fixed to a wall with a shallow air cavity between it and the wall. The function of this type of absorber is selectively to absorb low-frequency energy by resonance in its fundamental bending mode; the air cavity bulk stiffness, rather than the panel bending stiffness, often determines this frequency. The equation relevant to single-frequency diffuse-field excitation is the single-mode Eq. (5.35). The power dissipated by vibration at resonance in the fundamental mode is

$$\overline{P_d} = \eta_{int}^1\omega_1\bar{E}_1, \tag{5.60}$$

$$= \frac{4\pi\overline{p_0^2}c}{\rho_0\omega_1^2}\frac{R_{rad}^1 R_{int}^1}{(R_{rad}^1 + R_{int}^1)^2}, \tag{5.61}$$

and

$$\alpha = \frac{\overline{P_d}/A}{\overline{p_0^2}/4\rho_0 c} = \frac{16\pi c^2}{A\omega_1^2}\frac{R_{rad}^1 R_{int}^1}{(R_{rad}^1 + R_{int}^1)^2}. \tag{5.62}$$

The absorption coefficient is maximised by making $R_{rad}^1 = R_{int}^1$; then

$$\alpha_{max} = 4\pi c^2/A\omega_1^2 = \lambda_1^2/\pi A. \tag{5.63}$$

This result is analogous to that for a Helmholtz resonator in a wall in that the maximum absorption is obtained by matching the radiation and internal resistances; also the maximum absorption $(\alpha_{max}A)$ m^2 is independent of panel or neck area. It is the necessity of matching the two resistances to achieve optimum performance that gives membrane absorbers their characteristic single low-frequency peak, corresponding to fundamental-mode resonance. As discussed in Chapter 2, the fundamental mode of an edge supported panel has a much higher radiation efficiency than any of the higher-order modes within the next few octaves above f_1, because no volume cancellation occurs. The radiation resistance corresponds to that of a baffled piston of equivalent volume velocity at low ka, and is approximately

$$R_{rad}^{1} = \rho_0 A^2 \omega_1^2 / 8\pi c, \tag{5.64}$$

where A is the panel area. This is the value to which the damping of the membrane absorber should be matched for optimum absorption at resonance.

Problems

5.1 Why is the field scattered by a rigid circular cylinder located in a plane-wave field propagating in a direction normal to the cylinder axis not exactly equal to that produced by uniform translation of the cylinder at a velocity equal to the plane-wave particle velocity at the position of the axis of the rigid cylinder, but in its absence? Why does the parameter ka influence the error incurred by equating the two fields?

5.2 Under what conditions would you expect the acoustically induced response of the fundamental mode of a flat panel to be rather insensitive to the directions of an incident plane wave? Why is the fundamental mode unique in this respect?

5.3 Would you expect an array of small, independently mounted, flexible panels to scatter an incident plane wave more "diffusely" than one large single panel of the same overall area? Explain the physical basis of your answer.

5.4 A flat plate steel structure is excited into vibration in a 200 m^3 reverberation chamber by mechanical excitation in the one-third octave band centred on 500 Hz. The spatial-average mean-square acceleration is measured to be 10^6 m^2 s^{-4}, the average sound pressure level in the reverberation chamber is 106 dB, and its reverberation time in the 500 Hz band is 4 s. The area of the plate is 2 m^2 and its thickness is 3 mm. Estimate the average mean-square velocity of response of the structure to a 500-Hz one-third-octave band random sound field of 100-dB average sound-pressure level produced in the chamber by loudspeakers. Assume that the average internal loss factor of the structure in this band is 5×10^{-3}. Equations (2.22) and (5.37) apply.

5.5 It is desired to make an estimate of the rate at which a lightweight sheet roof of a factory absorbs sound when a large number of different types of machine operate simultaneously, so that an effective roof absorption coefficient can be approximately determined. Assuming that you have access to the roof via scaffolding, suggest how this objective might be achieved. Hint: In addition to the appropriate transducers and measurement equipment, you are provided with a large hammer and information concerning the material and thickness of the roof.

5.6 A nuclear reactor designer is concerned that the vibrational stresses in the steel casing of a heat exchanger may be unacceptably high because of acoustic excitation by the sound field generated by a circulator that drives the coolant gas through the exchanger. The coolant is helium gas, which is at an average temperature of 800°C, and a static pressure of 20 atm in the heat exchanger. Tests are made on a heat exchanger–circulator combination in atmospheric air at an average pressure of 1.3 atm and an average temperature of 45°C. Measurements are made of sound pressures in the heat exchanger and vibration strains in the casing.

Assuming that the sound fields in both cases can be considered to be broad band and diffuse, use the appropriate response equation to assess whether the results of such a test could reasonably be extrapolated to operating conditions. Indicate specifically, in qualitative terms, where the major uncertainties are expected to exist.

5.7 Assuming that the radiation loss factor of a structure is much smaller than the internal loss factor, show that the sound power dissipated by the structure when excited by a broad-band acoustic field is nearly independent of the actual internal loss factor. Is this true for pure tone excitation at the resonance frequency of one of the structural modes? What are the implications for the design of platelike sound absorbers for application in building acoustics?

6 Acoustic Coupling between Structures and Enclosed Volumes of Fluid

6.1 Practical Importance of the Problem

In our earlier considerations of sound radiation by vibrating structures and of the associated fluid-loading effects, we made the implicit assumption that the geometry of the boundaries of the fluid were such that sound waves could be radiated away to infinity and were not reflected back on the surface of the structure to produce added fluid loading. An exception to this general assumption was made in the analysis of a vibrating piston loaded by a column of fluid of finite length, in which the wave motion was restricted to the one-dimensional, plane-wave case.

There are, however, many systems of practical interest in which a structure is in contact with a fluid that is contained within a finite volume by physical boundaries that may wholly or only partly comprise the surface of the structure under consideration. The most significant difference between the acoustic behaviour of fluid contained within physical boundaries and that of unconstrained fluid is the existence in the former of natural modes of vibration and associated natural frequencies; these are called *acoustic modes*. As in solid structures, acoustic modes arise from interference between intersecting waves, and the natural frequencies are associated with the correspondence between the spatial characteristics of the interference pattern and the geometry of the physical boundaries of the fluid volume. An enclosed fluid exhibits resonant acoustic behaviour, the effects of which are to produce a strongly frequency-dependent response to vibration by contiguous structures, together with fluid-loading effects that also exhibit strong dependence upon frequency. Practical examples of interest include machinery noise control enclosures, double-leaf partitions, vehicle cabin spaces, tympanic and bodied musical instruments, loudspeaker enclosures, and fluid transport ducts. The skin drum is an example of a system in which solid–fluid interaction is fundamental to its vibrational behaviour.

Where the influence of the fluid loading has a significant effect on the vibration of the enclosing structure, the problem of analysis of the vibrational behaviour of the resulting coupled system is rather involved, except in certain rather idealised cases, such as the one-dimensional case already mentioned. It is, however, by no means invariably the case that the presence of an enclosed fluid greatly changes the structural vibration characteristics: the "art of the game" is to be able to judge whether a fully coupled vibration analysis is necessary or not. To some extent the necessity for such analysis depends upon the objective sought. For example, it is known that the sound power radiated by a vibrating structure into an enclosed volume of gas is often largely insensitive to the detailed form of fluid behaviour and its influence on structural vibration, provided that the bandwidth of excitation is broad and that the power is evaluated in frequency bands encompassing more than about ten natural frequencies of the coupled system. In fact, it may often be assumed that the fluid exhibits no resonant behaviour at all. However, if discrete frequency response of the structure and/or fluid is required, it will probably be necessary to account for resonant behaviour of the coupled system.

6.2 Fundamentals of Fluid–Structure Interaction

The process of interaction between a structure and an enclosed volume of fluid can be expressed in terms of functions that represent the response of each medium to simple harmonic-point excitation. For a structure this is customarily the point force transfer mobility $\tilde{Y}(\mathbf{r}_s, \mathbf{r}_f, \omega)$, where $\tilde{Y}$ is the velocity amplitude produced at $\mathbf{r}_s$ by unit force of frequency ω applied at $\mathbf{r}_f$: $\tilde{Y}$ is strictly a six-component quantity because both velocity and force are vector quantities. However, we shall restrict the definition to components of velocity and force normal to the surface of the local structure, since it is the normal motion of plates and shells that determines the fluid–structure interaction, and acoustic pressures in inviscid fluids can exert only normal forces on contiguous structures.

The appropriate function for a fluid is the acoustic Green's function, which is a solution to the inhomogeneous scalar Helmholtz equation (wave equation for simple harmonic time dependence):

$$(\nabla^2 + k^2)\tilde{G}(\mathbf{r}, \mathbf{r}_0) = -\delta(\mathbf{r} - \mathbf{r}_0), \tag{6.1}$$

where $\delta(\mathbf{r} - \mathbf{r}_0)$ is the three-dimensional Dirac delta function representing a point source at $\mathbf{r}_0$. The Dirac delta function is a generalised function (Lighthill, 1958) having the following properties:

$$\int_V f(\mathbf{r})\,\delta(\mathbf{r} - \mathbf{r}_0)\,dV(r) = \begin{cases} f(\mathbf{r}_0), & \mathbf{r}_0 \text{ within } V, \\ \frac{1}{2}f(\mathbf{r}_0), & \mathbf{r}_0 \text{ on the boundary of } V, \\ 0, & \mathbf{r}_0 \text{ outside } V. \end{cases} \tag{6.2}$$

Clearly, if $\tilde{\mathbf{G}}$ is the solution to the inhomogeneous wave equation (6.1), it must represent some form of physical quantity that depends upon the physical nature of the point source. Since $\int_V A\,\delta(\mathbf{r} - \mathbf{r}_0)\,dV(\mathbf{r}) = A$, where A is a constant, the dimensions of the delta function are L^{-3}, and since the dimensions of k^2 are L^{-2}, the dimensions of $\tilde{G}$ in Eq. (6.1) are L^{-1}. Any connection with a physical quantity seems somewhat obscure, especially since dimensions do not uniquely define any physical quantity, e.g., consider the quantities work and moment.

Since we are concerned primarily with acoustic pressures, it would seem sensible to ask what form the right-hand side of Eq. (6.1) would take if $\tilde{G}$ had the dimensions of pressure. Then the dimensions of each term must be those of k^2p, or $\mathrm{ML}^{-3}\mathrm{T}^{-2}$. Since $\delta(\mathbf{r} - \mathbf{r}_0)$ is of dimensions L^{-3}, multiplication of the Dirac delta source term in Eq. (6.1) by a quantity having dimensions MT^{-2} produces an equation for pressure: a likely candidate (with hindsight!) is rate of change of mass flux or, since we are mainly concerned with vibrating surfaces, (density) × (acceleration) × (area), which is (density) × (volume acceleration). This choice is clearly not the only one since the product MT^{-2} can be produced by a large variety of combinations of physical quantities; fortunately only a few actually correspond to physical sources of sound. Another quantity that has these dimensions and that represents a physical source of sound is the first spatial derivative of a force acting on a fluid medium, such as that due to a solid body in turbulent flow.

So far we have not mentioned any boundary conditions on Eq. (6.1). In free space, the Green's function must satisfy the condition that only waves traveling outward from the point source are allowed, and that the pressure tends to zero at an infinite distance from the source. Advanced acoustics texts (e.g., Pierce, 1981) show by application of Gauss' integral theorem that the solution of Eq. (6.1) is

$$\tilde{G}(\mathbf{r}, \mathbf{r}_0, \omega) = (1/4\pi|\mathbf{r} - \mathbf{r}_0|)\exp(-jk|\mathbf{r} - \mathbf{r}_0|), \tag{6.3}$$

a solution that corresponds to Eq. (2.2) for the pressure field produced by a small pulsating sphere ($ka \ll 1$).

Applications of Gauss' integral theorem to any volume of fluid bounded by any physical interface, or by any imaginary boundary surface in the homogeneous fluid, produces the Kirchhoff–Helmholtz integral equation, already encountered in a particular form in Eq. (3.1):

$$\tilde{p}(\mathbf{r}) = \int_S [\tilde{p}(\mathbf{r}_s)\,\partial\tilde{G}(\mathbf{r}, \mathbf{r}_s)/\partial n + j\omega\rho_0\tilde{v}_n(\mathbf{r}_s)\tilde{G}(\mathbf{r}, \mathbf{r}_s)]\,dS$$
$$+ j\omega\rho_0 \int_V \tilde{q}(\mathbf{r}_0)\tilde{G}(\mathbf{r}, \mathbf{r}_0)\,dV, \tag{6.4}$$

in which $\tilde{q}(\mathbf{r}_0)$ is the distribution of volume velocity source strength per unit volume, and $\partial\tilde{G}/\partial n$ the derivative of the Green's function with respect to the outward-going normal to the local surface: v_{n} is directed into the fluid. The first integral is evaluated over all bounding surfaces and the second over the entire volume. The Green's function in this equation has only to satisfy Eq. (6.1) and the radiation condition at infinity, the condition on the bounding surface being automatically satisfied by the application of Gauss' integral equation. The second term in the surface integral is actually a fluctuating mass-flux term, which is consistent with our earlier discussion. In Eq. (3.1), the first integral is expressed explicitly in terms of the free-space Green's function, the second (volume) integral not being appropriate in that case since the vibrating surface is the only source of disturbances in the fluid. It should be carefully noted that Eq. (6.4) is a special form of a more general integral equation in which the time dependence is arbitrary and phase $k|\mathbf{r}|$ is replaced by time difference $(t - |\mathbf{r}|/c)$.

A physical interpretation of the pressure term in Eq. (6.4) is that it represents the effect of the presence of the boundaries on the otherwise freely spreading fields of the distributed elemental volume and surface sources of fluctuating mass flux. The form of Green's function used in Eq. (6.4) in any practical case may be chosen so as to conveniently suit the geometry of the boundaries of the fluid. Clearly it has only to satisfy the wave equation and the far-field radiation conditions; but if it can also be chosen so as to satisfy the condition $\tilde{G} = 0$ or $\partial\tilde{G}/\partial n = 0$ on the boundaries of a given system, then one of the surface integral terms in Eq. (6.4) disappears. In these cases, only the surface pressure or surface normal velocity distributions, respectively, need to be known to solve for the whole pressure field. In the special case of the plane surface a choice of a Green's function consisting of a sum of the form in Eq. (6.3) and an image source form produces the Rayleigh equation (2.4), in which only velocity distributions over the plane need to be known. If the sign of the image source Green's function is changed so that the combination is zero on the plane, then only the pressure distribution on the plane needs to be known.

In a volume of fluid completely bounded by solid surfaces it is relatively easy to appreciate the physical interpretation of the Green's function satisfying one of these two homogeneous boundary conditions. If a small source of fluctuating volume, a point monopole, is placed in a fluid that is contained within boundaries, then, as we have seen, the pressure is a solution to Eq. (6.1) with a suitable form of mass fluctuation source term on the right-hand side. If the boundaries are rigid, then $\partial p/\partial n = 0$ on the boundaries, and therefore a form of Green's function that satisfies (6.1) and the boundary condition $\partial \tilde{G}/\partial n$ on the walls is acceptable. In the absence of sources in the fluid, the acoustic pressure satisfies the homogeneous wave equation

$$\nabla^2 p + k^2 p = 0, \tag{6.5}$$

subject to the rigid-wall boundary condition.

Let a solution to Eq. (6.5) be written as

$$p(\mathbf{r}) \exp(j\omega_n t) = \tilde{A}_n \psi_n(\mathbf{r}) \exp(j\omega_n t) \tag{6.6}$$

where ψ_n is the acoustic-pressure mode shape corresponding to the natural frequency ω_n of the rigid-walled space. Hence

$$\nabla^2 \psi_n(\mathbf{r}) + k_n^2 \psi_n(\mathbf{r}) = 0, \tag{6.7}$$

where $k_n = \omega_n/c$. These functions ψ_n satisfy the condition $\partial\psi_n/\partial_n = 0$ on the walls and therefore are candidates for incorporation in a Green's function satisfying the same condition. Hence it is reasonable to try to express $\tilde{G}$ as

$$\tilde{G}(\mathbf{r}, \mathbf{r}_0, \omega) = \sum_{n=0}^{\infty} \tilde{B}_n \psi_n(\mathbf{r}), \tag{6.8}$$

and

$$\nabla^2 \tilde{G} = \sum_n \tilde{B}_n \nabla^2 \psi_n(r). \tag{6.9}$$

From Eq. (6.7)

$$\nabla^2 \psi_n(\mathbf{r}) = -k_n^2 \psi_n(\mathbf{r}), \tag{6.10}$$

giving

$$\nabla^2 \tilde{G}(\mathbf{r}, \mathbf{r}_0, \omega) = -\sum_n k_n^2 \tilde{B}_n \psi_n(\mathbf{r}). \tag{6.11}$$

Equation (6.1) can now be written as

$$-\sum_n k_n^2 \tilde{B}_n \psi_n(\mathbf{r}) + k^2 \sum_n \tilde{B}_n \psi_n(\mathbf{r}) = -\delta(\mathbf{r} - \mathbf{r}_0). \tag{6.12}$$

It is a condition of natural modes of closed elastic systems that they are mutually orthogonal; since the mean fluid density is assumed to be uniform,

$$\int_V \psi_m(\mathbf{r})\psi_n(\mathbf{r})\,dV = \begin{cases} 0, & m \neq n, \\ \Lambda_n, & m = n, \end{cases}$$

where $\Lambda_n = \int_V \psi_n^2(\mathbf{r})\,dV$. If Eq. (6.12) is multiplied by $\psi_m(\mathbf{r})$ and integrated over the fluid volume, the summations disappear by virtue of the orthogonality condition to yield

$$\tilde{B}_n \Lambda_n (k^2 - k_n^2) = -\psi_n(\mathbf{r}_0), \tag{6.13a}$$

or

$$\tilde{B}_n = \psi_n(\mathbf{r}_0)/\Lambda_n(k_n^2 - k^2). \tag{6.13b}$$

Hence

$$\tilde{G}(\mathbf{r}, \mathbf{r}_0, \omega) = \sum_n \frac{\psi_n(\mathbf{r})\psi_n(\mathbf{r}_0)}{\Lambda_n(k_n^2 - k^2)}. \tag{6.14}$$

This function is reciprocal in $\mathbf{r}$ and $\mathbf{r}_0$ as it should be, and $\partial\tilde{G}/\partial_n = 0$ on the walls because $\partial\psi_n/\partial n = 0$ there also.

Now, if it is desired to evaluate the pressure field created in a fluid enclosed in a rigid boundary by a volume distribution of sources of volume acceleration, the volume integral of Eq. (6.4) may be used, with $\tilde{G}$ expressed by Eq. (6.14): the surface integral disappears. In the less likely case of a volume enclosed by boundaries on which the pressure vanishes, Eq. (6.14) takes the same form, but the mode shape functions ψ_n and frequencies ω_n are those appropriate to this boundary condition.

6.3 Interaction Analysis by Green's Function

We are primarily concerned with fluid–structure interaction, and therefore we now proceed to apply Eqs. (6.4) and (6.14) to the case of a fluid volume bounded by a thin-plate or shell structure that is excited by mechanical forces.

The equation of motion of the structure is

$$L[w(\mathbf{r}_s)] + m(\mathbf{r}_s)\,\partial^2 w(\mathbf{r}_s)/\partial t^2 = f(\mathbf{r}_s) + p(\mathbf{r}_s), \tag{6.15}$$

where w is the normal surface displacement of the structure, L the operator governing the elastic forces in the structure, f the distribution of mechanically applied force per unit area, and p the distribution of surface pressures;

f and w are directed outwards from the fluid volume, and fluid pressures external to the enclosed volume are ignored for simplicity. The surface pressure distribution is equal to twice the value given by Eq. (6.4). Hence, for simple harmonic excitation,

$$L[\tilde{w}(\mathbf{r}_s)] - \omega^2 m(\mathbf{r}_s)\tilde{w}(\mathbf{r}_s) = \tilde{f}(\mathbf{r}_s) + 2\omega^2\rho_0 \int_{S'} \tilde{w}(\mathbf{r}_s')\tilde{G}(\mathbf{r}_s, \mathbf{r}_s', \omega)\, dS', \quad (6.16)$$

where

$$\tilde{G}(\mathbf{r}_s, \mathbf{r}_s', \omega) = \sum_n [\psi_n(\mathbf{r}_s)\psi_n(\mathbf{r}_s')/\Lambda_n(k_n^2 - k^2)].$$

Hence, if the structural motion is expressed as a summation over the motions in the *in vacuo* natural modes as $\tilde{w}(\mathbf{r}_s) = \sum_p \tilde{w}_p\phi_p(\mathbf{r}_s)$, which satisfies

$$L[\phi_p(\mathbf{r}_s)] - \omega_p^2\, m(\mathbf{r}_s)\phi_p(\mathbf{r}_s) = 0, \quad (6.17)$$

Eq. (6.16) becomes

$$m(\mathbf{r}_s)\left[\sum_p \omega_p^2\tilde{w}_p\phi_p(\mathbf{r}_s) - \omega^2\sum_p \tilde{w}_p\phi_p(\mathbf{r}_s)\right]$$
$$= f(\mathbf{r}_s) + 2\omega^2\rho_0 \int_{S'}\left[\sum_q \tilde{w}_q\phi_q(\mathbf{r}_s')\sum_n [\psi_n(\mathbf{r}_s)\psi_n(\mathbf{r}_s')/\Lambda_n(k_n^2 - k^2)]\right] dS'. \quad (6.18)$$

Multiplication of Eq. (6.18) by $\phi_r(\mathbf{r}_s)$ and integration over the surface of the structure yields

$$\tilde{w}_p\Lambda_p(\omega_p^2 - \omega^2) - 2\omega^2\rho_0\tilde{w}_p\sum_n \alpha_{ppn}$$
$$= \int_S f(\mathbf{r}_s)\phi_p(\mathbf{r}_s)\, dS + 2\omega^2\rho_0 \sum_{q\neq p}\sum_n \tilde{w}_q\alpha_{pqn}, \quad (6.19)$$

where

$$\alpha_{ppn} = \frac{\left[\int_S \psi_n(\mathbf{r}_s)\phi_p(\mathbf{r}_s)\, dS\right]^2}{\Lambda_n(k_n^2 - k^2)},$$

$$\alpha_{pqn} = \frac{\left[\int_S \psi_n(\mathbf{r}_s)\phi_p(\mathbf{r}_s)\, dS\right]\left[\int_{S'} \psi_n(\mathbf{r}_s')\phi_q(\mathbf{r}_s')\, dS'\right]}{\Lambda_n(k_n^2 - k^2)},$$

and

$$\Lambda_p = \int_S \phi_p^2(\mathbf{r}_s)m(\mathbf{r}_s)\, dS,$$

which is the modal-generalised mass.

In Eq. (6.19) the modal self (direct)-fluid-loading terms α_{ppn} have been separated from the modal mutual (cross)-fluid-loading terms α_{pqn}. It is seen from the presence of all the modal displacements $\tilde{w}_{q \neq p}$ in the equation for $\tilde{w}_p$ that the action of the fluid is to couple the *in vacuo* structural modes. In fact, such a closed elastic system does possess orthogonal natural modes and associated natural frequencies, but they are different from those of the uncoupled-fluid and structural systems.

It is clear from the form of the denominators of α_{ppn} and α_{pqn} that the resonant behaviour of the fluid in the closed volume has a significant influence on the fluid loading. The physical nature of α_{ppn} can be appreciated by considering the term $-\omega^2 \rho_0 \tilde{w}_p/(k_n^2 - k^2)$. At frequencies well below the acoustic mode natural frequency ω_n, $k \ll k_n$, and this term represents inertial fluid loading, proportional to ω^2; at frequencies well above ω_n, $k \gg k_n$, and the term represents frequency-independent elastic stiffness, the numerator containing the fluid adiabatic bulk modulus $\rho_0 c^2$. The physical influence of α_{pqn} depends upon the sign of α_{pqn}, which varies with p, q, and n, and therefore the corresponding fluid loading may be masslike or stiffnesslike.

Equation (6.19) suggests infinite fluid loading at the acoustic natural frequencies of the rigid-walled enclosure. This condition corresponds to the complete rigidity of the fluid column discussed in Section 3.5, at frequencies where the length of the fluid column equals an integer number of half-wavelengths; that is, at the acoustic natural frequencies of the duct when terminated at each end by a rigid plug. In practice, the fluid loading is limited by dissipative mechanisms in the fluid and at its boundaries; this can be modelled in the Green's function of Eq. (6.14) by an ad hoc complex fluid bulk modulus which leads to a small imaginary term in the denominator. Similarly, it is customary to incorporate damping into the structural model by means of a complex elastic modulus. Although these dissipative mechanisms limit the fluid-loading effects, they do not change their basic characteristics. Near a natural frequency of the fluid volume with pressure release ($p = 0$) boundary conditions, the fluid loading produced by the associated acoustic mode disappears, but of course the contributions of other modes still exist.

As already mentioned, it is not common to attempt to evaluate the modes and response of the fully coupled system and it is usual to employ approximations; the most common one is to neglect the mutual (cross)-fluid-loading terms α_{pqn}. In many cases of reasonably large volumes of a gas, such as air, at static pressure close to atmospheric, it is generally unnecessary to account for fluid loading at all. Then the *in vacuo* structural response to mechanical excitation is evaluated, and Eq. (6.4) is used to estimate the pressure field produced by that given vibration. Fairly frequently only a relatively small proportion of the surface integral terms of the form $\int_S \psi_n(\mathbf{r}_s)\phi_p(\mathbf{r}_s)\, dS$ are significant, and further simplifications can be made. This feature is particularly useful in statistical analyses of fluid–structure interaction (Fahy, 1970).

6.4 Modal-Interaction Model

There is an alternative to the integro-differential equation formulation of the interaction between structures and enclosed fluids. The essential difference from the aforegoing analysis is that the fluid field is expressed directly in terms of its uncoupled modes and its behaviour is described by means of a differential equation for each mode. The fluid pressure on the surface of the structure is the agent by which the fluid influences structural motion, and the normal surface acceleration of the structure is the agent by which the structure influences the fluid field.

We may write the inhomogeneous wave equation for pressure as

$$\nabla^2 p - (1/c^2)\,\partial^2 p/\partial t^2 = -\rho_0\,\partial q/\partial t, \tag{6.20}$$

where q represents the distribution of source volume velocity per unit volume. As discussed earlier, the right-hand side represents a rate of change of mass flux per unit volume. The vibration of a bounding surface may be represented as a volume velocity distribution q_b confined to an infinitesimally thin layer situated on a surface S_0 just inside a rigid boundary, by the use of a one-dimensional Dirac delta function in a coordinate normal to surface. Since we have already assumed the normal structural displacement w to be directed outwards from the fluid volume, we may write

$$q_b(\mathbf{r}) = -2(\partial w(\mathbf{r}_s/\partial t)\,\delta(\xi - \xi_0). \tag{6.21}$$

Since δ is one dimensional it has the dimensions $(\text{length})^{-1}$, and therefore $q(\mathbf{r})$ has the correct dimensions of volume velocity per unit volume. Hence

$$\nabla^2 p - (1/c^2)\,\partial^2 p/\partial t^2 = 2\rho_0[\partial^2 w(\mathbf{r}_s)/\partial t^2]\,\delta(\xi - \xi_0) - \rho_0\,\partial q(\mathbf{r}_0)/\partial t, \tag{6.22}$$

where $q(\mathbf{r}_0)$ is the distribution of sources other than those representing boundary motion. We now express the acoustic pressure as a sum of the pressure distributions in the acoustic modes of the fluid volume with rigid boundaries:

$$p(\mathbf{r}, t) = \sum_{n=0}^{\infty} p_n(t)\psi_n(\mathbf{r}). \tag{6.23}$$

Using Eqs. (6.7) and (6.23) in Eq. (6.22) we obtain

$$\sum_n - k_n^2 p_n(t)\psi_n(\mathbf{r}) - (1/c^2)\sum_n \ddot{p}_n(t)\psi_n(\mathbf{r})$$
$$= 2\rho_0[\partial^2 w(\mathbf{r}_s)/\partial t^2]\,\delta(\xi - \xi_0) - \rho_0\,\partial q(\mathbf{r}_0)/\partial t. \quad (6.24)$$

Multiplying by $\psi_m(\mathbf{r})$, integrating over the fluid volume, and applying the orthogonality condition, Eq. (6.24) yields a differential equation for the modal-pressure coordinate $p_n(t)$:

$$\ddot{p}_n + \omega_n^2 p_n = -\frac{\rho_0 c^2}{\Lambda_n}\int_S \psi_n(\mathbf{r}_{s0})[\partial^2 w(\mathbf{r}_s)/\partial t^2]\,dS + (\rho_0 c^2/\Lambda_n)\dot{Q}_n, \quad (6.25)$$

where Q_n is a generalised volume velocity source strength given by $\int_V q(\mathbf{r}_0)\psi_n(\mathbf{r}_0)\,dV$ and $\mathbf{r}_{s0}$ refers to the surface at an infinitesimally small distance from the surface of the structure: ω_n is the natural frequency of the acoustic mode with rigid boundaries. It should be noted that, because of a mathematical condition associated with infinite sums, termed Gibb's phenomenon, the summation of rigid-wall acoustic modes does not converge to the correct boundary normal velocity, but does converge correctly to the surface pressure, which is all that is needed for a correct formulation of the coupled equations.

Substitution into Eq. (6.15) of an expression for the structural displacement in terms of a summation over the *in vacuo* normal modes,

$$w(\mathbf{r}_s t) = \sum_{p=1}^{\infty} w_p(t)\phi_p(\mathbf{r}_s), \quad (6.26)$$

together with the application of Eq. (6.17), followed by multiplication by $\phi_q(\mathbf{r}_s)$ and integration over the structural surface, yields the coupled modal equations of motion for the structure,

$$\ddot{w}_p + \omega_p^2 w_p = \frac{S}{\Lambda_p}\sum_p p_n C_{np} + F_p/\Lambda_p, \quad (6.27)$$

where F_p is a generalised force given by $\int_S f(\mathbf{r}_s)\phi_p(\mathbf{r}_s)\,dS$ and $\Lambda_p = \int_S m(\mathbf{r}_s)\phi_p^2(\mathbf{r}_s)\,dS$. (Check the derivation of the equation yourself.) For the fluid,

$$\ddot{p}_n + \omega_n^2 p_n = -(\rho_0 c^2 S/\Lambda_n)\sum_p \ddot{w}_p C_{np} + (\rho_0 c^2/\Lambda_n)\dot{Q}_n. \quad (6.28)$$

C_{np} is a dimensionless coupling coefficient given by the integral of the product of the structural and acoustic mode shape functions over the surface of the structure:

$$C_{np} = \frac{1}{S}\int_S \psi_n(\mathbf{r}_s)\phi_p(\mathbf{r}_s)\,dS, \quad (6.29)$$

where S is the total surface area of the structure. Note that Sp_n has the dimensions of force and $S\ddot{w}_p$ the dimensions of volume acceleration, as comparison with the applied generalised force F_p and source strength Q_n shows they should. Equations (6.27) and (6.28) together actually represent a doubly infinite set of simultaneous differential equations; these sets have to be limited in order to obtain solutions.

These coupled equations of motion have been developed in terms of the acoustic pressure p because it is the dynamical quantity most commonly measured and most familiar to readers. However, the equations take on a more obvious and aesthetically satisfying symmetry if pressure is replaced by the acoustic-field velocity potential Φ. This is a scalar quantity in terms of which an irrotational fluid field can be completely described by its spatial distribution. Books on fluid mechanics, such as that by Duncan *et al.* (1970), present complete treatments of the velocity potential. It suffices here to state that the pressure and Cartesian components of fluid particle velocity are related to it by the following equations:

$$p = -\rho_0 \, \partial\Phi/\partial t,$$
$$u = \partial\Phi/\partial x, \qquad v = \partial\Phi/\partial y, \qquad w = \partial\Phi/\partial z.$$

The acoustic-pressure mode shape functions ψ_n and frequencies ω_n are also appropriate to Φ. Equations (6.27) and (6.28) can now be written in terms of the structural displacement w and Φ as

$$\ddot{w}_p + \omega_p^2 w_p = -(\rho_0 S/\Lambda_p) \sum_n \dot{\Phi}_n C_{np} + F_p/\Lambda_p, \tag{6.30}$$

$$\ddot{\Phi}_n + \omega_n^2 \Phi_n = (c^2 S/\Lambda_n) \sum_p \dot{w}_p C_{np} - c^2 Q_n/\Lambda_n. \tag{6.31}$$

Note that Q_n and not $\dot{Q}_n$ now appears as the acoustic source term, the sign of the boundary velocity term being opposite because $\dot{w}_p$ is directed out of the volume and Q_n is directed into the volume. The aforementioned symmetry is now apparent in the dependence of the coupling terms on $\dot{\Phi}_n$ and $\dot{w}_p$. This particular form of coupling was termed "gyrostatic coupling" by Rayleigh; unlike viscous coupling, which is also a function of the first time derivative, energy dissipation is not involved in the coupling mechanism.

Up to this point no allowance has been made for internal dissipation mechanisms in the structure or the fluid. The most practical means of representing dissipation is to employ ad hoc viscous damping terms in the modal equations; these usually take the form $\beta_p \dot{w}_p$ and $\beta_n \dot{\Phi}_n$, where β_p and β_n are generalised modal-damping coefficients normally based upon empirical data. The corresponding modal loss factors can only strictly be related to β_p and β_n in terms of the energies of modal vibration at the corresponding resonance frequencies

$$\omega_p^2 \Lambda_p \beta_p \overline{w_p^2} = \eta_p \omega_p \bar{E}_p = \eta_p \omega_p^3 \Lambda_p \overline{w_p^2}$$

or

$$\eta_p = \beta_p / \omega_p. \tag{6.32}$$

Similarly,

$$\eta_n = \beta_n / \omega_n.$$

In using these models it is implicitly assumed that damping coupling between modes is neglected. This is generally a reasonable assumption and is adopted so as to simplify modal analysis.

6.5 Solutions of the Modal-Interaction Model

The natural frequencies and modes of a coupled system are usually different from those of the individual uncoupled systems. The total energy of vibration of a mode of a coupled system is divided between the structure and the fluid. In many cases the greater proportion of energy resides either in the fluid or in the structure; this fact leads us to talk loosely of "fluid"- and "structure"-(dominated) modes. Although the coupled modes differ from the uncoupled modes, the motions and energies of the two (or more) coupled components may still be expressed in terms of combinations of the uncoupled modes, as the previous analysis shows: in this case these modes may simply be thought of as convenient functions rather than having physical significance. In particular, the energy of vibration of the system is equal to the sum of the energies of the uncoupled modes, these energies being computed according to the proportional contribution of each uncoupled mode to the coupled mode under consideration. This useful fact arises from the orthogonality of the uncoupled modes: of course, the coupled modes are orthogonal over the whole fluid–structure region, but not over any one sub-region.

It is revealing to analyse the vibration of a coupled system modelled in terms of just one uncoupled fluid mode and one uncoupled structural mode. Consider Eqs. (6.30) and (6.31) in this case, with simplified coefficients and ad hoc damping terms:

$$\ddot{w}_p + \beta_p \dot{w}_p + \omega_p^2 w_p = -K_{np} \dot{\Phi}_n + K_p F_p, \tag{6.33}$$

$$\ddot{\Phi}_n + \beta_n \dot{\Phi}_n + \omega_n^2 \Phi_n = G_{np} \dot{w}_p - G_n Q_n. \tag{6.34}$$

Let us first consider free vibration. We wish to solve for the natural frequencies ω:

$$(-\omega^2 + j\omega\beta_p + \omega_p^2)\tilde{w}_p = -j\omega K_{np}\tilde{\Phi}_n, \tag{6.35}$$

$$(-\omega^2 + j\omega\beta_n + \omega_n^2)\tilde{\Phi}_n = j\omega G_{np}\tilde{w}_p, \tag{6.36}$$

yielding

$$[(-\omega^2 + j\omega\beta_p + \omega_p^2)(-\omega^2 + j\omega\beta_n + \omega_n^2) - \omega^2 K_{np}G_{np}]\tilde{\Phi}_n = 0. \tag{6.37}$$

The expression in brackets equals zero in the non-trivial case.

In the undamped case

$$2\omega^2 = \omega_n^2 + \omega_p^2 + K_{np}G_{np} \pm [(\omega_n^2 - \omega_p^2)^2 + 2K_{np}G_{np}(\omega_n^2 + \omega_p^2) + K_{np}^2G_{np}^2]^{1/2}. \tag{6.38}$$

The product $K_{np}G_{np}$ is always positive because it is proportional to C_{np}^2. Suppose $\omega_p = \omega_n$; then

$$2\omega^2 = 2\omega_n^2 + K_{np}G_{np} \pm [4K_{np}G_{np}\omega_n^2 + K_{np}^2G_{np}^2]^{1/2}. \tag{6.39}$$

Since the value of the term in brackets exceeds $K_{np}G_{np}$, one of the new natural frequencies is greater than ω_n and one is less than ω_n, an example of the well-known frequency-splitting phenomenon exhibited by coupled oscillators of the same uncoupled natural frequencies. The interaction of an uncoupled fluid mode with an uncoupled structural mode can be likened to the influence of an auxiliary mass–spring system on a primary oscillator, a phenomenon utilised in the design of "dynamic absorbers" or "vibration neutralisers." Suppose $\omega_n \gg \omega_p$: then

$$2\omega^2 \simeq \omega_n^2 + K_{np}G_{np} \pm (\omega_n^2 + K_{np}G_{np}), \tag{6.40}$$

and the two natural frequencies ω_1 and ω_2 of the coupled system satisfy the conditions $\omega_1^2 \ll \omega_2^2$, $\omega_2^2 \simeq \omega_n^2 + K_{np}G_{np}$.

The magnitudes of the frequency changes from the uncoupled state clearly depend upon the magnitude of $K_{np}G_{np}$. Reference to Eqs. (6.30) and (6.31) show this term to be given by

$$K_{np}G_{np} = \rho_0 c^2 S^2 C_{np}^2/\Lambda_n\Lambda_p, \tag{6.41}$$

in which $0 \leqslant C_{np} \leqslant 1$, $\frac{1}{8}V \leqslant \Lambda_n \leqslant V$, and $\frac{1}{4}M \leqslant \Lambda_p \leqslant \frac{1}{2}M$, assuming sinusoidal structural modes and cosinusoidal fluid modes in a rectangular box geometry. Hence

$$(K_{np}G_{np})_{\max} = 32\rho_0 c^2 S^2/MV = 32\rho_0 c^2 S/V\langle m\rangle, \tag{6.42a}$$

where $\langle m\rangle$ is an average structural mass per unit area. As indicated by the ratio S/V, this term is greatest for fluid volumes having one dimension considerably less than the other two: these are termed "disproportionate volumes" in room acoustics. For a rectangular volume of side lengths d, e, and f, with a uniform flexible plate forming one face of dimensions e and f, the coupling term becomes

$$(K_{np}G_{np})_{\max} = 32\rho_0 c^2/md = 32\omega_0^2, \tag{6.42b}$$

where ω_0 is the fundamental natural frequency of a plate of mass per unit area m on a fluid spring formed by a cavity of depth d. The bulk modulus of the fluid, $\rho_0 c^2$, is, as expected, a controlling parameter. The effect of coupling in altering natural frequencies from their uncoupled values is clearly a function of the ratios ω_n/ω_0 and ω_p/ω_0: in the case of large ratios of both, Eq. (6.38) indicates that little change is expected.

In fact, the maximum value of the coupling term, given in Eq. (6.42), is extremely unlikely to be achieved in practice because of the characteristics of the coupling coefficient C_{np}. The value of this coefficient is a measure of the degree of spatial matching of structural- and acoustic-mode shapes at the fluid–structure interface. Modes are standing waves formed by the interference between multiply reflected traveling waves, and it is more useful for the present purpose of discussing C_{np} to consider the natural wavenumbers in the structure and fluid. For simplicity, consider the rectangular panel–box system mentioned above: the panel is located in the xy plane.

In the panel

$$k_{x_\mathrm{p}}^2 + k_{y_\mathrm{p}}^2 = k_\mathrm{b}^2, \tag{6.43}$$

and in the fluid

$$k_{x_\mathrm{f}}^2 + k_{y_\mathrm{f}}^2 + k_{z_\mathrm{f}}^2 = k^2. \tag{6.44}$$

For the best matching to occur, $k_{x_\mathrm{p}} \simeq k_{x_\mathrm{f}}$ and $k_{y_\mathrm{p}} \simeq k_{y_\mathrm{f}}$. Hence

$$k_b^2 + k_{z_\mathrm{f}}^2 \simeq k^2. \tag{6.45}$$

Now, at frequencies below the critical frequency of the panel, $k_\mathrm{b} > k$. Hence, for fluid modes in which k_{z_f} is real, the best matching condition cannot occur, the nearest approach corresponding to acoustic modes with $k_{z_\mathrm{f}} = 0$, that is to say, two-dimensional acoustic modes that have component wavevectors parallel to the plate. Best coupling for a flat uniform panel can only occur at frequencies above the critical frequency, in which case ω_n, $\omega_p \gg \omega_0$, except for extremely shallow cavities. The fundamental mode of a panel is the sole exception to this general form of behaviour because it can couple efficiently with the zero-order, bulk-compression mode of the fluid for which $n = 0$ and $\omega_n = 0$. In this case, assuming a simply supported panel,

$$C_{np} = \frac{1}{S}\int_0^e \int_0^f \sin\left(\frac{\pi x}{e}\right)\sin\left(\frac{\pi b}{f}\right) 1\, dS = \frac{4}{\pi^2},$$

$$K_{np}G_{np} = (64/\pi^4)\omega_0^2,$$

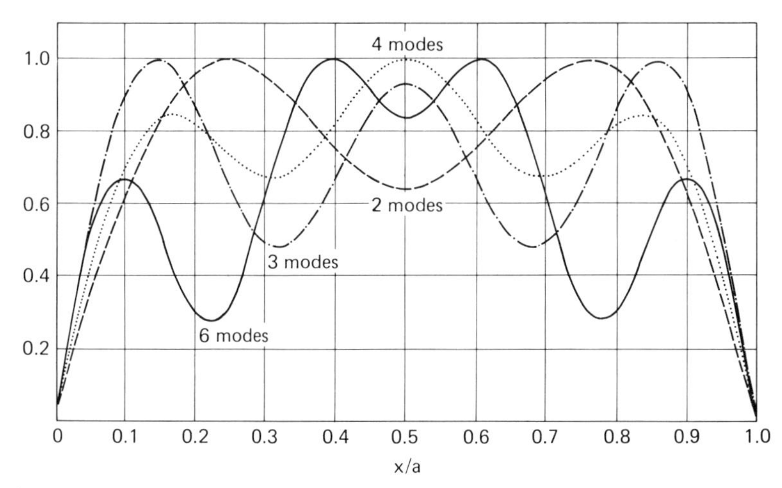

Fig. 107 Successive approximation to the fundamental modal deflection of the panel using an increasing number of *in-vacuo* panel modes in the analysis (Pretlove, 1965).

which is far less than $(K_{np}G_{np})_{\max}$. In this case, the mode shape of the panel in the fundamental coupled mode can be significantly altered by the influence of the fluid, as shown in Fig. 107 after Pretlove (1965).

The above argument based upon wavenumber matching must be modified for curved-shell structures because membrane stress effects can significantly increase flexural-wave phase velocities, and hence reduce flexural wavenumbers at given frequencies. This phenomenon has already been extensively discussed in Chapter 4 in relation to sound transmission through circular cylindrical shells. Wavenumber diagrams such as Fig. 102 show that natural flexural wavenumbers less than acoustic wavenumbers can exist in a circular cylinder below the critical frequency of a shell wall considered as a flat plate. Hence Eq. (6.45) can be satisfied in terms of the appropriate coordinate system, and good matching can occur, as indicated by the coincidence behaviour described in Chapter 4.

We now turn to the analysis of forced simple harmonic vibration of the coupled system. As already stated, if the coupling is fairly weak between the modes of interest (ω_n/ω_0, $\omega_p/\omega_0 \gg 1$), and direct excitation is applied either only to the fluid or only to the structure, then the relevant uncoupled modal equation [(6.31) or (6.30), respectively] is solved for the uncoupled modal response: ad hoc modal damping must be employed to limit the resonant responses. Then the other equation of the pair is solved with excitation arising only from the coupling term into which the uncoupled response of the other component is substituted.

Suppose the structure is directly excited by applied forces. Equation (6.30) becomes, with coupling neglected,

$$(-\omega^2 + j\omega\omega_p\eta_p + \omega_p^2)\tilde{w}_p = \tilde{F}_p/\Lambda_p. \tag{6.46}$$

Substituting for $\dot{w}_p = j\omega w_p$ in Eq. (6.31), modified to include acoustic-mode damping, and setting $Q_n = 0$, gives

$$\tilde{\Phi}_n = \frac{j\omega c^2 S}{\Lambda_n(-\omega^2 + j\omega\omega_n\eta_n + \omega_n^2)} \sum_p \frac{\tilde{F}_p C_{np}}{\Lambda_p(-\omega^2 + j\omega\omega_p\eta_p + \omega_p^2)}. \tag{6.47}$$

The total acoustic field is given by

$$\Phi(\mathbf{r}_s, t) = \exp(j\omega t) \sum_n \tilde{\Phi}_n \phi_n(\mathbf{r}_s). \tag{6.48}$$

The relative contributions to the acoustic field from the doubly infinite sum of modal terms depend upon the ratios ω/ω_p, ω/ω_n, the coupling coefficients C_{np}, and the generalised forces $\tilde{F}_p$. Maxima in the acoustic-frequency response would be expected when the excitation frequency is close to ω_p or ω_n, or to a pair of each, the magnitude of the maxima depending upon the relevant loss factors and the magnitude of the appropriate coupling coefficient.

The fact that this approximate method of response estimation yields maxima near the uncoupled-mode resonance frequencies is simply a result of ignoring the coupling of the two systems in Eqs. (6.46). Of course, the maxima in the frequency response of the coupled system should occur at frequencies near the resonance frequencies of the coupled system, obtained by solving the simple harmonic forms of Eqs. (6.30) and (6.31), with $F_p = Q_n = 0$. Conventional numerical eigenvalue techniques may be used to obtain solutions for the natural frequencies, once the infinite sets of equations are limited according to the frequency range of interest.

6.6 Power Flow Analysis

As an alternative to solving the integro-differential or modal equations of motion directly, a technique of estimating the time-average energy flow between the fluid and structure, called *Statistical Energy Analysis*, has been developed. It is most suitable for cases of broad-band excitation over a bandwidth encompassing many uncoupled-system natural frequencies. The reasons for this restriction are twofold: (1) the basic relationship between power flow and energy difference between oscillators (modes) takes a very simple form provided that the excitation spectrum extends over a frequency range encompassing the resonance frequencies of the uncoupled modes involved; (2) a statistical description of the parameters of the coupled system is only sensible and useful if a reasonably large population of modes is involved.

Each of the coupled systems is mathematically modelled as a set of oscillators corresponding to its uncoupled modes. As we have seen, the interaction between the systems can be represented exactly by coupled differential equations of motion in these modes, even though they do not form the actual orthogonal set of modes of the coupled system. The advantages of using the uncoupled modes are twofold: (1) knowledge of their spatial or wavenumber characteristics, and of their frequency-average distribution of natural frequencies, in terms of modal density, is more readily available than the equivalent information about the coupled system; and (2) the total time-average energies of each component system are exactly given by the sum of the energies expressed in terms of the uncoupled modes of that system.

The time-average power flow from a structure to a contiguous fluid may be expressed exactly as

$$\bar{P}_{12} = \frac{1}{T}\int_0^T\left[\int_S \dot{w}(\mathbf{r}_s)p(\mathbf{r}_s)\,dS\right]dt = \frac{\rho_0}{T}\int_0^T\left[\int_S \sum_p \dot{w}_p\phi_p(\mathbf{r}_s)\sum_n \dot{\Phi}_n\psi_n(\mathbf{r}_s)\,dS\right]dt, \tag{6.49}$$

where T is chosen to suit the frequency spectrum of the vibration.

The space and time integrals are independent, and therefore

$$\bar{P}_{12} = \rho_0\sum_p\sum_n \overline{\dot{w}_p\dot{\Phi}_n}\int_S \psi_p(\mathbf{r}_s)\phi_n(\mathbf{r}_s)\,dS = \rho_0 S\sum_p\sum_n C_{np}\overline{\dot{w}_p\dot{\Phi}_n}, \tag{6.50}$$

in which the overbar indicates time average and C_{np} is defined by Eq. (6.29). Referring back to Eq. (6.30), we can check Eq. (6.50) for energy conservation. Multiply Eq. (6.30), modified by a viscous damping term $\beta_p\dot{w}_p$, by $\dot{w}_p$ and consider the time average:

$$\Lambda_p(\overline{\ddot{w}_p\dot{w}_p} + \beta_p\overline{(\dot{w}_p)^2} + \omega_p^2\overline{w_p\dot{w}_p}) = -\rho_0 S\sum_n C_{np}\overline{\dot{w}_p\dot{\Phi}_n} + \overline{F_p\dot{w}_p}. \tag{6.51}$$

If the vibration is time stationary, the terms $\overline{\ddot{w}_p\dot{w}_p}$ and $\overline{w_p\dot{w}_p}$ are zero. (Why?) The term $\Lambda_p\beta_p\overline{\dot{w}_p^2}$ is the rate of dissipation of energy attributable to vibration in mode p, and the term $\overline{F_p\dot{w}_p}$ is the rate of work done by external forces attributable to motion in mode p. Note that the phrases "work done on mode p" and "energy dissipated by mode p" are avoided in order to reinforce the idea that these are not members of the set of actual orthogonal modes of the whole coupled system, but are mathematically useful functions in terms of which the motion of the structural portion of the coupled system may be expressed. Summation of the remaining terms in Eq. (6.51) over all p shows correctly that the total power injected into the structure is equal to the sum of that dissipated internally by the structure and that transferred to the fluid, and dissipated therein.

Of course, the power flow attributable to coupling between any two modes of the two uncoupled systems, expressed by Eq. (6.50) as $\rho_0 S C_{np} \dot{w}_p \dot{\Phi}_n$, cannot strictly be evaluated simply by considering the equations of motion of those two modes in isolation, since in principle all structural modes are coupled to all acoustic modes, as Eqs. (6.30) and (6.31) indicate. However, engineering analysis is largely about making reasonable assumptions that simplify the process of calculation and increase the ability of the analyst or user to understand the influence of the basic parameters, without reducing the accuracy of the results to an extent that renders them misleading or useless. Consequently, a commonly used simplifying assumption is that the total power flow between the structural and fluid systems is approximately equal to the sum of the power flows attributable to coupling between isolated pairs of modes.

Detailed analysis (Scharton and Lyon, 1968) shows that the time-average power flow between a pair of randomly excited, gyrostatically coupled, isolated oscillators is

$$\bar{P}_{12} = \eta_{12}\omega_c(\bar{E}_1 - \bar{E}_2), \tag{6.52}$$

where $\bar{E}_1$ and $\bar{E}_2$ are the oscillator energies, and the coupling loss factor η_{12} is given by

$$\eta_{12} = \frac{2\rho_0 c^2 S^2 C_{12}^2}{\Lambda_1 \Lambda_2}\left[\frac{(\beta_2\omega_1^2 + \beta_1\omega_2^2)(\omega_1 + \omega_2)^{-1}}{(\omega_1^2 - \omega_2^2)^2 + (\beta_2\omega_1^2 + \beta_1\omega_2^2)(\beta_1 + \beta_2)}\right], \tag{6.53}$$

in which the excitation band centre frequency ω_c has been replaced by the arithmetic mean of the modal natural frequencies, $(\omega_n + \omega_p)/2$; and C_{12}, Λ_1, and Λ_2 are equivalent to C_{np}, Λ_n, and Λ_p, respectively. It is significant that the coupling loss factor is symmetric in subscripts 1 and 2.

Equation (6.53) shows that the coupling loss factor is proportional to the square of the coupling coefficient and is extremely sensitive to the difference between the natural frequencies of the two oscillators. We may obtain approximate expressions for η_{12} that correspond either to conditions of close or proximate natural frequencies, or to conditions of distant or remote natural frequencies. In the first case we write $\omega_1 \simeq \omega_2 \simeq (\omega_1 + \omega_2)/2$. If, in addition, $2|\omega_1 - \omega_2| < \beta_1 + \beta_2$, then the first term in the denominator is small compared with the second and

$$\eta_{12} \simeq \frac{2\rho_0 c^2 S^2 C_{12}^2}{\Lambda_1 \Lambda_2}\left[\frac{1}{(\omega_1 + \omega_2)(\beta_1 + \beta_2)}\right]. \tag{6.54}$$

In this case the power flow produced by a given coupling coefficient between oscillators of given energy difference is inversely proportional to the sum of the modal loss factors.

In the case of remote natural frequencies, for which $2|\omega_1 - \omega_2| \ll \beta_1 + \beta_2$, the first term in the denominator is dominant. Since $\beta_1 = \eta_1\omega_1$ and $\beta_2 = \eta_2\omega_2$ are normally much less than ω_1 and ω_2, respectively, the coupling loss factor is, in this case, very small compared with that produced by proximate modal frequencies, given similar values of the coupling coefficient.

If it is assumed, as a first approximation, that the total power flow between a structure and a volume of fluid is given by the sum of the power flows attributable to coupling between isolated pairs of modes of the uncoupled components, then it is only necessary in practice to identify mode pairs that are well matched spatially and that have proximate natural frequencies; the other mode pairs can be discounted.

This approach was used by Fahy (1969, 1970) to evaluate the power flow between a rectangular plate and fluid in a box of which the plate formed one face, and also between a circular cylindrical shell and a contained fluid. It was found that the modal-average coupling loss factors so evaluated corresponded to those between the structures and the unbounded fluid external to the system, provided that the average difference between the natural frequencies of the modes of the two uncoupled components was not too large. This external radiation coupling loss factor is related to the radiation efficiency by $\eta_{rad} = \rho_0 c\sigma/\omega_c m$. The important conclusion to be drawn from these analyses is that, in cases in which the modal density of each component is sufficiently high, the broad-band power exchanged between a structure and an enclosed volume of fluid can be evaluated from a knowledge of the modal average free-space radiation characteristics. This convenient result does not apply to discrete-frequency or narrow-band excitation.

6.7 Wave Propagation in Structures Loaded by Fluid Layers

So far in this chapter we have assumed that the fluid is completely bounded, and consequently we have constructed modal models of the coupled structure–fluid system. As we have seen before, the modes of a system are formed by constructive interference between progressive waves, and consequently it is equally valid to formulate a wave propagation model for a coupled system. It transpires that the results of analyses of such models are particularly useful in cases where a plate or shell structure is coupled to a shallow "layer" of fluid confined within boundaries essentially parallel to the structural surface, in which case the close proximity of the fluid boundary can greatly modify the free-wave characteristics of the structure. Practical examples include fluid transport ducts, flooded sonar compartments in ships, double-leaf partitions, and the basilar membrane in the human auditory system. The latter, being a biological system, does not really come within the scope of this book, which is confined to linear, constant-parameter structures; however, like blood vessels, it shares some of the features of the systems dealt with below.

The most convenient model with which to illustrate the special features of wave propagation in structures loaded by shallow fluid layers is that of a fluid layer that separates two identical uniform elastic plates: a two-dimensional analysis is appropriate in this case. The system is shown in Fig. 108.

The acoustic pressure in the fluid satisfies the homogeneous wave equation subject to normal particle velocity boundary conditions on the surfaces of the plates. We shall seek a relationship between an assumed propagation wavenumber in the x direction and frequency; in other words, we wish to establish the dispersion characteristics of this system. Let the assumed wavenumber be k_x. The two-dimensional acoustic wave equation for the simple harmonic pressure $\tilde{p}(y)\exp[j(\omega t - k_x x)]$ yields

$$\partial^2 \tilde{p}(y)/\partial y^2 + (k^2 - k_x^2)\tilde{p}(y) = 0. \tag{6.55}$$

The solution for $\tilde{p}(y)$ is

$$\tilde{p}(y) = \tilde{A}\exp(-jk_y y) + \tilde{B}\exp(jk_y y), \tag{6.56}$$

with $k_y = (k^2 - k_x^2)^{1/2}$.

The equation of motion of the upper plate, neglecting external fluid, gives

$$(Dk_x^4 - \omega^2 m)\tilde{w}_1 = (\tilde{p})_{y=h}, \tag{6.57}$$

and for the lower plate gives

$$(Dk_x^4 - \omega^2 m)\tilde{w}_2 = -(\tilde{p})_{y=-h}. \tag{6.58}$$

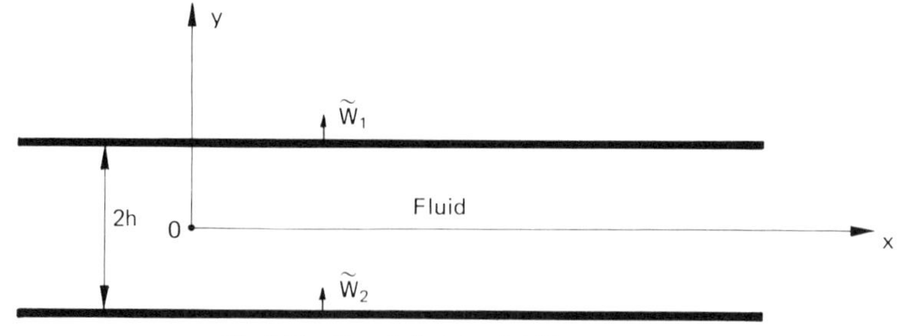

Fig. 108 Coupled plate–fluid-layer system.

The boundary conditions on the acoustic field are

$$[\partial\tilde{p}(y)/\partial y]_{y=h} = \rho_0\omega^2\tilde{w}_1, \quad (6.59)$$

$$[\partial\tilde{p}(y)/\partial y]_{y=-h} = \rho_0\omega^2\tilde{w}_2. \quad (6.60)$$

It is clear from considerations of symmetry that a solution must exist with $\tilde{w}_1 = -\tilde{w}_2 = \tilde{w}$. In this case, Eqs. (6.56), (6.59), and (6.60) yield $\tilde{A} = \tilde{B}$ and

$$\tilde{p}(y) = 2\tilde{A}\cos(k_y y), \quad (6.61)$$

giving a pressure amplitude $2\tilde{A}$ on the axis of symmetry. The solution for $\tilde{A}$ is

$$\tilde{A} = -\omega^2\rho_0\tilde{w}/2k_y\sin(k_y h). \quad (6.62)$$

Hence Eq. (6.57) becomes

$$D(k_x^4 - k_b^4)\tilde{w} = -\omega^2\rho_0\tilde{w}(k^2 - k_x^2)^{-1/2}\cot(k^2 - k_x^2)^{1/2}h, \quad (6.63)$$

where k_b is the *in vacuo*, free flexural wavenumber of the plate. This equation may be expressed in a non-dimensional form in terms of the ratio of axial to acoustic wavenumber k_x/k, the ratio of acoustic wavenumber to the acoustic (and flexural) wavenumber at the critical frequency $k/k_c = \omega/\omega_c$, a fluid-layer thickness parameter $k_c h$, and a plate mass parameter $m/\rho_0 h$:

$$\begin{aligned}(k_c h)(m/\rho_0 h)(k/k_c)^3[1 - (k_x/k)^2]^{1/2}[(k_x/k)^4 - (k_c/k)^2]\\ = -\cot[(k_c h)(k/k_c)(1 - (k_x/k)^2)^{1/2}]. \quad (6.64)\end{aligned}$$

This is the dispersion relationship for symmetric waves, which is, of course, independent of $\tilde{w}$ for the assumed linear system. It is effectively a fifth-order equation in k_x^2, but because of the presence of the trigonometric term it is an implicit equation in k_x/k, which must be solved by graphical or numerical techniques. Although numerical solutions are complete and accurate, the physical interpretation of equations such as (6.64) is often more clearly revealed by seeking approximate analytical solutions for various special cases. We do not know a priori whether k_x will be greater or less than k; that is to say, whether the axial phase velocity is smaller or greater than the speed of sound in the fluid. However, it is clear from examination of the equation that the forms of solution will be rather different in the two cases because of the terms $[1 - (k_x/k)^2]^{1/2}$.

We proceed by first considering cases with $k_x < k$.

(1) $k_x < k$: $k_x < k_b$. If $k_x < k_b$ then $(k_x/k)^4 \ll (k_c/k)^2$ and Eq. (6.64) may be approximated by

$$(m/\rho_0 h)(k/k_c)(k_c h)[1 - (k_x/k)^2]^{1/2} \simeq \cot[(k/k_c)(k_c h)(1 - (k_x/k)^2)^{1/2}], \quad (6.65)$$

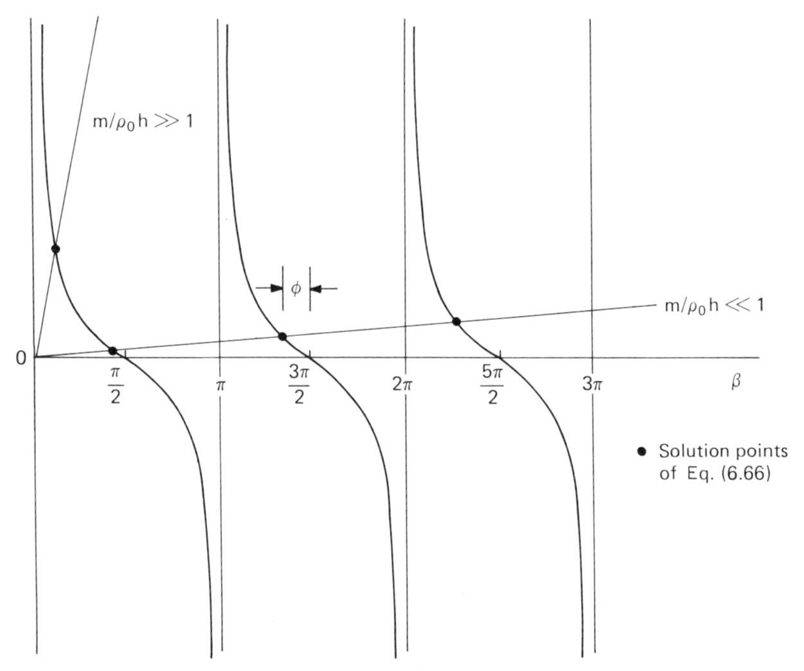

Fig. 109 Plots of cot β and $(m/\rho_0 h)$ vs. β.

or

$$(m/\rho_0 h)\beta \simeq \cot\beta, \tag{6.66}$$

where $\beta = (k/k_c)(k_c h)[1 - (k_x/k)^2]^{1/2}$. The two sides of Eq. (6.66) are plotted against β in Fig. 109.

When $m/\rho_0 h \gg 1$, the first intersection occurs at a value of β much less than unity. In this case, $\cot\beta \to (1 - \frac{1}{2}\beta^2)/\beta$ and

$$\beta^2 \simeq \rho_0 h/m$$

or

$$(k_x/k)^2 \simeq 1 - (\omega_0/\omega)^2, \tag{6.67}$$

where $\omega_0 = (\rho_0 c^2/mh)^{1/2}$ is the frequency of resonance of mass per unit area m on a rigidly terminated column of fluid of length h: it corresponds to the mass–air–mass frequency of double-leaf partitions. The ratio $(\omega_0/\omega_c)^2 = (1/12)(c_l'/c)^2(t/h)(\rho_0/\rho_s)$, where ρ_s is the density of the panel material. Since it was assumed initially that $k_x < k_b$, this solution is only valid if $k^2[1 - (\omega_0/\omega)^2] \ll k_b^2 = k/k_c$. This condition corresponds to a condition on frequency of $2(\omega/\omega_c) \ll 1 + [1 + (2\omega_0/\omega_c)^2]^{1/2}$. Propagating solutions corresponding to this first intersection (branch) do not exist for frequencies below ω_0 because k_x is then imaginary. When $\omega/\omega_0 \gg 1$, the wave corresponding to this branch propagates with an axial phase speed close to the speed of sound in the fluid.

Further intersections will occur at values of β close to $n\pi$. In these cases,

$$(k_x/k)^2 \simeq 1 - (n\pi/kh)^2, \qquad \text{integer } n \geqslant 1. \tag{6.68}$$

The condition $k_x < k_b$ is satisfied for $2(\omega/\omega_c) \ll 1 + [1 + (2n\pi/k_c h)^2]^{1/2}$ and propagation does not occur at frequencies ω less than $n\pi c/h$. The latter are the frequencies corresponding to n acoustic wavelengths between the plates. Figure 110a shows the behaviour of these various branches.

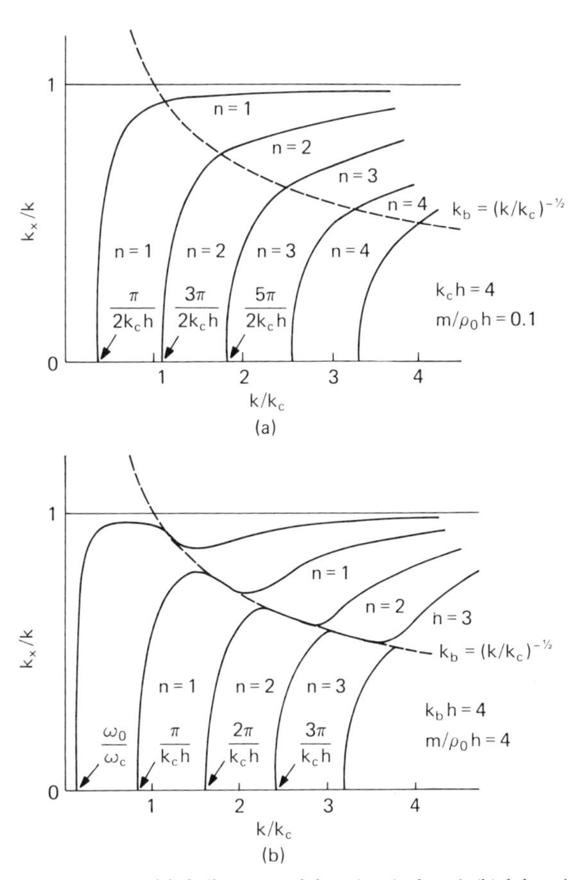

Fig. 110 Dispersion curves with $k_x/k < 1$. (a) $k_c h = 4$, $m/\rho_0 h = 4$; (b) $k_c h = 4$, $m/\rho_0 h = 0.1$.

If, on the other hand, $m/\rho_0 h \ll 1$ (Fig. 110b), intersections of the curves in Fig. 109 occur at values of β less than $(2n-1)\pi/2$ by a small value ϕ, which is much less than unity. Now $\cot[(2n-1)\pi/2 - \phi] \simeq \phi$. Hence intersection occurs when

$$\beta(1 + m/\rho_0 h) \simeq -(2n-1)\pi/2,$$

or

$$\left(\frac{k_x}{k}\right)^2 \simeq 1 - \left[\frac{(2n-1)\pi}{2kh(1 + m/\rho_0 h)}\right]^2, \qquad \text{integer } n \geqslant 1. \tag{6.69}$$

This wave does not propagate at frequencies below that for which

$$kh = \frac{(2n-1)\pi}{2(1 + m/\rho_0 h)}. \tag{6.70}$$

In the limit $m/\rho_0 h = 0$, these frequencies correspond to n half-wave-lengths between the "plates," which, of course, are the cutoff frequencies of acoustic modes in a two-dimensional waveguide with pressure-release ($p = 0$) boundaries. The assumed condition $k_x < k_b$ holds at frequencies satisfying

$$2(\omega/\omega_c) \ll 1 + \{1 + [(2n-1)\pi/k_c h(1 + m/\rho_0 h)]^2\}^{1/2}. \tag{6.71}$$

The transition of the wave characteristics from a similarity to those in a duct with rigid boundaries ($m/\rho_0 h \to \infty$), to a similarity to those in a duct with pressure-release boundaries ($m/\rho_0 h \to 0$), is seen clearly from Fig. 109 by following the intersection of straight lines of slope varying from infinity to zero with any one cotangent curve.

We now consider another special case:

(2) $k_x < k$: $k_x = k_b$. Equation (6.63) indicates that this will occur when $(k^2 - k_b^2)^{1/2} h = (2n-1)\pi/2$. Solving for frequency yields

$$2(\omega/\omega_c) = 1 + \{1 + [(2n-1)\pi/k_c h]^2\}^{1/2}. \tag{6.72}$$

Hence equality of the axial phase speed of the coupled wave with that of the *in vacuo* flexural wave in a plate can only occur above the critical frequency. At these frequencies the impedance of the fluid layer, as seen by the plate, is zero and therefore has no influence on the plate. Figure 110 shows how the branches revealed above for $k_x < k$ cross the k_b curve at frequencies given by Eq. (6.72). Although Eq. (6.72) indicates that the value of $\omega/\omega_c = k/k_c$ at which $k_x = k_b$, for a given n, is independent of $m/\rho_0 h$, it is clear from Fig. 110 that the tendency for the dispersion curves in this region to follow that of the *in vacuo* bending wavenumber k_b, increases as $m/\rho_0 h$ increases.

We now seek branches representing waves travelling with axial phase speeds greater than that of sound in the fluid, but less than that of the *in vacuo* flexural waves in the plate. This can obviously only occur at frequencies greater than the critical frequency.

(3) $k_x < k$: $k_x > k_b$. Equation (6.64) becomes

$$(k_c h)(m/\rho_0 h)(k/k_c)^3(k_x/k)^4[1 - (k_x/k)^2]^{1/2} = -\cot\{(k_c h)(k/k_c)[1 - (k_x/k)^2]^{1/2}\},$$

or

$$(m/\rho_0 h)(k_x/k)^4(k/k_c)^2\beta \simeq -\cot \beta. \tag{6.73}$$

It is difficult to generalise about this equation because, although k_x/k is less than unity, k/k_c, which must be greater than unity, may be indefinitely large. When $m/\rho_0 h \to 0$, solutions correspond to $\beta \to (2n - 1)\pi/2$, and when $m/\rho_0 h \gg 1$, solutions correspond to $\beta \to n\pi$.

Consequently it is necessary to resort to numerical solution, the form of which is shown in Fig. 110. Of course, if $m/\rho_0 h$ is very small, the branches correspond closely with pressure-release boundary modes except at very high frequencies, when $(k/k_c)^2$ is very large and there is a transition to rigid-wall duct behaviour. If $m/\rho_0 h$ is much greater than unity, the modes correspond closely to rigid-walled duct modes except close to the frequencies of intersection with k_b, given by Eq. (6.72).

So far we have assumed that k_x/k is less than unity. We now relax this condition, but in so doing, we should return to the original derivation of Eqs. (6.63) and (6.64) to check the physics.

(4) $k_x > k$: $k_x < k_b$. Here we seek waves that have axial phase speeds less than the speed of sound in the fluid but greater than *in vacuo* flexural waves in the plates. Returning to Eq. (6.56) we must write instead

$$\tilde{p}(y) = \tilde{A}\exp(-k_y y) + \tilde{B}\exp(k_y y), \tag{6.74}$$

where $k_y = (k_x^2 - k^2)^{1/2}$. Application of the boundary conditions that the plate displacements be in anti-phase yields $\tilde{A} = \tilde{B}$ and

$$\tilde{p}(y) = 2\tilde{A}\cosh(k_y y). \tag{6.75}$$

The solution for $\tilde{A}$ is

$$\tilde{A} = \omega^2\rho_0\tilde{w}/2k_y \sinh(k_y y). \tag{6.76}$$

Hence Eq. (6.57) becomes

$$D(k_x^4 - k_b^4)\tilde{w} = \omega^2\rho_0\tilde{w}(k_x^2 - k^2)^{-1/2}\coth(k_x^2 - k^2)^{1/2}k,$$

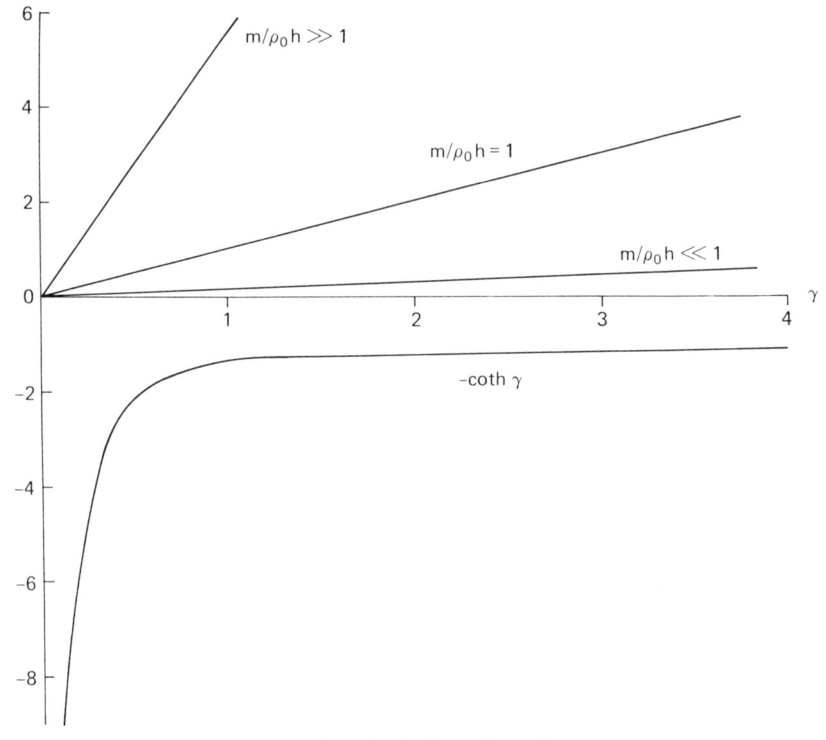

Fig. 111 Plot of $(m/\rho_0 h)\gamma$ and $-\coth\gamma$ vs. γ.

or

$$(k_c h)(m/\rho_0 h)(k/k_c)^3[(k_x^2/k)^2 - 1]^{1/2}[(k_x/k)^4 - (k_c/k)^2]$$
$$= \coth\{(k_c h)(k/k_c)[(k_x/k)^2 - 1]^{1/2}\}. \tag{6.77}$$

Applying the condition $k_x < k_b$ gives an equation equivalent to (6.66):

$$(m/\rho_0 h)\gamma \simeq -\coth\gamma, \tag{6.78}$$

where $\gamma = (k_c h)(k/k_c)[(k_x/k)^2 - 1]^{1/2}$. The two sides of this equation are plotted against γ for values of $m/\rho_0 h$ greater than, equal to, and less than unity in Fig. 111. It is seen that no solution exists for this combination of conditions. Let us see if waves can exist that have axial phase speeds less than both the speed of sound in the fluid and *in vacuo* flexural waves in the plates.

(5) $k_x > k$: $k_x > k_b$. Equation (6.77) reduces to

$$(k/k_c)^2(m/\rho_0 h)(k_x/k)^4\gamma \simeq \coth\gamma.$$

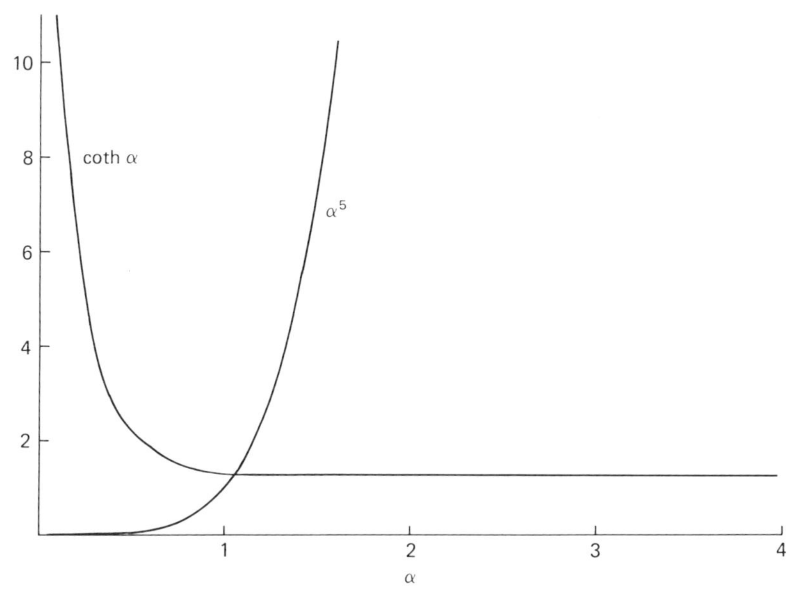

Fig. 112 Plots of coth α and α^5 vs. α.

This may also be written

$$(k/k_c)^{-2}(m/\rho_0 h)(k_c h)^{-4}\alpha^5 \simeq \coth\alpha, \tag{6.79}$$

where $\alpha = k_x h$ and it has been assumed that $[(k_x/k)^2 - 1]^{1/2} \simeq k_x/k$. The positive sign on the right-hand side allows this equation to possess a solution for any chosen values of the parameters k/k_c, $m/\rho_0 h$, and $k_c h$, as Fig. 112 shows. However, there is only one intersection, only one branch, and therefore only one type of wave. Numerical analysis shows that this branch corresponds to flexural plate waves that are inertially loaded by the evanescent fields in the fluid, and hence have phase speeds below their *in vacuo* values. The fact that no waves having $k_x < k_b$ and $k_x > k$ exist indicates that the fluid does not exert a stiffness-type loading on this branch. Above the critical frequency, this branch asymptotes to a wave with a constant-phase velocity just below that of sound in the fluid, as shown in Fig. 113.

The analyses of anti-symmetric wave motion, in which $\tilde{w}_1 = \tilde{w}_2$, and of axially evanescent fields are left as exercises for the (keen) student.

Analysis of the wave propagation characteristics of fluid-filled, thin-walled, circular cylindrical shells (Fuller and Fahy, 1981) shows them to be substantially more complex than those of the two-dimensional flat-plate–cavity system analysed above. One of the complicating factors is the existence of complex axial wavenumbers, even in the undamped *in vacuo* shell.

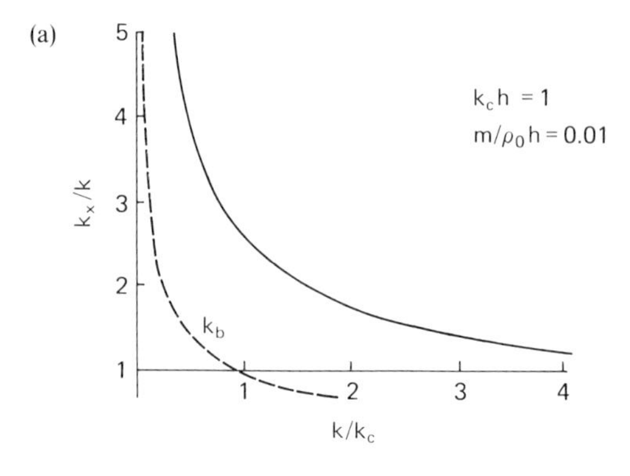

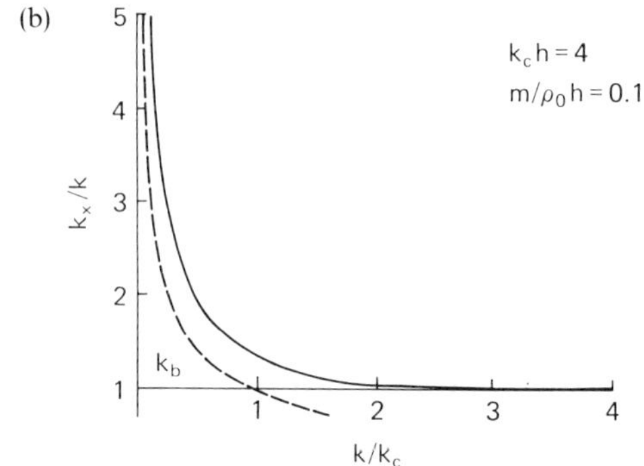

Fig. 113 Dispersion curves with $k_x/k > 1$. (a) $k_c h = 1$, $m/\rho_0 h = 0.01$; (b) $k_c h = 4$, $m/\rho_0 h = 0.1$.

Problems

6.1 Obtain explicit expressions for the modal fluid-loading terms α_{ppn} and α_{pqn} in Eq. (6.19) in the case of a rectangular, simply supported uniform panel that forms one wall of an otherwise rigid rectangular box. Hence demonstrate that inter-modal coupling may vary widely in magnitude and in dynamic form, i.e., analogous to stiffness or mass.

6.2 Why are the acoustic-pressure mode shapes and natural frequencies of an enclosure also appropriate to the velocity potential?

6.3 Check Eqs. (6.27) and (6.28) for dimensional consistency.

6.4 Prove that gyrostatic coupling of the form appearing in Eqs. (6.30) and (6.31) represents energy exchange between fluid and structural modes, and not energy dissipation in the coupling, unlike mode coupling through a viscous dashpot. Hint: Write the equations of motion of two mass–spring systems coupled through a dashpot.

6.5 Can you think of a simple coupled-beam system of which at least some of the modes of the coupled system have the same natural frequencies and mode shapes as the uncoupled systems?

6.6 Explain the ranges of Λ_n and Λ_p in Eq. (6.41).

6.7 Evaluate the maximum change from the *in vacuo* natural frequency of the fundamental panel mode of a rectangular panel on a rectangular box according to Eq. (6.38), when the box measures 300 × 400 mm and the panel is 1 mm thick, measures 500 × 400 mm, and is made of aluminium alloy. Assume the fluid medium is air at atmospheric pressure and 20°C. How would this frequency change if the temperature of the air were increased to 200°C, assuming that the pressure was maintained constant? What reasons would you have for doubting the validity of this model?

6.8 A thin square panel is simply supported across one end of a square section duct, the other end of which is open to the atmosphere. Assuming that the resulting fundamental natural frequency of the panel, loaded by the fluid in the duct, is sufficiently low for only plane-wave propagation to occur in the duct, and that the small-ka expressions for circular piston radiation impedance provide reasonable approximations to the open end impedance, derive an equation of free vibration of the panel. Evaluate the damped natural frequency and loss factor of the lowest "panel" mode for the following conditions: aluminium alloy panel of dimensions 300 × 300 × 2 mm; duct length 3 m; air in the duct. Numerical or graphical solutions may be employed. Obtain an equivalent piston radius by equating areas.

6.9 Write computer programs to check the accuracy of the approximate solutions to Eq. (6.64) for axial wavenumber k_x given in Section 6.7.

7 Introduction to Numerically Based Analyses of Fluid–Structure Interaction

7.1 The Role of Numerical Analysis

So far we have mainly considered analytical techniques for solving the equations that express relationships between the quantities we choose for our mathematical models of the physical world. The great advantage of analytic solutions is that they often display the dependence of the model behaviour on its non-dimensional parameters in an explicit fashion, and they lend themselves well to "large-parameter" and "small-parameter" asymptotic approximations, so that physical interpretation of the solutions may readily be achieved. The other side of the coin is that it is very often not possible to obtain analytic solutions to the model equations because of the geometric, dynamic, and/or kinematic complexity of the system modelled. The prime example of this problem is that of fluid turbulence; the governing equations are well known, but a general analytic solution cannot be found. As seen in the previous chapter, the vibrational behaviour of bounded fluid and solid systems may generally be expressed in terms of series solutions to the governing equations; however, it is generally not possible to derive analytic expressions for the terms in the series, unless the systems and their boundaries are particularly simple.

As an alternative to seeking approximate analytical means of solving equations, we may substitute numbers into the various terms of our equations and see how well the equations and their boundary conditions are satisfied. We may then employ progressive adjustment (iterative) techniques to "travel" toward a solution that satisfies the equations and boundary conditions to an acceptable degree of accuracy: the choice of the acceptability criteria depends upon the requirements of the particular applications. We hope that the iteration process converges and that the result converged upon is physically valid. Many systematised routines for performing numerical analysis have been introduced and the development of efficient computational techniques for this implementation is an important area of research.

It cannot be emphasised too strongly that the first priority in the mathematical analysis of the behaviour of physical systems is to generate a mathematical model appropriate to the desired objective of the analysis and to the likely degrees of influence and ranges of variation of the physical quantities involved. In other words, a model must incorporate representations of all the physical quantities thought to exert significant influences on the behaviour of the primary quantities of interest, and the selection of the quantities must be made on the basis of a reasonably thorough qualitative understanding of the physics of a problem, and of the physical inter-relationships between the quantities involved. The desired degree of correspondence between the behaviour of the mathematical model and that of the physical system modelled necessarily influences the detail to which the modelling process extends. The reader may like to consider the latter point in relation to the choice of reference coordinate system for the analysis of dynamic behaviour of systems on the earth's surface. When and why should account be taken of the rotation of the earth? Closer to home, the question of how to represent the boundaries of a distributed system, which is part of a much more extensive system, is a problem frequently encountered in the fields of vibration and acoustic analysis. The reason for emphasising the paramount importance of giving very careful thought to the process of generating a mathematical model is that, however sophisticated the numerical-analysis techniques used, an inadequate model yields inadequate equations, and the subsequent solution is of little value. The technique of solution must never take pride of place over the model to which it is applied.

Numerical analysis can, of course, be performed to any degree of precision desired, subject to the limitations of the precision to which numbers can be held and manipulated in the computer. Although this is generally very high, significant errors can accumulate as a result of successive arithmetic operations, and vigilance in this respect should always be maintained. Although very precise numerical solutions can usually be obtained, the influence of individual system parameters may not be easy to evaluate from a welter of computer output, and much tedious interpretive work may often be avoided by employing a judicious combination of approximate analytical techniques and sensible parametric studies on the computer. Naturally the acid test is whether the theoretical results compare favourably with observed physical behaviour, and a healthy scepticism regarding the results of theoretical studies is the hallmark of the mature analyst.

Having noted the traps awaiting the unwary, let us now proceed with caution to consider various approaches to the numerical analysis of vibrating systems.

7.2 Numerical Analysis of Sound Fields

We concentrate here on sound fields rather than vibrational fields of solid structures because numerical analysis of structural vibration forms a complete specialist field in its own right, possessing a vast literature (e.g., Cook, 1981). The problems of practical interest divide into three main categories: (1) analysis of the free- and forced-vibration behaviour of fluid in enclosed spaces; (2) analysis of the acoustic fields generated in essentially unbounded regions of fluid by vibrating surfaces, and (3) the interaction between flexible structures and sound fields incident upon them.

We have already seen that fluid-loading effects may sometimes significantly affect structural motion, in which case a fully coupled analysis is necessary. Examples of such cases include underwater sonar transducers and liquid containment vessels such as fuel tanks and nuclear reactor pressure vessels. Whatever form of numerical analysis is used, the essential feature common to all is that field quantities that are actually continuous throughout a region of an elastic continuum are represented by their numerical values at discrete points in space, together, in some cases, with assumed interpolation functions, which approximate the distribution of these quantities between the points. A sound field in a fluid has to satisfy the wave equation, together with certain conditions on bounding surfaces and, in the case of transient problems, certain initial conditions. In the case of steady-state radiation into a fluid region that is infinitely extended, certain conditions at infinity (the Sommerfeld condition), which ensure that the field is physically valid, also have to be applied.

Analysis of sound fields in closed or nearly closed volumes is most usually accomplished by using *Finite-Element* (FE) *analysis*, in which the fluid space is theoretically divided into contiguous elements of linear dimension substantially smaller than an acoustic wavelength at the highest frequency of interest.

This requirement is analogous to the Shannon criterion for the digital representation of signals, by which the sampling rate must exceed twice the highest frequency present in a signal. Field-variable distributions are assumed, and expressions for potential and kinetic energies, and work done by applied forces, are derived in terms of variables assigned as degrees of freedom at nodal points on element boundaries. Then Lagrange's equations are applied and solved for the degrees of freedom that define the field. This technique is more useful for enclosed volumes than for unbounded fluid volumes because the necessary number of degrees of freedom is limited. It has been used most successfully for the analysis of sound fields in enclosures of arbitrary boundary geometry (Petyt *et al.*, 1976), and for studying the behaviour of sound in ducts (Eversman and Astley, 1981).

It is in principle possible to apply *Finite-Difference* (FD) *analysis* to sound fields (Shuku, 1975; Cabelli, 1982). Again, the fluid region is divided up by a line grid and field values are assigned to the grid intersection points. In a rectangular Cartesian coordinate system the grid is square, whereas FE analysis can take diverse geometric forms. The reason is that in FD analysis the derivatives in the partial differential wave equation are represented by finite-difference approximations, those pertaining to any given point involving values at surrounding grid points. An initial set of field values is assumed, and systematic techniques exist for iterating the values to a stable solution. The main practical problem with FD analysis in application to sound fields in volumes of arbitrary geometry is that it does not readily accommodate boundaries that do not conform to the grid line pattern. In addition, it is much more sensitive to local errors of field representation than the FE method, in which intermediate energy expressions, which are the results of integration over the whole region, are effectively formed.

For the purpose of evaluating the field radiated by a vibrating surface into an infinitely extended volume of fluid, it is usual to apply numerical techniques to the evaluation of the field on the surface of the radiator, through the approximate solution of the Kirchhoff–Helmholtz integral Eqs. (3.1) and (6.4); the radiated field can easily be determined once the surface field is known. Numerical solutions are usually required because the surface geometry of the radiator is not one of the small family for which the Green's function can be expressed in closed, or series, form and satisfies either zero pressure or zero normal particle velocity on the surface. Hence it is usual to use the free-space form of the Green's function, which for simple harmonic time variation is given by Eq. (6.3).

The radiating surface is divided up into discrete elements and the aim of the analysis is to obtain an estimate of the surface pressure distribution that corresponds to the assumed normal velocity distribution. As we shall see, there are certain fundamental analytical problems to be overcome before numerical analysis can be successfully applied to this problem. In the fully coupled case, where the surface is driven by some form of applied forces, or is subject to an incident sound field, it is of course necessary to solve simultaneously for the surface field and for the motion, which both generates and is affected by the fluid loading. In the following sections the principles of various numerical techniques are introduced, and some of the practical aspects of their application are discussed.

7.3 Finite-Element Methods

Finite-element analysis may be applied to any physical system of which the behaviour is described by a set of partial differential equations, the independent variables usually being space and time. The approach consists in the subdivision of the material space into a finite set of contiguous elements defined by a gridwork of lines (mesh), which may be straight or curved. At discrete points (nodes) on this grid, which may lie solely at intersections, or also at intermediate points, certain field variables, plus their spatial derivatives, which are pertinent to the problem under consideration, are selected as the nodal degrees of freedom. The distributions of the variables between the nodes are assumed to take simple forms that satisfy certain continuity conditions at the nodes; the actual continuity conditions depend upon the physical nature of the field under investigation.

In cases of analysis of vibration of a fluid or solid system, or combination thereof, use is made of a fundamental principle of mechanics known as Hamilton's principle, which states that "among all the displacements that satisfy the presented (geometric) boundary conditions and the prescribed conditions at $t = t_1$ and $t = t_2$, the actual solution renders the integral $\int_{t_1}^{t_2} (T - U + W)\,dt$ stationary." In this integral, T denotes the total kinetic energy of the system, U the total potential energy of the system, and W the work done by any non-conservative forces acting on or within the system. (How would you define a non-conservative force?) The application of this principle is often referred to as the "variational approach" because the condition may also be stated as

$$\delta \int_{t_1}^{t_2} (T - U + W)\,dt = 0, \tag{7.1}$$

in which δ denotes the first variation of the integral. The potential (strain) energy of an element of a linear elastic system can always be expressed in terms of the square of the deformation displacement of the element, and therefore in terms of the square of any other variable linearly related to this displacement, such as elastic stress or strain. The kinetic energy is expressed in terms of the element speed relative to an appropriate frame of reference: for simple harmonic motion, speed is linearly related to displacement. The work done on a system by applied forces is a function of the displacement of the point of action relative to an appropriate frame of reference. Hence, knowledge of the applied forces and the elastic displacement field allows the quantities in Eq. (7.1) to be evaluated. (What frame of reference would you choose for the analysis of vibration of a blade on a manoeuvring helicopter, and why?)

Suppose that the state of a vibrating distributed system is expressed in terms of the values of field variables and their derivatives (degrees of freedom) at the chosen set of discrete nodal points. The energy and work associated with the vibration of each element of the system can be evaluated only if expressions for the distributions of the appropriate field variable in the region between the nodal points are available, because the energy and work are expressed as integrals over the continuous extent of the element. Therefore, the spatial variation of the particular field variable must be prescribed by the analyst. In practice, what is done is to choose prescribed functions such that each one corresponds to unit value of a particular degree of freedom at one nodal point in the element, while corresponding to zero value for all other degrees of freedom at all the nodal points of the element. Hence the value of the prescribed variable at any point in an inter-nodal region can be expressed as a simple summation of the individual influences of each degree of freedom at each node, each contribution being weighted by the value of the appropriate prescribed function at the interior point. Finite elements differ in their geometry, number of nodes, number and natures of degrees of freedom, and forms of prescribed function.

As a simple example, let us consider a one-dimensional elastic system in the form of a uniform beam in which the transverse bending displacement is w and the spatial coordinate is x. We imagine the system to be divided up into lengthwise elements with nodes at the element junctions. As degrees of freedom we arbitrarily choose w and its first space derivative w'.* Since the values of w and w' at each end of the element are to be allowed, through the prescribed functions, to exercise four independent influences on the variation of w with x, it should be possible to find a polynomial expression for w having four independent coefficients, which are determined by the unit value requirement on the prescribed functions described above.

Let

$$w(x) = w(0)f_1(x) + w'(0)f_2(x) + w(l)f_3(x) + w'(l)f_4(x), \tag{7.2}$$

* As we shall see later, this choice may not be entirely arbitrary.

in which l is the element length and f_1, f_2, f_3, and f_4 are the prescribed functions. Also let $w(x)$ be described by the polynomial expression

$$w(x) = a_0 + a_1 x + a_2 x^2 + a_3 x^3. \tag{7.3}$$

Then

$$\begin{aligned} w(0) &= a_0, \qquad w'(0) = a_1, \\ w(l) &= a_0 + a_1 l + a_2 l^2 + a_3 l^3, \qquad w'(l) = a_1 + 2a_2 l + 3a_3 l^2. \end{aligned} \tag{7.4}$$

Solving for the functions f gives

$$\begin{aligned} f_1(x) &= 1 - 3x^2/l^2 + 2x^3/l^3, \qquad f_2(x) = x - 2x^2/l + x^3/l^2, \\ f_3(x) &= 3x^2/l^2 - 2x^3/l^3, \qquad f_4(x) = -x^2/l + x^3/l^2. \end{aligned} \tag{7.5}$$

Note that $f_1(0) = f'_2(0) = f_3(l) = f'_4(l) = 1$, $f_1(l) = f'_2(l) = f_3(0) = f'_4(0) = 0$, as required.

Let the beam have cross-sectional area A, cross-sectional second moment of area I, elastic modulus E, and density ρ. The potential energy of each element is

$$U_e = \tfrac{1}{2}EI \int_0^l (\partial^2 w/\partial x^2)^2 \, dx, \tag{7.6}$$

in which, from Eq. (7.2),

$$\partial^2 w/\partial x^2 = w(0)f''_1(x) + w'(0)f''_2(x) + w(l)f''_3(x) + w'(l)f''_4(x).$$

Thus by using Eq. (7.5), the potential energy may be expressed in terms of the degrees of freedom at the element nodes. Similarly, the kinetic energy may be expressed in terms of the time derivatives of the quantities defined as the degrees of freedom as

$$T_e = \tfrac{1}{2}\rho A \int_0^l (\dot{w})^2 \, dx, \tag{7.7}$$

in which, from Eq. (7.2),

$$\dot{w}(x) = \dot{w}(0)f_1(x) + \dot{w}'(0)f_2(x) + \dot{w}(l)f_3(x) + \dot{w}'(l)f_4(x).$$

It should be noted carefully that the prescribed functions must be continuous at each node in derivatives of order up to $p - 1$, where the highest derivative in the energy expressions is of order p: in the case above, only the displacement w and its spatial derivative w' need to be continuous. Having obtained expressions for the quantities in Eq. (7.1) for one element, the equivalent expression may now be obtained for the whole system by addition.

The application of Eq. (7.1) to a system in which the energies are expressed in terms of a set of independent displacement degrees of freedom q_i, and their time derivatives $\dot{q}_i$, produces Lagrange's equations

$$\frac{d}{dt}\left(\frac{\partial T}{\partial \dot{q}_i}\right) + \frac{\partial D}{\partial \dot{q}_i} + \frac{\partial U}{\partial q_i} = f_i, \tag{7.8}$$

in which D is an energy dissipation function that represents the dissipative components of the internal non-conservative forces and for which the instantaneous rate of dissipation of energy associated with the degree of freedom q_i is equal to $D\dot{q}_i^2$. The generalised forces f_i are defined as the work done during unit displacement q_i.

It is customary to express the various equations used in FE analysis in matrix form, which is concise and well adapted to computational procedures. In the case of the elastic beam discussed above, the set of degrees of freedom of each element may be written as a column matrix $\{q\}_e$, where $\{q\}_e^{\mathrm{T}} = \lfloor w(0)\, w'(0)\, w(l)\, w'(l) \rfloor$. Equation (7.2) may be written as

$$w(x) = \lfloor F \rfloor \{q\}_e, \tag{7.9}$$

where $\lfloor F \rfloor = \lfloor f_1\ f_2\ f_3\ f_4 \rfloor$. Hence the potential energy of an element becomes, from Eq. (7.6),

$$U_e = \tfrac{1}{2}\{q\}_e^{\mathrm{T}}[K]_e\{q\}_e, \tag{7.10}$$

where

$$[K]_e = EI \int_0^l \lfloor F'' \rfloor^{\mathrm{T}} \lfloor F'' \rfloor \, dx, \tag{7.11}$$

and $\lfloor F'' \rfloor = \lfloor f_1''\ f_2''\ f_3''\ f_4'' \rfloor$. The kinetic energy of an element becomes, from Eq. (7.7),

$$T_e = \tfrac{1}{2}\{\dot{q}\}_e^{\mathrm{T}}[M]_e\{\dot{q}\}_e, \tag{7.12}$$

where

$$[M]_e = \rho A \int_0^l \lfloor F \rfloor^{\mathrm{T}} \lfloor F \rfloor \, dx. \tag{7.13}$$

The integrals in Eqs. (7.11) and (7.13) may easily be evaluated using the expressions for the prescribed functions in Eq. (7.5). (Try it.) The matrices $[K]_e$ and $[M]_e$, which are square symmetric (check this statement), may be termed "stiffness" and "mass" matrices, although the dimensions of only some of their elements actually have the dimensions of simple spring stiffness (MT^{-2}) or mass (M) because not all the degrees of freedom share the form of linear physical displacements.

Up to this point we have not considered the geometric boundary conditions of the beam, any applied forces or non-conservative internal forces. Only those geometric boundary conditions involving derivatives of order up to $p - 1$ need to be explicitly satisfied, by simply eliminating those prescribed functions in the appropriate bounding elements that do not satisfy them.

Expressions for the work done by external forces may be obtained in terms of integrals over the elements of products of the applied force distributions and the displacements w, that is, in terms of $\{q\}_e$ and integrals of products of the applied-force distribution and $[F]$: hence a work matrix will be formed. The dissipation function matrix may be evaluated in a similar fashion once a model for the dissipation mechanism has been chosen.

Once all the element matrices have been formed, the equivalent system matrices are formed by addition. Since only a subset of the system degrees of freedom applies to any one element, the matrix relating the system degrees of freedom $\{q_i\}$ to the element degrees of freedom $\{q\}_e$ must first be formed, thus

$$\{q\}_e = [\alpha]_e\{q_i\}, \tag{7.14}$$

where, in this case,

$$[\alpha]_1 = \begin{bmatrix} 1 & 0 & 0 & 0 & 0 & 0 & 0 & \cdots \\ 0 & 1 & 0 & 0 & 0 & 0 & 0 & \cdots \\ 0 & 0 & 1 & 0 & 0 & 0 & 0 & \cdots \\ 0 & 0 & 0 & 1 & 0 & 0 & 0 & \cdots \end{bmatrix},$$

since each element has four degrees of freedom. Summing the energies over all elements gives

$$T = \tfrac{1}{2}\{\dot{q}_i\}^T[M]\{\dot{q}_i\}, \tag{7.15}$$

where

$$[M] = \sum_{e=1}^{n} [\alpha]_e^T [M]_e[\alpha]_e,$$

and there are n elements, and

$$U = \tfrac{1}{2}\{q_i\}^T[K]\{q_i\}, \tag{7.16}$$

where

$$[K] = \sum_{e=1}^{n} [\alpha]_e^T[K]_e[\alpha_e].$$

Application of Lagrange's equation yields $4n$ coupled equations of motion of the form

$$[M]\{\ddot{q}_i\} + [C]\{\dot{q}_i\} + [K]\{q_i\} = \{f_i\}, \tag{7.17}$$

where it has been assumed that the damping may be modelled by a linear viscous model. These equations may be solved directly, but in practice a large number of degrees of freedom makes a direct solution costly. Hence it is common to use a modal substitution method, by which the undamped, free-vibration problem (eigenvalue problem) with $f_i = 0$ is solved by one of a number of standard methods (see Jennings, 1977), for a limited number of the lowest natural frequencies ω_n and modes ψ_n, which very often dominate the total response. The system degree-of-freedom matrix is then transformed as follows:

$$\{q_i\} = [\psi_n]\{P_n\}. \tag{7.18}$$

The columns of matrix $[\psi_n]$ represent the natural modes of vibration of the structure; if only a small number are included $\{P_n\}$ has fewer elements than $\{q_i\}$. If, in addition, it is valid to model the dissipation process in terms of a mass- and/or stiffness-proportional viscous damping, the resulting equations for P_n are completely uncoupled and the usual single-degree-of-freedom solutions apply. Such a procedure is not valid in the case of significant fluid loading of a structure, unless the fluid is completely bounded by nondissipative boundaries, because the modes of structures loaded by an infinitely extended fluid are not orthogonal (Mkhitarov, 1972).

In the case of a two-dimensional rectangular finite element with four corner nodes, the distribution of any variable within the element may be described by a product of prescribed functions so that the unit dependence requirement is maintained at each node.

The energy functional appropriate to any particular problem depends upon the physical nature of the system and the form of vibration modeled. For instance, the energy expressions for bending vibrations of a uniform thin flat plate are

$$U = \frac{1}{2} D \int_S \left\{ (\nabla^2 w)^2 - 2(1 - \nu)\left[\frac{\partial^2 w}{\partial x^2}\frac{\partial^2 w}{\partial y^2} - \left(\frac{\partial^2 w}{\partial x\,\partial y} \right)^2 \right] \right\} dS, \tag{7.19}$$

$$T = \frac{1}{2} m \int_S (\dot{w})^2\, dS, \tag{7.20}$$

where $\nabla^2 = (\partial^2 w/\partial x^2) + (\partial^2 w/\partial y^2)$ and D and m have their other usual meanings.

Having seen how FE analysis can be applied to a simple elastic structure, we can now extend the principles to a bounded volume of fluid. The potential energy density of an acoustic field is related to the acoustic pressure by

$$\varepsilon_U = \tfrac{1}{2} p^2/\rho_0 c^2. \tag{7.21}$$

The kinetic energy density is most obviously expressed in terms of the acoustic particle velocity as

$$\varepsilon_T = \tfrac{1}{2}\rho_0 |\mathbf{u}|^2 \tag{7.22}$$

where $\mathbf{u} = u\mathbf{i} + v\mathbf{j} + w\mathbf{k}$.

It is clearly sensible to express both forms of energy in terms of the same field variable. There are a number of possibilities including pressure, displacement potential, velocity potential, particle displacement, and particle velocity. The one that seems most suitable at first sight is the pressure because it is a scalar and a "forcelike" quantity, which will complement the displacement commonly used for structural analysis. In particular, it is useful to have this combination when considering coupling work or energy exchange. The choice of variable is discussed in detail by Gladwell (1966).

If we restrict our attention to simple harmonic problems or to the equivalent Fourier components of stationary vibration, we can readily convert Eq. (7.22) into a pressure form, using the fluid momentum Eqs. (1.5):

$$\varepsilon_T = (\tfrac{1}{2}\rho_0\omega^2)[(\partial p/\partial x)^2 + (\partial p/\partial y)^2 + (\partial p/\partial z)^2], \tag{7.23}$$

where the pressure gradient terms represent the real magnitudes of the time-varying quantities. The potential energy density remains in the same form as Eq. (7.19). The boundary condition for a rigid-walled cavity is that of zero normal particle velocity, or

$$\partial p/\partial n = 0. \tag{7.24}$$

Because the highest-order spatial derivative of pressure in Eq. (7.23) is of order one, it is only necessary to satisfy continuity of pressure at the nodes of the finite elements.

Selection of node points and prescribed functions, which in principle may be simply linear, enables the derivatives in Eq. (7.23) to be evaluated in terms of the nodal pressures. Hence the energies of elements may be found in terms of integrals of Eqs. (7.21) and (7.22) over the element volume, and then the total energies of the system obtained by a matrix assembly procedure similar to that demonstrated in the case of the beam. Hence the problem of finding the acoustic natural frequencies and mode shapes of a cavity reduces to an application of Rayleigh's principle when these (maximum) energy expressions are equated. (What difference do you notice between the ratio that gives ω^2 and that for an oscillating mechanical system when written in terms of displacements?)

The shapes and sizes of the finite elements must be chosen to suit the particular geometry and frequency range being investigated. Although it is quite acceptable to employ spatial gradients of pressure, or velocity potential, as degrees of freedom, it appears that the most versatile finite elements use only the scalar quantity as degrees of freedom, but employ polynomial prescribed functions and additional nodes at mid-points of the element sides; these are known as iso-parametric elements (e.g., Petyt *et al.*, 1976). A one-dimensional example of such an element of total length l would have the distribution

$$P(x) = a_0 + a_1 x + a_2 x^2 \tag{7.25}$$

or

$$P(x) = p(0) f_1 + p(l/2) f_2 + p(l) f_3, \tag{7.26}$$

where

$$f_1 = 1 - 3x/l + 2x^2/l^2, \qquad f_2 = 4x/l - 4x^2/l^2, \qquad f_3 = -x/l + 2x^2/l^2.$$

These functions satisfy the necessary conditions

$$\begin{aligned} f_1(0) &= 1, & f_2(0) &= 0, & f_3(0) &= 0, \\ f_1(l/2) &= 0, & f_2(l/2) &= 1, & f_3(l/2) &= 0, \\ f_1(l) &= 0, & f_2(l) &= 0, & f_3(l) &= 1. \end{aligned}$$

In this case, the derivative $\partial p/\partial x$ at one end, say $x = l$, is given by

$$(\partial p/\partial x)_l = [p(0) - 4p(l/2) + 3p(l)]/l,$$

so that the over-restrictive condition produced by the linear prescribed function is relaxed, giving generally more accurate results, at the expense of increasing the size of the computational task.

Where structural and fluid systems are coupled together, energy expressions can be written for the whole dynamical system as the sum of the energies of each in terms of appropriate field quantities. The coupling forces acting at the interface between the systems are non-dissipative internal forces as far as the complete system is concerned and hence will not appear explicitly. Nodes on the interface are common to both systems, and boundary conditions of equality of normal particle velocity can be satisfied at these nodes. Because different forms of prescribed function are usually used for structures and fluids it is not usually possible, or actually necessary, to satisfy the boundary conditions in internodal regions, provided that the element mesh is adequately fine.

The coupling between the two components of the system will be manifested in the presence in the structural-energy expressions of acoustic degrees of freedom for the fluid elements adjacent to the interface; and in the fluid-energy expressions there will appear structural displacement degrees of freedom for the same reason. (Why is this so?)

As an alternative to writing a single energy expression for the whole system, separate expressions may be written for the two components, but then the actions of the coupling forces have to be explicitly incorporated in the work component of the functional, which is to be made stationary by Eq. (7.1). The generalised force distribution on the structure is rather easily determined since the work done on element δS by acoustic pressure p during normal displacement δw into the fluid is

$$\delta W = -p\,\delta w\,\delta S. \tag{7.27}$$

The work associated with incremental normal structural displacement degree of freedom δw_s of any one structural element, due to fluid pressure distribution over the interface between that element and the associated fluid element is given by

$$\delta W_s = -\int_{S_e} p(S_e)\,\delta w_s(S_e) f_s(S_e)\,dS_e, \tag{7.28}$$

where $f_s(S_e)$ is the prescribed function associated with w_s and the integration extends over the element interface S_e. The pressure distribution $p(S_e)$ is associated with all the degrees of freedom of the acoustic finite element on the interface; let us call them P_a. With each P_a is associated a prescribed function $f_a(S_e)$. Hence the total work done by the fluid field in incremental displacement of one structural degree of freedom δw_s is given formally by

$$\delta W_s = -\delta w_s \sum_a P_a \int_{S_e} f_a(S_e) f_s(S_e)\,dS_e. \tag{7.29}$$

The generalised force associated with each combination of P_a and w_s is

$$f_{as} = P_a G_{as}, \tag{7.30}$$

where

$$G_{as} = \int_{S_e} f_a(S_e) f_s(S_e)\,dS_e.$$

Equation (7.30) can be cast in matrix form as

$$\{f\}_{as} = -[G_{as}]\{P_a\}, \tag{7.31}$$

with

$$[G_{as}] = \int_{S_e} \lfloor f_s(S_e) \rfloor^{\mathrm{T}} \lfloor f_a(S_e) \rfloor\,dS_e.$$

The equivalent generalised force for the degrees of freedom of the complete structural system can be obtained by transforming the structural and fluid element coordinates to system coordinates q, as in Eq. (7.14). Note that only the fluid coordinates local to the interface will appear. Finally, equations similar to Eq. (7.17), with $\{f_i\}$ replaced by $\{f\}_{as}$, are formed:

$$[\mathbf{K} + j\omega\mathbf{C} - \omega^2\mathbf{M}]\{q\} = -[G]\{p\}. \tag{7.32}$$

The work done by the fluid on the structure is equal in magnitude to the work done by the structure on the fluid, but is of opposite sign. However, because the kinetic and potential energy expressions in Eqs. (7.21) and (7.23) are in terms of pressures and pressure gradients, it is not quite so simple to see how the appropriate coupled equations for the fluid are derived. It is helpful here to consider the Helmholtz equation for a simple harmonic acoustic pressure field, which in rectangular Cartesian coordinates is

$$\frac{\partial^2 p}{\partial x^2} + \frac{\partial^2 p}{\partial y^2} + \frac{\partial^2 p}{\partial z^2} + k^2 p = 0. \tag{7.33}$$

If this equation is multiplied by the acoustic pressure p and integrated over the fluid volume, integration by parts yields

$$\int_V \{[(\partial p/\partial x)^2 + (\partial p/\partial y)^2 + (\partial p/\partial z)^2] - k^2 p^2\}\, dV = \int_S (\partial p/\partial n) p\, dS = \rho_0 \omega^2 \int_S wp\, dS, \tag{7.34}$$

where the term on the right-hand side is an integral over all bounding surfaces. If the pressure field is represented by finite elements having pressure degrees of freedom and associated prescribed functions, the volume and surface integrals can be evaluated to yield

$$\{p\}^{\mathrm{T}}[A]\{p\} - k^2\{p\}^{\mathrm{T}}[B]\{p\} = \rho_0 \omega^2 \{q\}^{\mathrm{T}}[G]\{p\}. \tag{7.35}$$

If the variation of Eq. (7.34) is set equal to zero, the following set of equations results:

$$[\mathbf{A} - k^2 \mathbf{B}]\{p\} = \rho_0 \omega^2 [G]^{\mathrm{T}}\{q\}. \tag{7.36}$$

These are the complementary equations to those of Eq. (7.32). Together they can be solved by standard eigenvalue solution techniques for the damped natural frequencies and modes of the coupled system. The matrices are unsymmetric in the pressure formulation but may be made symmetric by employing the velocity potential as the fluid degree of freedom (Everstine, 1981).

If the structure is subject to applied forces, a column matrix of generalised applied forces must be added to the right-hand side of Eq. (7.32). If the fluid contains sources of volume velocity Q, a column of generalised sources of the form $-j\rho_0\omega\{Q\}$ must be added to the right-hand side: see Eq. (6.20). An example of the application of the FE technique is illustrated by Fig. 114, which shows the acoustic field isobars associated with coupling of the fluid in two rooms through an intervening panel (Craggs and Stead, 1976).

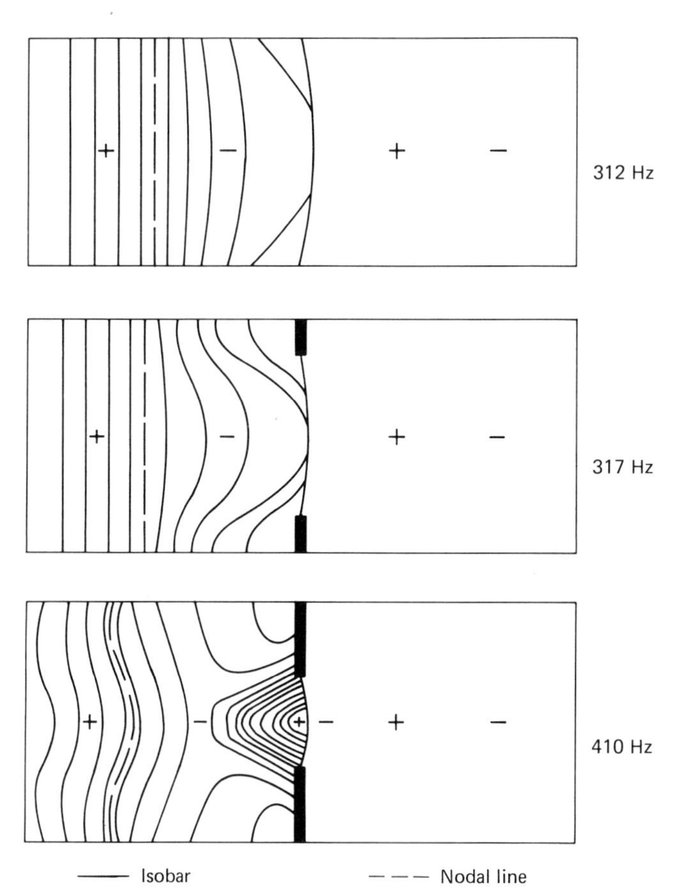

Fig. 114 Sound pressure isobars in a room partially bounded by a flexible panel (Craggs and Stead, 1976).

7.4 Integral Equation Analysis

It has already been mentioned that FE analysis, in which an acoustic field is described by means of the values of various field quantities at a finite number of nodal points distributed throughout the volume, together with assumed internodal distributions, is most practicable when the extent of the region of fluid concerned is not too large in terms of an acoustic wavelength. It is, however, often necessary to determine the characteristics of acoustic fields that are created in virtually infinite expanses of fluid by the action of sound sources located in a fluid. In this book we are concerned, in particular, with sound fields associated with the mechanically induced vibration of structures in contact with a fluid, or with the interaction of incident sound fields and structures immersed in a fluid.

In principle, it is possible to determine such sound fields from a knowledge of the distribution of the acoustic pressure and normal component of acoustic particle velocity on any bounding surfaces of a fluid, by the application of the Kirchhoff–Helmholtz integral equation, which is given by Eq. (6.4) for simple harmonic motion:

$$\tilde{p}(\mathbf{r}) = \int_S [\tilde{p}(\mathbf{r}_s)\,\partial\tilde{G}(\mathbf{r},\mathbf{r}_s)/\partial n + j\omega\rho_0\tilde{v}_n(\mathbf{r}_s)\tilde{G}(\mathbf{r},\mathbf{r}_s)]\,dS + j\omega\rho_0\int_V \tilde{q}(\mathbf{r}_0)\tilde{G}(\mathbf{r},\mathbf{r}_0)\,dV, \tag{7.37}$$

in which we may use the free-field Green's function

$$\tilde{G}(\mathbf{r},\mathbf{r}_0,\omega) = (1/4\pi|\mathbf{r}-\mathbf{r}_0|)\exp(-jk|\mathbf{r}-\mathbf{r}_0|).$$

Even where the distribution of surface normal velocity $\tilde{v}_n(\mathbf{r}_s)$ is known a priori, it is not generally possible to evaluate this integral, except where the surface geometry is particularly simple: the surface pressure distribution $\tilde{p}(\mathbf{r}_s)$ is not independent of $\tilde{v}_n(\mathbf{r}_s)$, but essentially determined by it.

The problem of developing analytical and numerical mathematical techniques to solve the integral equation has exercised the skills of applied mathematicians for many years, and much of the material is beyond the scope of this book. However, it is appropriate to outline the bases of the various approaches and to indicate the strength and limitations of each from a practical point of view.

In cases where the distribution of normal particle velocity is known, or can be assumed to be independent of the surface pressures, as in the case of solid structures vibrating in fluids of low density, it would appear at first sight that it is only necessary to let the field observation point $\mathbf{r}$ approach the surface and to evaluate the resulting pressure distribution: once this is known Eq. (7.37) can be solved for any field point. With $\mathbf{r}$ on the surface, and with no volume-distributed sources, Eq. (7.37) becomes

$$\tilde{p}(\mathbf{r}'_s) = 2\int_S^* [\tilde{p}(\mathbf{r}_s)\,\partial\tilde{G}(\mathbf{r}'_s,\mathbf{r}_s)/\partial n + j\omega\rho_0\tilde{v}_n(\mathbf{r}_s)\tilde{G}(\mathbf{r}'_s,\mathbf{r}_s)]\,dS, \tag{7.38}$$

where $\int^*$ indicates that the principal values of the integrals are to be taken, since they are singular when $\mathbf{r}'_s$ and $\mathbf{r}_s$ coincide. The principal value is obtained by taking the limit of the integrals, evaluated over a surface excluding an area $\pi\varepsilon^2$ around the point of singularity, as $\varepsilon \to 0$ (Courant and Hilbert, 1962).

For the purposes of numerical analysis, the radiating surface may be divided into a set of small discrete surface elements, over which it may reasonably be assumed that the pressure and the normal particle velocity are constant. If there are n elements, Eq. (7.38) can be written as n simultaneous algebraic equations and solved for $\tilde{p}_n$. Unfortunately, in the case of a surface that encloses an interior volume, there is no unique solution at frequencies that equal the natural frequencies of the enclosed fluid with pressure release ($p = 0$) boundaries. This is, of course, only an analytical problem, since physically a unique solution always exists. However, these frequencies are not known unless an auxiliary analysis is performed to determine them. One apparent solution to this difficulty is to require the pressure at certain points inside the closed volume to vanish, by overdetermining the system of equations. The effectiveness of this strategy is dependent upon the number and location of the interior points chosen; they must not, for instance, be at a point of zero pressure in the interior acoustic mode, of which the pressure distribution is actually unknown. Another difficulty is that the discretisation of the integral operator can produce ill-conditioned equations over bands of frequency, rather than at single frequencies (Seznec, 1983). However, in some cases of practical interest, this approach has been found to give accurate results for the radiated field (Schenck, 1968; Hodgson and Sadek, 1983; Koopmann and Benner, 1982). In the latter work, the amount of numerical integration necessary in the evaluation of Eq. (7.38) for each elemental area has been reduced by using large kr, and small kr, approximations to the free-space Green's function and its derivative, and deriving closed-form analytic expressions for the resulting contributions to the integral. The large-kr approximation is used for elements separated by a large non-dimensional distance kr, and the small-kr approximation is used for closely spaced elements.

Alternative means of overcoming the problem of non-uniqueness of the solution at characteristic frequencies of the interior space have been proposed. Burton and Miller (1971) propose the use of the differentiated form of Eq. (7.38), which gives the value of the normal pressure gradient on the surfaces of the body in linear combination with the original equation for surface pressure. Unfortunately the differential form contains a strongly singular term in the form of the second derivative of the Green's function. Filippi (1977) proposes the use of unknown surface distributions of a layer potential, which requires the computation of intermediate functions, but appears to be a promising approach from a practical point of view. Sayhi *et al.* (1981) develop this approach further.

In a modified version of this direct approach, known as the *boundary-element technique*, the vibrating surface is divided into a set of finite elements to which nodal normal velocities and surface pressures are allocated, the former being known and the latter unknown in the radiation problem considered here. Prescribed distributions of these variables over the elements are then chosen, as in the FE method described in the previous section, and the contributions to the integral of Eq. (7.38) are evaluated. The singular nature of the integrands is accommodated by various analytical devices. Finally, a matrix equation linking the normal velocity distribution and the surface pressure distribution is generated, but this technique suffers from the problem of non-uniqueness of the solution at certain frequencies.

As mentioned in Chapter 2, it is possible to make direct use of the Rayleigh integral for calculating the sound power radiated by a single structural mode of a plane, baffled structure provided that the mode is purely real, i.e., all points move in phase or in antiphase with each other.

The surface pressure is, from Eq. (2.4),

$$\tilde{p}(\mathbf{r}'_s) = \frac{j\omega\rho_0}{2\pi}\int_S \frac{\tilde{v}_n(\mathbf{r}_s)e^{-jkR}}{R}\, dS, \tag{7.39}$$

where $R = |\mathbf{r}_s - \mathbf{r}'_s|$. In pure modal vibration, $\tilde{v}_n(\mathbf{r}_s) = \tilde{v}_n\phi(\mathbf{r}_s)$. The radiated sound power is given by

$$\begin{aligned}\bar{P} &= \int_{S'} \frac{1}{2}\,\mathrm{Re}\,[\tilde{p}(\mathbf{r}'_s)\tilde{v}_n^*(\mathbf{r}'_s)]\, dS' \\ &= \frac{\omega\rho_0|\tilde{v}_n^2|}{4\pi}\int_{S'} \phi(\mathbf{r}'_s)\left[\int_S \phi(\mathbf{r}_s)\frac{\sin kR}{R}\, dS\right] dS'.\end{aligned} \tag{7.40}$$

There is no problem of singularity with the integrand because $(\sin kR)/R \to k$ as $R \to 0$. If a mode shape is calculated or measured, Eq. (7.40) provides a very simple means of evaluating the sound power by numerical integration. Recently, modern FFT techniques have been applied with considerable success to the evaluation of sound fields radiated by baffled and unbaffled plane panels (Williams, 1983).

In problems involving the interaction between structures and fluids, it would clearly be convenient to combine structural finite-element analysis with some form of acoustic finite-element analysis. A combination of FE and boundary-element techniques has been proposed by Zienkiewicz *et al.* (1977), among others. As a means of overcoming the problem presented by infinite fluid domains, so-called infinite (really semi-infinite) elements have been developed (Bettes, 1980; Ton and Rassetos, 1977). It is not really possible to generalise about the advantages and disadvantages of these various methods: in fact, the only way really to evaluate their effectiveness is to apply them to a range of problems. Unfortunately very few readers will have the opportunity, the time, the money, or the stamina to perform a comprehensive set of comparative tests. What can be stated with confidence, however, is that in the numerical analysis of the vibrational behaviour of coupled fluid–structural systems, it is the structure, and not the fluid, that will present the greatest difficulty in the modelling process.

Problems

7.1 Write a finite-element program to evaluate the lowest natural frequency and mode shape of a thin, simply supported panel coupled to a rectangular cavity containing a compressible fluid. Restrict the problem to two dimensions and use prescribed functions of parabolic form, with pressure or velocity potential as your acoustic degree of freedom. Investigate the convergence of your solution as you reduce the mesh size.

7.2 Use Eq. (7.40) to evaluate the sound power radiated per unit length of a baffled-plate strip of infinite length and width 300 mm, vibrating in the second mode at 500 Hz. Assume that the strip is simply supported at the edges.

References

Argyris, J. H. (1954). *Aircr. Eng.* **36**(302), 102–112.

Bank, G., and Hathaway, G. T. (1981). *J. Audio Eng. Soc. Am.* **29**(5), 314–319.

Belousov, Yu. I., and Rimskii-Korsakov, A. V. (1975). *Sov. Phys. Acoust.* **21**(2), 103–109.

Bettes, P. (1980). *Int. J. Numer. Methods Eng.* **15**, 1613–1626.

Bies, D. L. (1971). *In* "Noise and Vibration Control" (L. L. Beranek, ed.), p. 246. McGraw-Hill, New York.

Bijl, L. A. (1977). "Measurement of Vibrations Complementary to Sound Measurements," CONCAWE Rep. No. 8/77. The Oil Companies International Study Group for Conservation of Clean Air and Water in Europe, Den Haag, The Netherlands.

Bishop, R. E. D., and Mahalingham, S. (1981). *J. Sound Vibration* **77**, 149–163.

Blevins, R. D. (1979). "Formulas for Natural Frequency and Mode Shape." Van Nostrand-Reinhold, Princeton, New Jersey.

Bull, M. K., and Norton, M. P. (1982). On coincidence in relation to prediction of pipe wall vibration and noise radiation due to turbulent pipe flow disturbed by pipe fittings. *Proc. Int. Conf. Flow Induced Vibrations Fluid Eng.* 1982 Paper H2. Reading Univ., England.

Burton, A. J., and Miller, G. F. (1971). *Proc. R. Soc. London, Ser. A* **323**, 201–210.

Cabelli, A. (1982). *J. Sound Vibration* **85**, 423–434.

Cederfeldt, L. (1974). "Sound Insulation of Corrugated Plates. A Summary of Laboratory Measurements," Rep. 55. Division of Building Technology, Lund Institute of Technology, Sweden.

Cook, R. D. (1981). "Concepts and Applications of Finite Element Analysis," 2nd ed. Wiley, New York.

Courant, R., and Hilbert, D. (1962). "Methods of Mathematical Physics," Vol. 2, p. 256. Wiley (Interscience), New York.

Craggs, A., and Stead, G. (1976). *Acustica* **35**(2), 89–98.

Cremer, L. (1942). Theorie der Schalldämmung dünner Wände bei Schrägen einfall. *Akust. Ztg.* **7**, 81.
Cremer, L. (1968). Schallschutz in Gebäuden. "Berichte aus der Bauforschung," No. 56. Von Wilhelm Erst und Sohn, Berlin.
Cremer, L. (1981). "Physik der Geige." Hirzel, Stuttgart.
Cremer, L., Heckl, M., and Ungar, E. E. (1973). "Structure-borne Sound." Springer-Verlag, Berlin and New York.
Crighton, D. G. (1972). *J. Sound Vibration* **20**(2), 209–218.
Crighton, D. G. (1977). *J. Sound Vibration* **54**(3), 389–391.
Davies, H. G. (1971). *J. Sound Vibration* **15**(1), 107–126.
Duncan, W. J., Thom, A. S., and Young, A. D. (1970). "Mechanics of Fluids," 2nd ed., pp. 81–84 and 132–137. Edward Arnold, London.
Einarsson, S., and Söderquist, J. (1982). *Proc. Inter-Noise 82* pp. 467–470.
Eversman, W., and Astley, R. J. (1981). *J. Sound Vibration* **74**, 103–122.
Everstine, G. C. (1981). *J. Sound Vibration* **79**(1), 157–160.
Fahy, F. J. (1969). *J. Sound Vibration* **10**(3), 490–512.
Fahy, F. J. (1970). *J. Sound Vibration* **13**(2), 171–194.
Filippi, P. T. (1977). *J. Sound Vibration* **54**, 473–500.
Fuller, C. R., and Fahy, F. J. (1981). *J. Sound Vibration* **81**(4), 501–518.
Gladwell, G. M. L. (1966). *J. Sound Vibration* **4**(2), 172–186.
Gomperts, M. C. (1977). *Acustica* **37**(2), 93–102.
Gu, Q., and Wang, J. (1983). *Chinese J. Acoust.* **2**(2), 113–126.
Heckl, M. (1960). *Acustica* **10**(2), 109–115.
Heckl, M. (1962a). *J. Acoust. Soc. Am.* **34**, 803–808.
Heckl, M. (1962b). *J. Acoust. Soc. Am.* **34**(10), 1553–1557.
Hodgson, D. C., and Sadek, M. M. (1983). *Proc. Inst. Mech. Eng., Part C* **197**, 189–197.
Holmer, C. I., and Heymann, F. J. (1980). *J. Sound Vibration* **70**(3), 275–301.
Ingemansson, S., and Kihlman, T. (1959). Sound insulation of frame walls. *Trans. Chalmers Univ. Technol., Gothenburg* No. 222.
Jennings, A. (1977). "Matrix Computation for Engineers and Scientists." Wiley, London.
Jones, R. E. (1981). *Noise Control Eng.* **16**(2), 90–105.
Josse, R., and Lamure, C. (1964). *Acustica* **14**, 266–280.
Junger, M. C., and Feit, D. (1972). "Sound, Structures and Their Interaction." M.I.T. Press, Cambridge, Massachusetts. (2nd ed. due 1985).
Kamkar, H. (1981). "The Effect of Fluid Loading on the Vibration Response of a Planar Arbitrary Shaped Baffled Plate," Rep. RD/B/5046N81. Central Electricity Generating Board, Berkeley Nuclear Laboratories, Berkeley, Gloucestershire, England.
Kihlman, T. (1967a). "Transmission of Structure-borne Sound in Buildings," Rep. 9. National Swedish Institute for Building Research, Stockholm. 1967.
Kihlman, T. (1967b). *J. Sound Vibration* **11**(4), 435–445.
Kinsler, L. E., Frey, A. R., Coppens, A. B., and Sanders, J. V. (1982). "Fundamentals of Acoustics," 3rd ed. Wiley, New York.
Koopmann, G. H., and Benner, H. (1982). *J. Acoust. Soc. Am.* **71**(1), 78–89.
Kurtze, G., and Watters, B. G. (1959). *J. Acoust. Soc. Am.* **31**(6), 739–748.
Lagrange, J. L. (1788). "Mechanique Analytique."
Leissa, A. W. (1969). "Vibration of Plates," NASA SP-160. Office of Technology Utilization, National Aeronautics and Space Administration, Washington, D.C.
Leissa, A. W. (1973). "Vibration of Shells," NASA SP-288. Office of Technology Utilization National Aeronautics and Space Administration, Washington, D.C.
Liamshev, L. M. (1958). *Sov. Phys. Acoust.* **4**(1), 50–58.

Liamshev, L. M. (1960). *Sov. Phys. Acoust.* **5**(4), 431–438.
Lighthill, M. J. (1958). "Introduction to Fourier Analysis and Generalised Functions," Cambridge Univ. Press, London and New York.
Lin, G. F., and Garrelick, J. M. (1977). *J. Acoust. Soc. Am.* **61**(4), 1014–1018.
Lomas, N. S., and Hayek, S. I. (1977). *J. Sound. Vibration* **52**(1), 1–25.
Lord Rayleigh, Strutt (1896). "The Theory of Sound," 2nd ed. (reprinted by Dover, New York, 1945).
Macadam, J. A. (1976). *Appl. Acoust.* **9**, 103–118.
Maidanik, G. (1962). *J. Acoust. Soc. Am.* **34**, 809–826.
Maidanik, G., and Kerwin, E. M. (1966). *J. Acoust. Soc. Am.* **40**(5), 1034–1038.
Mead, D. J. (1975). *J. Sound Vibration* **40**(1), 1–18.
Mindlin, R. D. (1951). *J. Appl. Mech.* **18**, 31–38.
Mkhitarov, R. A. (1972). *Sov. Phys. Acoust.* **18**(1), 123–126.
Morse, P. M. (1948). "Vibration and Sound," 2nd ed. McGraw-Hill, New York (reprinted by the Acoustical Society of America, 1981).
Muller, P. (1983). *J. Sound Vibration* **87**(1), 115–141.
Nikiforov, A. S. (1981). *Sov. Phys. Acoust.* **27**(1), 87–89.
Northwood, T. D. (1970). "Transmission loss of plasterboard walls," Build. Res. Note No. 66. National Research Council of Canada, Ottawa.
Pagliarini, G., and Pompoli, R. (1983). *Acustica* **52**(5), 296–299.
Pallett, D. S. (1972). Applications of statistical methods to the vibrations and acoustic radiation of fluid-loaded cylindrical shells. Ph.D. Thesis, Pennsylvania State University, State College.
Petyt, M., Lea, J., and Koopmann, G. H. (1976). *J. Sound Vibration* **45**, 495–502.
Pierce, A. D. (1981). "Acoustics: An Introduction to its Physical Principles and Applications." McGraw-Hill, New York.
Pierri, R. A. (1977). Study of a dynamic absorber for reducing the vibration and noise radiation of plate-like structures. M.Sc. dissertation, University of Southampton, England.
Pretlove, A. J. (1965). *J. Sound Vibration* **2**(3), 197–209.
Quirt, J. D. (1982). *J. Acoust. Soc. Am.* **72**(3), 834–844.
Quirt, J. D. (1983). *J. Acoust. Soc. Am.* **74**(2), 534–542.
Randall, R. B. (1977). "Frequency Analysis." Brüel & Kjaer, Naerum, Denmark.
Rennison, D. C. (1977). The vibrational response of, and the acoustic radiation from, thin-walled pipes. Ph.D. thesis, University of Adelaide, Australia.
Sayhi, M. N., Ousset, Y., and Verchery, G. (1981). *J. Sound Vibration* **74**, 187–204.
Scharton, T. D., and Lyon, R. H. (1968). *J. Acoust. Soc. Am.* **43**(6), 1332–1343.
Schenck, H. A. (1968). *J. Acoust. Soc. Am.* **44**(1), 41–58.
Sewell, E. C. (1970). *J. Sound Vibration* **12**(1), 21–32.
Seznec, R., (1983). Les methodes de calcul numerique en acoustique. *Proc. 11th Int. Congr. on Acoustics, Paris.* Vol. 8, 75–88.
Sharp, B. (1978). *Noise Control Eng.* **11**(2), 53–63.
Shuku, T. (1975). *J. Acoust. Soc. Jpn.* **28**, 5–10.
Smith, P. W. (1962). *J Acoust. Soc. Am.* **34**(5), 640–647.
Stepanishen, P. E. (1982). *J. Acoust. Soc. Am.* **71**(4), 813–823.
Strawderman, W. A., Ko, S. H., and Nuttall, A. N. (1979). *J. Acoust. Soc. Am.* **66**(2), 579–585.
Szechenyi, E. (1971). *J. Sound Vibration* **19**(1), 65–82.
Ten Wolde, T., Verheij, J. W., and Steenhoek, H. F. (1975). *J. Sound Vibration* **42**(1), 49–56.
Timoshenko, S. P., and Goodier, J. N. (1951). "Theory of Elasticity." McGraw-Hill, New York.
Ton, P., and Rassetos, J. N. (1977). "Finite Element Method: Basic Technique and Implementation." M.I.T. Press, Cambridge, Massachusetts.
Tweed, L. W., and Tree, D. R. (1978). *Noise Control Eng.* **10**(2), 74–79.

Vér, I. L., and Holmer, C. I. (1971). *In* "Noise and Vibration Control" (L. L. Beranek, ed.), pp. 287—296. McGraw-Hill, New York.
von Venzke, G., Dämmig, P., and Fischer, H. W. (1973). *Acustica* **29**(1), 29–40.
Wallace, C. E. (1972). *J. Acoust. Soc. Am.* **51**(3), Part 2, 946–952.
Warburton, G. B. (1976). "The Dynamical Behaviour of Structures," 2nd ed. Pergamon, Oxford.
Warburton, G. B. (1978). *J. Sound Vibration* **60**(3), 465–469.
Watson, G. N. (1966). "A Treatise on the Theory of Bessel Functions," 2nd ed. Cambridge Univ. Press, London and New York.
Williams, E. G. (1983). *J. Acoust. Soc. Am.* **74**(1), 343–347.
Zienkiewicz, O. C., Kelly, D. W., and Bettes, P. (1977). *Int. J. Numer. Methods Eng.* **11**, 355–375.

Answers

Chapter 1

1.1 $c^2 = \dfrac{E_1}{\rho_1}\left[\dfrac{1 + [(d_2/d_1)^2 - 1](E_2/E_1)}{1 + [(d_2/d_1)^2 - 1](\rho_2/\rho_1)}\right]$

1.3 Central force per unit axial length $= 2\pi[Eh/(1 - \nu^2)](w/a)$

$\omega_r = c_l'/a$

1.4 Impedance of mounted system "seen" by floor $= \tilde{Z}_I$

$$\tilde{Z}_I = \frac{j\omega m(s + j\omega r)}{-\omega^2 m + s + j\omega r}$$

Velocity of floor in presence of system $= \tilde{v}_1$
Velocity of floor in absence of system $= \tilde{v}_0$

$\tilde{v}_1/\tilde{v}_0 = [1 + \tilde{Z}_I/\tilde{Z}_F]^{-1}$, where $\tilde{Z}_F$ is the point impedance of floor

Velocity of mass $= \tilde{v}_m$
$\tilde{v}_m/\tilde{v}_0 = (\tilde{Z}_1/j\omega m)(1 + \tilde{Z}_I/\tilde{Z}_F)^{-1}$

$\tilde{Z}_F$ purely real: real part of $\tilde{Z}_I$ positive
Maximum response of mass when $Im\{\tilde{Z}_I\} = 0$: $\omega^2 = s/(m - r^2/s)$

1.5 $\bar{P} = \frac{1}{2}|\tilde{F}|^2 \operatorname{Re}\{1/Z^*\}$

$Z = A\omega^{1/2}(1 + j) - jK/\omega$

$$\operatorname{Re}\{1/Z^*\} = \frac{A\omega^{5/2}}{A^2\omega^3 + (A\omega^{3/2} - K)^2}$$

$\operatorname{Re}\{1/Z^*\}$ is maximum when $\omega = 0.91(K/A)^{2/3}$

1.6 Tip mass M: beam mass per unit length m

$$\tilde{B} = -\tilde{A}\left[\frac{1 + 2k_b M/m + j}{1 + 2k_b M/m - j}\right]$$

$$\tilde{D} = -\tilde{A}\left[\frac{2j}{1 + 2k_b M/m - j}\right]$$

$M = 0$: free end

$\tilde{B} = -\tilde{A}j$: $\quad |\tilde{B}/\tilde{A}|^2 = 1$

$\tilde{D} = \tilde{A}(1 - j)$

$M = \infty$: simple support

$\tilde{B} = -\tilde{A}$: $\quad |\tilde{B}/\tilde{A}|^2 = 1$

$\tilde{D} = 0$

Spring stiffness K

$$\tilde{B} = -\tilde{A}\left[\frac{2K/EIk_b^3 - 1 - j}{2K/EIk_b^3 - 1 + j}\right]$$

$$\tilde{D} = \tilde{A}\left[\frac{2j}{(2K/k_b^3 - 1) + j}\right]$$

$K = 0$: free end

$\tilde{B} = -\tilde{A}j$: $\quad |\tilde{B}/\tilde{A}|^2 = 1$

$\tilde{D} = \tilde{A}(1 - j)$

$K = \infty$: simple support

$\tilde{B} = -\tilde{A}$: $\quad |\tilde{B}/\tilde{A}|^2 = 1$

$\tilde{D} = 0$

1.7 In free wave propagation the vibration amplitude of each mass is the same, but the phase decreases by a constant increment from mass to adjacent mass, i.e., $x_{n+1} = x_n \exp(-\mu)$. Solve for μ from the equation of motion of any mass. $c_g \propto \partial\omega/\partial\mu = 0$ at the cut-off frequency.

$\cosh \mu = 1 - m\omega^2/2k$

$\omega_c^2 = 4k/m$

1.8 $\bar{P} = \frac{1}{2}\,\mathrm{Re}\{\tilde{M}(-j\omega\beta^*)\} + \frac{1}{2}\,\mathrm{Re}\{\tilde{S}(-j\omega\eta^*)\}$
$\bar{P} = \frac{1}{2}EI\omega k_b^3|\tilde{\eta}|^2$
$\tau = 0.5$

1.9 Strip thickness = b: strip width = h
Torsional stiffness = GKb^3h (Table 2)

$k_t = \omega[\rho_s bh^3/12Kb^3hG]^{1/2} = (\omega h/b)[\rho_s/12GK]^{1/2}$
$k_b = (\omega/b)^{1/2}[12\rho_s(1 - \nu^2)]^{1/4}$

At the critical frequency $k_t = k_b$

$\omega_c = 12\sqrt{3}bKc_l''(1 - \nu^2)^{1/2}/h^2(1 + \nu)$

Chapter 2

2.1 $\overline{p^2} = \omega^2\rho_0^2|\tilde{Q}|^2/32\pi^2r^2 = 4 \times 10^{-4}\quad \mathrm{Pa}^2$
$|\tilde{Q}| = 3.7 \times 10^{-3}\quad \mathrm{m^3s^{-1}}$

2.2 Piston radius a: axial distance R_0

$\tilde{p} = -(\omega\rho_0\tilde{v}_n/k)[\exp(-jkR') - \exp(-jkR_0)]$,

where R' is the distance from the piston edge to the observer position

$$\frac{\tilde{p}_{\mathrm{approx}}}{\tilde{p}_{\mathrm{exact}}} = -(ja^2k/2R_0)\{\exp[-jk(R' - R_0)] - 1\}^{-1}$$

$R_0 = 1$ m, $\quad f = 1$ kHz, $\quad$ magnitude error = 13%

2.5 Water: $\eta_{\mathrm{rad}}/\sigma = 20.4$
Air: $\eta_{\mathrm{rad}}/\sigma = 5.6 \times 10^{-3}$

2.6 $\bar{P} = \omega^2\rho_0|\tilde{Q}|^2/4\pi c\qquad$ (one side)
$\tilde{Q} = 4(a/\pi)^2\tilde{v}_0$
$\eta_{\mathrm{rad}} = \bar{P}/\omega M\langle\overline{v^2}\rangle = 8\bar{P}/\omega M|\tilde{v}_0|^2$
$= 32\omega\rho_0a^2/\pi^5c\rho_sh$
Water: $\eta_{\mathrm{rad}} = 0.41$
Air: $\eta_{\mathrm{rad}} = 2 \times 10^{-3}$

2.11 $\tilde{v}_0/\tilde{v}_i = [1 + \tilde{Z}_p/\tilde{Z}_c]^{-1}$
$|\tilde{v}_i|^2 = 0.195\ \mathrm{m^2\,s^{-2}}$
$\tilde{Z}_c = j\omega m = 314.6j$
$\tilde{Z}_p = 860.6\ \mathrm{Nm^{-1}s}$
$|\tilde{v}_0/\tilde{v}_i|^2 = 0.118$

$|\tilde{v}_0|^2 = 2.3 \times 10^{-2}$

$$\bar{P} = \frac{\rho_0 c^2 |\tilde{v}_0|^2}{2\omega_c}$$
$$= 5.9 \times 10^{-2}\ \text{W}$$

2.12 $I_z/\text{unit length} = \frac{1}{\lambda}\int_0^\lambda A^2 h \sin^2(\pi x/\lambda)\, ds,$

where $ds^2 = dx^2 + dy^2$

Can approximate ds by dx because y^2 is greatest where dy/dx is least

$I_z/\text{unit length} \approx hA^2/2, \quad I_x/\text{unit length} = h^3/12$

$D_x^{1/2}k_x^2 + D_z^{1/2}k_z^2 = \omega m^{1/2}$ and $k_x^2 + k_z^2 = k^2$ at intersection

Hence $k_x^2 = \dfrac{\omega m^{1/2} - D_z^{1/2}k^2}{D_x^{1/2} - D_z^{1/2}}$

$D_x = 3500$ Nm, $\quad D_z = 64.1$ Nm

Hence $k_x = 12.4\ \text{m}^{-1}$, $\quad k_z = 34.5\ \text{m}^{-1}$, $\quad \theta = 19.8°$ from z axis

Chapter 3

3.1 $m = 8\rho a_0^3/3$

$\omega_0 - \omega_0(M/M + m)^{1/2}$

$f_0 = 31.06$ Hz

3.2 100 Hz, $\quad k_b = 9.05\ \text{m}^{-1}$, $\quad m = 39\ \text{kg m}^{-2}$, $\quad c_b = 69.4\ \text{m s}^{-1}$

$m_{air} \approx 0.27\ \text{kg m}^{-2}$, $\quad m_{water} \approx 221\ \text{kg m}^{-2}$

$c_{b(water)} \approx 43.2\ \text{m s}^{-1}$

1 kHz: $\quad k_b = 28.63\ \text{m}^{-1}$, $\quad c_b = 219.4\ \text{m s}^{-1}$

$c_{b(water)} \approx 169.8\ \text{m s}^{-1}$

2,2 mode

$f_0 = 289.8$ Hz: $\quad f_{0(air)} \approx 288.7$ Hz, $\quad f_{0(water)} \approx 139.4$ Hz

10,10 mode

$f_0 = 7243$ Hz: $\quad f_{0(air)} \approx 7243$ Hz, $\quad f_{0(water)} \approx 5614$ Hz

3.3 Equation (3.26) gives

$$\tilde{p}_1(a)/\tilde{v}_0 = \frac{-j\rho_0 c H_0(ka)}{H_0'(ka)}.$$

For $ka \ll 1$, this may be written approximately as

$$\tilde{p}_1(a)/\tilde{v}_0 \approx j\rho_0 cka \ln(ka).$$

Hence the effective inertial loading per unit area m' is $\rho_0 a \ln(ka)$, which is added to the cylinder mass per unit area $\rho_s h$. The ring frequency is reduced in the ratio $(1 + m'/\rho_s h)^{-1/2}$.

$f_r = c_l'/\pi d = 2274$ Hz, $\quad f_r' = 659$ Hz

3.4 Let the pressure difference across hole equal $R\tilde{u}$, where $\tilde{u}$ is the particle velocity through the hole. Continuity of particle velocity at the piston and volume velocity at the termination yield the following expression for the total impedance:

$$\tilde{z}_t = \tilde{z}_m + \rho_0 c \left[\frac{1 - \beta^2 - 2j\beta \sin 2kl}{(1 + \beta^2) - 2\beta \cos 2kl} \right],$$

where $\tilde{z}_m$ is the *in-vacuo* piston impedance, $\beta = (R - \alpha\rho_0 c)/(R + \alpha\rho_0 c)$, and α is the hole/tube cross-sectional area ratio. The real part of the expression in the square brackets represents dissipation (damping); the imaginary part represents stiffness or inertia, depending upon sign. The damping is zero if $R = \infty$ (closed end).

3.5 $\tilde{Z}_0 = R_0 + jX_0, \quad \tilde{Z}_l = R + jX$
$\bar{P} = \frac{1}{2}S^2|\tilde{u}|^2 \operatorname{Re}\{\tilde{Z}_0\} = \frac{1}{2}S^2|\tilde{u}|^2 R_0$

$$R_0 = \frac{R(\rho_0 c/S)^2(1 + \tan^2 kl)}{(\rho_0 c/S - X \tan kl)^2 + R^2 \tan^2 kl}$$

Putting $\partial R_0/\partial(kl) = 0$ yields stationary values given by

$$\tan 2kl = \frac{2(\rho_0 c/S)X}{|\tilde{Z}_l|^2 - (\rho_0 c/S)^2},$$

or $X(1 - \tan^2 kl)(\rho_0 c/S) = (|\tilde{Z}_l|^2 - (\rho_0 c/S)^2) \tan kl$.

$$\text{Therefore } \tan kl = \frac{-b \pm \sqrt{b^2 - 4ac}}{2a},$$

where $a = -c = X/(\rho_0 c/S)$ and $b = |\tilde{Z}_l|^2/(\rho_0 c/S)^2$.
$R/(\rho_0 c/S) = 5, \quad X/(\rho_0 c/S) = -10, \quad |\tilde{Z}_l|^2/(\rho_0 c/S)^2 = 125$

$$\tan kl = \frac{-125 \pm \sqrt{1.56 \times 10^4 + 400}}{-20}$$

$$= 12.57 \quad \text{or} \quad -7.45 \times 10^{-2}$$

$kl = 1.49$ or 7.44×10^{-2}
$f_1 = 8.12$ Hz, $\quad f_2 = 162.7$ Hz
Maximum power at f_1, minimum power at f_2

Chapter 4

4.1 $\tilde{v} = \dfrac{\tilde{p}_{bl}S}{\tilde{Z}_m + \tilde{Z}^m_{rad}}, \qquad \tilde{p}_{bl} = 2\tilde{p}_i, \qquad Z^m_{rad} = S^2\tilde{Z}_{rad} \qquad$ (Eq. 3.10)

$\bar{P}_{rad} = \frac{1}{2}|\tilde{v}|^2 \operatorname{Re}\{\tilde{Z}^m_{rad}\}$
$\bar{P}_{inc} = S|\tilde{p}_i|^2 \cos\phi/2\rho_0 c$
$\tau = 4\rho_0 cS \operatorname{Re}\{\tilde{Z}_{rad}\}/|\tilde{Z}_m + \tilde{Z}^m_{rad}|^2 \cos\phi$
$\tilde{Z}_m = j(\omega m - s/\omega) + r$
$\tilde{Z}_{rad} = 2\rho_0 cS[(ka)^2/2 + 8ka/3\pi]$
$\omega_0 = 314 \text{ rad s}^{-1}, \qquad s = \omega_0^2 m = 1972 \text{ N m}^{-1}$
At resonance $\tilde{Z}_m + \tilde{Z}^m_{rad} = \operatorname{Re}\{\tilde{Z}^m_{rad}\}$
$\tau_{res} = 4\rho_0 cS/\operatorname{Re}\{\tilde{Z}^m_{rad}\} \cos\phi = 4/(ka)^2$
$\tau_{res} = 1909!$

At frequencies much greater than ω_0, the inertial impedance of the piston controls transmission:

$\tau \approx 4(\rho_0 cS)^2(ka)^2/\omega^2 m^2 \cos\phi.$

τ is frequency independent and varies as $\sec\phi$ as long as $ka \ll 1$.

4.2 To reduce the transmission coefficient by 10, the mechanical loss factor must be given by $\eta_m = 10\rho_0 cS(ka)^2/\omega_0 m = 10^{-2}$.

4.3 Normal particle velocity $\tilde{v}$

$\tilde{v} = 2\tilde{p}_i[j\rho_2 c_2[(k_1 \sin\phi_i/k_2)^2 - 1]^{-1/2} + \rho_1 c_1 \sec\phi_i]^{-1}$
$\tilde{p}_2 = 0.312\tilde{p}_i$

Incidence angle beyond critical: no energy transmission

4.4 Refraction causes effective incidence angle to increase
$f < f_c$: transmission increased
$f > f_c$: stiffness-controlled transmission decreased
: mass-controlled transmission increased
: overall effect is to increase diffuse field transmission

4.5 Eq. (4.56): $\tau_\infty/\tau_r \approx 200(A/P\lambda_c)^2\eta_{tot} = 1$
$\lambda_c = 2.63 \times 10^{-2}, \qquad \eta_{tot} = 4.6 \times 10^{-5}$
Eq. (4.65): $3/2 + \ln(2\omega/\Delta\omega) = 2.52, \qquad \omega/\omega_c = 3.83 \times 10^{-2}$
Field incidence $R = R(0) - 5$ dB
Therefore, $2.52 + 2.42 \times 10^{-3}/\eta = 8$, or $\eta = 4.4 \times 10^{-4}$
The difference lies in approximation for σ in Eq. (4.55)

4.6 $\omega_1 = 219 \text{ rad s}^{-1}, \qquad \delta_1 = 9.4 \times 10^{-2}$
$\omega_0 = 3212 \text{ rad s}^{-1}, \qquad f_0 = 511$ Hz

4.7 $\tilde{\xi}_1/\tilde{\xi}_2 = -m_2/m_1$

4.8 $$\frac{\partial^2 u}{\partial x^2} - \frac{1}{c^2}\frac{\partial^2 u}{\partial t^2} - \frac{r}{\rho_0 c^2}\frac{\partial u}{\partial t} = 0$$

Let $u = \tilde{u}\exp[j(\omega t + \lambda x)]$, where $\lambda = \alpha + j\beta$,
$\alpha^2 - \beta^2 + 2j\alpha\beta = j\omega r/\rho_0 c^2 - k^2$,
$\alpha^2 - \beta^2 = -k^2$, normally $\alpha \ll \beta$.
Hence $\alpha \approx r/2\rho_0 c$

4.10 Power radiated by enclosure per unit area $= \frac{1}{2}\rho_0 c|\tilde{v}|^2$
Power dissipated by enclosure per unit area $= \frac{1}{2}r|\tilde{v}|^2$
Power radiated by source/power radiated by enclosure $= 1 + r/\rho_0 c$
At frequencies far above ω_1, the enclosure mass controls transmission. Maxima occur when $\sin kl = 1$. Therefore, $\bar{P}_e/\bar{P} \approx (\rho_0 c/\omega m)^2$, IL $\approx R(0) + 6$ dB

4.11 $f_c = c^2/1.8hc_l'$ $\quad f_r = c_l'/\pi d$
$f_c/f_r = (\pi/1.8)(d/h)(c/c_l')^2$
Air at 20°C: $d/h = 122$
Air at 100°C: $d/h = 96$

Chapter 5

5.1 Incident field varies slightly in phase over surface of cylinder

5.2 $ka \ll \pi$: a is larger panel dimension
All other modes have a direction of maximum response because the peak in the wavenumber spectrum is not at $k_x = 0$

5.4 $\bar{P}_{rad} = \rho_0 c S\sigma\langle\overline{v^2}\rangle = \langle\overline{p^2}\rangle a/4\rho_0 c$,
where $a = 0.16V/T$
$\sigma = 4.73 \times 10^{-4}$
$\eta_{rad} = 2.6 \times 10^{-6}$, $\quad \eta_{rad} \ll \eta_{int}$
$n_s(\omega) = 3.67 \times 10^{-2}\,\text{rad}^{-1}\,\text{s}$
$\langle\overline{v^2}\rangle = 10^{-7}\,\text{m}^2\,\text{s}^{-2}$

5.5 $\eta_{int} = 2.2/f\,T_{mech}$
$\bar{P}_{diss} = \eta_{int}\omega M\langle\overline{v^2}\rangle$
$\bar{P}_{inc} = S\langle\overline{p^2}\rangle/4\rho_0 c$
$\alpha = 4\eta_{int}\omega m\rho_0 c\langle\overline{v^2}\rangle/\langle\overline{p^2}\rangle$

5.6 $$\frac{\langle\overline{v^2}\rangle}{\langle\overline{p^2}\rangle} = \frac{2\pi^2 c n_s(\omega)}{M\rho_0\omega^2}\left[\frac{(\rho_0 c/\omega m)\sigma}{(\rho_0 c/\omega m)\sigma + \eta_{int}}\right]$$

Assume $\eta_{int} \gg \eta_{rad}$ in test $\rightarrow n_s(\omega)\sigma/\eta_{int}$
Assume $\langle\overline{\varepsilon^2}\rangle \propto \langle\overline{v^2}\rangle$

Assume dependence of σ on f/f_c and $P\lambda c/A$ [see Eq. (4.55)]: extrapolate to operating conditions
Major uncertainty in form of σ

Chapter 6

6.1 For $\psi_n = \cos(p\pi x/a)\cos(q\pi z/b)\cos(r\pi y/c)$,
$\phi_p = \sin(m\pi x/a)\sin(n\pi z/b)$,
and $\phi_q = \sin(m'\pi x/a)\sin(n'\pi z/b)$,

$$\Lambda_n(k_n^2 - k^2)\alpha_{ppn} = \left[\frac{(-1)^{m+p} - 1}{1 - (m/p)^2}\right]^2 \left[\frac{(-1)^{n+q} - 1}{1 - (n/q)^2}\right]^2 \left[\frac{mnab}{p^2q^2\pi}\right]^2$$

$$\Lambda_n(k_n^2 - k^2)\alpha_{pqn} = \left[\frac{(-1)^{m+p} - 1}{1 - (m/p)^2}\right]\left[\frac{(-1)^{n+q} - 1}{1 - (n'/q)^2}\right]\left[\frac{(-1)^{m'+p} - 1}{1 - (m'/p)^2}\right] \times \left[\frac{(-1)^{n'+q} - 1}{1 - (n'/q)^2}\right]\left[\frac{mnm'n'a^2b^2}{p^4q^4\pi^2}\right]$$

6.7 $\omega_p = 146.6$ rad s^{-1}, $f_1 = 23.3$ Hz
$\omega_n = 0$ (bulk compression) because C_{np} is greatest for this mode
$C_{np} = (2/\pi)^2$, $K_{np} = 16\rho_0 ab/M\pi^2$, $G_{np} = 4c^2S/\pi^2 v$
$\Lambda_n = V$, $\Lambda_p = M/4$
$\omega = 401.9$ rad s^{-1}, $f = 64$ Hz
No change of frequency will occur because $\rho_0 c^2$ is unchanged

6.8 $(M_p/4)(-\omega^2 + \omega_0^2) + 4j\omega b^2 \tilde{Z}_0 S/\pi^2 = 0$

$$\tilde{Z}_0 = \frac{\rho_0 c}{S}\left[\frac{\tilde{Z}_l + (\rho_0 c/S)\tan kl}{(\rho_0 c/S) + j\tilde{Z}_l \tan kl}\right] = \frac{\tilde{p}}{\tilde{Q}}$$

$\tilde{Q} = 4b^2\tilde{v}_0/\pi^2$: b is the panel side length
$a = b/\sqrt{\pi} = 0.17$ m

Chapter 7

7.2 Second mode: $v_n = \tilde{v}_n \sin(2\pi z/b)\exp(j\omega t)$
$\tilde{p}$ is uniform along lines parallel to the edges: x axis along one edge
For any point on the z axis, $R^2 = x^2 + (z - z')^2$
Divide surface into suitably sized squares and evaluate inner integral numerically for given value of z'. Then integrate numerically over z'. Investigate the convergence of the solution as the grid size is decreased.

Index

A

B

C

Y